Macroscopic Quantum Tunneling

Macroscopic quantum phenomena are particularly important when considering the problem of Schrödinger's cat. This book contains a coherent and self-contained account of such phenomena, focusing on the central role played by macroscopic quantum tunneling.

Beginning with an explanation of the nature and significance of the cat problem, Shin Takagi introduces the concept of macroscopic quantum tunneling. He deals with typical examples in detail, elucidating how quantum mechanical coherence may be lost (so-called 'decoherence') or how it may be maintained despite the effects of environment and measurement processes. Recent experimental and theoretical advances are discussed, and the remaining problems described. The final chapter describes an experiment to decide between quantum mechanics and macrorealism in the light of Einstein's moon.

Assuming only a knowledge of elementary quantum mechanics, this book emphasizes conceptual aspects rather than technical details. It provides a firm introduction to the subject for graduate students and researchers.

SHIN TAKAGI obtained his Ph.D. from the University of Tokyo in 1974 and is probably best known for the so-called Leggett–Takagi equation which explains the spin dynamics of superfluid helium 3. He has worked in several fields of theoretical physics including low-temperature physics, quantum field theory in curved space and the problem of time and history in quantum mechanics. Currently Professor of Physical Sciences at Fuji Tokoha University, Shin Takagi has contributed to two other books in Japanese: *Quantum Phenomena in the Macroscopic World* (Shokabo Publishers, 1999) and *The World of the Quantum* (Iwanami Shoten Publishers, 1999).

MACROSCOPIC QUANTUM TUNNELING

SHIN TAKAGI

Fuji Tokoha University,
Fuji, Japan

CAMBRIDGE
UNIVERSITY PRESS

CAMBRIDGE UNIVERSITY PRESS
Cambridge, New York, Melbourne, Madrid, Cape Town, Singapore, São Paulo

Cambridge University Press
The Edinburgh Building, Cambridge CB2 2RU, UK

Published in the United States of America by Cambridge University Press, New York

www.cambridge.org
Information on this title: www.cambridge.org/9780521800020

Originally published in Japan, in Japanese
by Iwanami Shoten, Publishers, Tokyo
© Shin Takagi 1997

First published in English by Cambridge University Press 2002
as *Macroscopic Quantum Tunneling*
This digitally printed first paperback version 2005

A catalogue record for this publication is available from the British Library

ISBN-13 978-0-521-80002-0 hardback
ISBN-10 0-521-80002-1 hardback

ISBN-13 978-0-521-67571-0 paperback
ISBN-10 0-521-67571-5 paperback

Contents

Preface   *page* viii
Acknowledgements   x

1   Introduction   1
    1.1   The cat and the moon   1
    1.2   Leggett program   4
    1.3   What is meant by "macroscopic"?   6
          1.3.1   Intuitive consideration   6
          1.3.2   S-Cattiness   7
    1.4   Macroscopic quantum tunneling   9
          1.4.1   Leggett program and macroscopic quantum tunneling   9
          1.4.2   Classification of macroscopic quantum tunneling:
                  MQC and MQT   13
    Exercises   16

2   Overview of macroscopic quantum tunneling   17
    2.1   Standard form of Hamiltonian and Schrödinger equation:
          dimensional analysis   17
    2.2   Overview of MQC   20
    2.3   Overview of MQT   25
    2.4   Formulas for ground-state energy splitting and decay rate   30
    2.5   Quantum decay and irreversible processes   32
    2.6   Euclidean action: bounce and instanton   34
    Exercises   38

3   Some candidate systems for macroscopic quantum tunneling   41
    3.1   SQUID   41
          3.1.1   Josephson junction and equivalent circuit for SQUID   41
          3.1.2   Phase difference and magnetic flux   44

	3.1.3	Equation of motion for magnetic flux	46
	3.1.4	Naive quantum theory	47
	3.1.5	Friction coefficient	52
3.2		Liquid ^{3}He–^{4}He mixture	53
	3.2.1	Phase separation in liquid ^{3}He–^{4}He mixture	53
	3.2.2	Rayleigh–Plesset model	56
	3.2.3	Naive quantum theory	58
	3.2.4	Field theory of nucleation	62
3.3		Single-domain magnet	72
	3.3.1	Naive quantum theory	74
	3.3.2	Longitude basis and longitude-represented wavefunction	77
	3.3.3	Longitude representation of spin operator	82
	3.3.4	Longitude-represented Schrödinger equation and its consequences	85
	3.3.5	Symmetry and interference effect	88
	Exercises		91

4	Environmental problems		95
4.1	Coherence and decoherence		95
4.2	General discussion of dephasing in MQC		96
4.3	Dissipative environment		102
4.4	Non-dissipative environment		106
4.5	Quantum Zeno effect		110
	Exercises		114

5	Harmonic environment		117
5.1	Linearly-coupled harmonic-environment model		117
5.2	Heisenberg equation of motion: part 1		120
5.3	Environmental frequency distribution function		123
5.4	Retarded resistance function		124
5.5	Heisenberg equation of motion: part 2		126
5.6	Equation of motion in the classical regime		128
	Exercises		131

6	Quantum resonant oscillation in the harmonic environment		133
6.1	Preliminary consideration on perturbation theory		133
6.2	Perturbation theory of time evolution		138
6.3	Macroscopic quantum resonant oscillation		143
6.4	Estimation of quantum-resonance frequency		147
	Exercises		149

| 7 | Quantum decay in the harmonic environment | | 151 |
| 7.1 | Conjectures on decay rate | | 151 |

7.2 Euclidean action functional and bounce 152
    7.2.1 Effective Euclidean action functional and
        a stationary point 153
    7.2.2 Effective Euclidean action functional in the
        harmonic-environment model 154
7.3 Bounce and decay rate in a bilinear model 157
7.4 Remark on decay rate at finite temperatures 159
Exercises 160

8 General versus harmonic environments 161
8.1 Modified Born–Oppenheimer basis 161
8.2 Modified Born–Oppenheimer representation for Hamiltonian 164
8.3 Caldeira–Leggett assumption 166
Exercises 170

9 The cat in the moonlight 171
9.1 Macrorealism 171
9.2 Leggett–Garg inequality 172
9.3 Measurable correlation functions 174
9.4 Quantum-mechanical correlation function 175
9.5 Thought experiment to test Leggett–Garg inequality 177
9.6 Concluding remarks 184
Exercises 185

Appendix A. Euclidean space and Hilbert space 187
A.1 Three-dimensional Euclidean space 187
A.2 Hilbert space 187

Appendix B. Virtual ground state of a system of a single degree
          of freedom and its decay 189
B.1 Energy eigenfunction 189
B.2 Quasi-stationary wave 191
B.3 Resonance and virtual bound states 192
B.4 Decay of the virtual ground state 193

Appendix C. Functional derivative 195

Appendix D. Miscellanea about spin 199
D.1 Spin operator 199
D.2 Spin-1/2 decomposition of spin S 200
D.3 Spin disentanglement theorem 201
D.4 Spin coherent state 203

Bibliography 207
Index 209

Preface

The phrase *tunneling*, which has classical-mechanical overtones, is a way of viewing certain dynamical processes within the paradigm of quantum mechanics. *Macroscopic quantum tunneling*, in particular, offers an indispensable viewpoint from which to ponder the cat.[1] This viewpoint, which proposes to free the cat from the confines of the Gedanken world and put it to an experimental test, has become a realistic one by virtue of recent remarkable advances in experimental technologies.

The readership anticipated for this book consists of those graduate students, both experimental and theoretical, who have just finished their undergraduate courses. Assuming only a knowledge of elementary quantum mechanics, this book surveys *macroscopic quantum tunneling* and its significance. The emphasis is laid on conceptual aspects rather than on technical details. The aim of this book is not so much to elaborate on established and systematized theories as to locate *macroscopic quantum tunneling* in the broad perspective and invite students to one of the most important and fascinating footpaths of physics in the twenty-first century.

Linguistic policy

(i) The language used is <u>physical mathematics</u>, that is, an intuitive mathematics and reasoning; highbrow theoretical techniques are relegated to appropriate literature.

(ii) In order to distinguish generally accepted statements from the personal opinions of the author, he adopts the following convention:

A sentence of the type "A is called α" is meant to be a generally accepted statement, whereas a sentence of the type "A is to be called α" is to be understood to imply

---

[1] If one regarded quantum mechanics as merely the way to compute transition probabilities, there would be no logical necessity to single out quantum tunnelings, which constitute merely a class of quantum-mechanical transitions. However, it is often the case that the essence of something is recognizable only from a certain viewpoint; a problem may look different not only if the paradigm itself is changed (see e.g. T. S. Kuhn, *The Structure of Scientific Revolutions*, University of Chicago Press, 1962) but also if a viewpoint is changed while retaining the same paradigm.

that "the present author has decided to call A α in the absence of or regardless of common practice".[2]

(iii) For brevity and logical clarity, four kinds of equality symbols are employed; the usual symbol $=$ is to be used in an equation which follows from what precedes it, and $\equiv$ is to rephrase a mathematical symbol with words or another symbol, whereas $:=$ and $=:$ are to be used in definitions.[3] Here are three examples:

(a) The equation

$$\hat{H}_S \equiv \text{Hamiltonian for the } \underline{\text{macrosystem}} := \cdots$$

implies that "the symbol $\hat{H}_S$ denotes the Hamiltonian for the underline{macrosystem}, and this Hamiltonian is defined as ...".

(b) The phrase "by use of $B := S[\mathcal{Q}]$" implies "by use of B, which is defined by the equation $B = S[\mathcal{Q}]$".

(c) The equation $A =: B$ is to be understood to imply "A defines B", that is, "B is defined by the equation $B = A$".

(iv) The set consisting of all the real numbers is denoted by $\mathbf{R}$, and those of the natural and the complex numbers by $\mathbf{N}$ and $\mathbf{C}$, respectively.[4] For example, the sentence "x is a real number" is sometimes abbreviated as $x \in \mathbf{R}$.

(v) In some cases the absence of appropriate or concise technical terms has required the author to introduce new words, which are underlined. The underlining is to warn the reader that they are not, at least as yet, commonly used nor authorized in academic circles. Here are two examples of such underlined words:

(a) S-Cat is to be used as the abbreviation for Schrödinger's cat.

(b) The sentence "this potential is to be called the bumpy slope" is to be understood as "in the absence of an appropriate authorized word, the present author calls this potential the bumpy slope".

Whenever a strange phrase is encountered, the reader is advised to regard it as a kind of (mathematical) symbol; after all, words are nothing but symbols.

(vi) More than half a century ago, Casimir advocated the use of Broken English by the world scientific community.[5] The present author follows Casimir's spirit to write this book in Beinglish, which, being a revised version of the Broken English, is a common tool of communication among global scientific beings; its grammar, which vaguely resembles that of English, is "customizable" and may be improvised freely.[6]

---

[2] The sentence of the latter type, therefore, is liable to contain errors and may not necessarily be generally accepted.

[3] Of course, the boundary between rephrasing and defining is somewhat fuzzy.

[4] We follow the modern practice to include 0 in $\mathbf{N}$.

[5] H. B. G. Casimir, *Broken English* in *J. Jocular Phys.* **III** (1955) 14 [Reprinted in R. L. Weber, *More Random Walks in Science* (The Institute of Physics, 1982), p.2.]. JJP were irregularly published by the Niels Bohr Institute. The present author is grateful to Professor Chris Pethick for a copy of the collection volume *FAUST and JOURNAL of JOCULAR PHYSICS Volumes I, II and III Reprinted on the Occasion of NIELS BOHR'S CENTENARY October 7, 1985.*

[6] It is unfortunate that spoken Beinglish cannot be reproduced here; it may or may not at all sound like English depending on the speaker.

Acknowledgements

This book is based on lectures given by the author in several graduate courses in Japan, especially at Tohoku University where he taught a regular course. He is grateful to the staff and students of the following universities for inviting him as a lecturer of short-term intensive courses and making valuable comments: Ochanomizu University, Kyoto University, University of Tsukuba, Osaka City University and Nagoya University. Many thanks are due to the former graduate students of Tohoku University: Norifumi Yamada, Kaoru Hiyama, Taichi Hiraoka, Shigenobu Suzuki, Takashi Nakamura, Yusuke Kanno, Takashi Isozaki and Yoshihiro Yamamoto, from whom the author learned much while acting as their thesis supervisor; many of the results of collaboration with these students have been incorporated into this book without explicitly giving them credit. Valuable comments by yet another former graduate student Kousuke Shizume on the Japanese edition are much appreciated; they were of great help to the author in revising some of the subtle points overlooked therein. It should be mentioned that this book would not have been written if the author had not met Professor Tony Leggett more than 25 years ago. During this quarter of a century the author has learned physics in its essence from Tony, both directly through conversation and collaboration on specific problems and indirectly through reading his writings. Tony also kindly answered many of the author's elementary questions on macroscopic quantum tunneling on the commencement of the writing of the original Japanese edition of this book. For all this, the author cannot thank Tony too much. Finally, thanks are due also to Professors Yosuke Nagaoka and Keiji Kikkawa for providing the author with an opportunity of writing the Japanese edition of this book, to Dr. Nobuaki Miyabe of Iwanami Publishers and Dr. Simon Capelin of Cambridge University Press for their effort in publishing Japanese and English editions, to John B. Laing, a non-physicist colleague of the author, for his generous and time-consuming help of correcting and improving the author's Beinglish, and to poet Inge and relativist Werner for an amusing correspondence about S-Cat, namely the Fitzgerald (not FitzGerald) contracted Schrödinger cat.

1

Introduction

We begin by surveying what *macroscopic quantum phenomena* are, and what is the significance of searching for such phenomena, thereby locating *macroscopic quantum tunneling* in the broad perspective of physics in the new century.

1.1 The cat and the moon

It should not be necessary to elaborate on a Young-type interference experiment, which has by now been realized not only with electrons or neutrons but also with atoms such as He, Ne and Na. In a typical experiment, a particle of a given kinetic energy is sent through a double slit to a planar array of particle counters. What happens is that one and only one of the counters fires and is marked by a bright spot. As many particles of the same kind and the same kinetic energy as the first particle are sent one by one successively, bright spots accumulate and eventually emerge as an almost smooth interference pattern. This impressive emergence of the pattern may best be appreciated by watching a movie that records such an experiment in real time. In view of recent remarkable advances in experimental technology, one cannot but be curious about the prospect in the not-unforeseeable future: can the Young-type experiment be realized with an even bigger object, and how big an object can one deal with? Here is a dialogue between Weizsäcker[1] and Glauber[2] at a meeting on quantum mechanics in the early 1990s.

W: However far the technology should advance, one would not be able to see an interference pattern with tennis balls.
G: It might be possible with soccer balls, though.[3]

---

[1] C. F. von Weizsäcker is a theoretician known for his contribution to nuclear physics, etc. He is a brother of the former president of Germany.
[2] R. J. Glauber is a theoretician known for his contribution to quantum optics, etc.
[3] The molecule C_{60} consisting of 60 carbon atoms is often called a soccer ball because of its shape. **Note added in the English edition**: In 1999, only 2 years after the Japanese edition was published, an interference pattern was observed successfully with C_{60} (see Ref. [10] in the bibliography).

The fundamental equation of quantum mechanics[4] is the Schrödinger equation[5] which describes the time evolution of a given system. Its most important property is linearity. Let $|\Psi(t)\rangle$ be the state of the system at time t. (Hereafter a *state* is to be understood as a *quantum state* unless otherwise mentioned.) The Schrödinger equation may be written generally in the following (integrated) form:

$$|\Psi(t)\rangle = \hat{U}(t)|\Psi(0)\rangle. \tag{1.1.1}$$

The symbol $\hat{U}(t)$ denotes a unitary operator determined by the Hamiltonian $\hat{H}$ of the system. For the present purpose, it is sufficient to regard $\hat{U}(t)$ as a sort of a linear black box; given a state at time 0, (i) the state at an arbitrary time t is determined uniquely by the above equation, and (ii) the following equality holds for arbitrary $|\Psi_1\rangle$, $|\Psi_2\rangle$ and t:

$$\hat{U}(t)(|\Psi_1\rangle + |\Psi_2\rangle) = \hat{U}(t)|\Psi_1\rangle + \hat{U}(t)|\Psi_2\rangle. \tag{1.1.2}$$

This relationship embodies the above-mentioned linearity, which is supported by interference effects demonstrated in various experiments, especially ones of the Young-type.

In the microscopic world,[6] a variety of superpositions (namely, linear combinations) of states have been confirmed experimentally. One of the familiar examples is a superposition of spin-up and spin-down states. By virtue of the linearity (1.1.2), a superposition in the microscopic world can in principle be magnified to one in the macroscopic world. A mechanism of such a magnification is Schrödinger's linear theater, of which a simplified version would run as follows. On the stage is a box containing a cat. The play is so designed that a radioactive nucleus is thrown into the box and is swallowed by the cat at time 0, and that there are two possible states $|\phi_\pm\rangle$ for the nucleus at time 0 such that the nucleus has not yet decayed if it is in $|\phi_+\rangle$, but has decayed already if in $|\phi_-\rangle$. The cat will remain intact if the nucleus is in $|\phi_+\rangle$ at time 0, but will eventually die due to radiation hazard if the nucleus is in $|\phi_-\rangle$ at time 0. Let $|\psi\rangle$ be the state of the cat at time 0, and $|\Psi_\pm\rangle$ be the state describing the cat either remaining intact (subscript $+$) or being dead (subscript $-$), at some time T, after swallowing the nucleus which was in $|\phi_\pm\rangle$. Note that $|\Psi_\pm\rangle$ are states of the entire system composed of the cat and the nucleus:

$$|\Psi_\pm\rangle := \hat{U}(T)|\psi, \phi_\pm\rangle, \quad |\psi, \phi_\pm\rangle \equiv |\psi\rangle|\phi_\pm\rangle. \tag{1.1.3}$$

The curtain is to be closed at the time T. So much for the setting of the stage. Now, immediately before the curtain is opened at time 0, the nucleus is to be prepared

---

[4] Quantum mechanics here includes quantum field theory as well.

[5] Some of the readers might associate the Schrödinger equation with the equation "$\hat{H}|\Psi\rangle = E|\Psi\rangle$". The latter, however, being a special case of the former, is appropriate only if "$|\Psi\rangle$ is a stationary state".

[6] The word "world" here is meant to represent vaguely the whole collection of various physical phenomena.

in neither of the two states $|\phi_\pm\rangle$, but in the superposition $|\phi_+\rangle + |\phi_-\rangle$.[7] Hence, the initial state of the entire system is of the following form:

$$|\Psi(0)\rangle = |\psi\rangle(|\phi_+\rangle + |\phi_-\rangle) = |\psi, \phi_+\rangle + |\psi, \phi_-\rangle. \qquad (1.1.4)$$

Combining this with (1.1.1), (1.1.2) and (1.1.3), one finds the state at the time T:

$$|\Psi(T)\rangle = \hat{U}(T)(|\psi, \phi_+\rangle + |\psi, \phi_-\rangle) = |\Psi_+\rangle + |\Psi_-\rangle. \qquad (1.1.5)$$

This equation shows that the superposition in the microscopic world ($|\phi_+\rangle + |\phi_-\rangle$) can be magnified to that in the macroscopic world ($|\Psi_+\rangle + |\Psi_-\rangle$):

MMM: Magnification {microscopic $\longrightarrow$ macroscopic}

Given the state $|\Psi(T)\rangle$, which is a superposition of $|\Psi_+\rangle$ and $|\Psi_-\rangle$, the cat can neither be said to be alive ($|\Psi_+\rangle$) nor dead ($|\Psi_-\rangle$); it may only be said to be in the state of $|\Psi_+\rangle$ AND $|\Psi_-\rangle$.

In view of the fact that superposition of distinct states (e.g. $|\phi_\pm\rangle$ in the above example) in the microscopic world has been confirmed experimentally, the appearance of *macroscopic superposition* (or, equivalently, *macroscopic linear combination*) such as (1.1.5) cannot be avoided so long as linearity of the Schrödinger equation is taken for granted. (Here, "macroscopic superposition" is meant to imply "superposition of *macroscopically distinct* states".) However, this sort of strange state is incompatible with the *macrorealism*,[8] according to which the cat, exposed to the radiation, must necessarily be either in the state $|\Psi_+\rangle$ or in $|\Psi_-\rangle$ ($|\Psi_+\rangle$ OR $|\Psi_-\rangle$); a cat in the state $|\Psi_+\rangle$ AND $|\Psi_-\rangle$ is totally incomprehensible.[9]

During one of his walks, Einstein is said to have asked his colleague[10] "Do you really believe that the moon is there only when you look at it?" What lies at the core of the discussion is the problem of the transition[11] from AND of quantum mechanics to OR of macrorealism,

TAO: Transition {AND $\longrightarrow$ OR}.

---

[7] A radioactive nucleus evolves into a state of this type even if it was originally in $|\phi_+\rangle$.

[8] Loosely speaking, a naive everyday-life realism. See Chapter 9 for details.

[9] According to Schrödinger as translated into English, it is 'ridiculous'; he ridiculed it by creating his linear theater but without forgetting to mention '...the living and the dead cat (pardon the expression)...'. Would Schrödinger have said 'Scat!' to the S-Cat ($\equiv$ Schrödinger's cat)? Note that the audience are not allowed to enter the theater before the closing time T! They are invited to examine the cat only somewhat later. Then, some of them will find the cat alive and the others will find it dead.

[10] A. Pais, *Rev. Mod. Phys.* **51** (1979), 863–914, Section X. Although this question may not have addressed specifically to the problem concerning the macroscopic world, it is undoubtedly an eloquent representative of the macrorealism.

[11] Also called Collapse, Objectification or REalisation (put together as CORE).

If a measuring apparatus replaces the cat, this problem reduces to the "problem of measurement in quantum mechanics", which has been debated since the birth of quantum mechanics.

It should be noted that the state (1.1.5) is not a simple product of a state of the nucleus and a state of the cat but a sum (linear combination) of such products. In this state, the nucleus and the cat cannot be separated; rather they are as it were inseparably entangled. In general such a state is called an *entangled state* (see Chapter 4 for details).

1.2 Leggett program

Let us take a look at the "quantum measurement problem" or the "*S-Cat* (Schrödinger's cat) paradox" from a laboratory-rooted point of view. This paradox presupposes the universal validity of quantum mechanics even in the macroscopic world. This premise, however, lacks in experimental evidence. If it were not valid, the paradox would either disappear or change its character. If, on the other hand, the premise is valid and a macroscopic superposition is realized, one should expect *QIMDS*,[12] namely, *quantum interference of macroscopically distinct states*. The question then is this: how macroscopic can an object be for a laboratory experiment to be able to detect QIMDS, thereby confirming a macroscopic superposition with the object? Note here a traditional opinion against QIMDS:

> Even if quantum mechanics was valid in the macroscopic world, it would be impossible in practice to detect QIMDS.

The argument runs as follows:[13]

> A macroscopic system has a large number of degrees of freedom. Accordingly, QIMDS must result as a sum of a large number of interference effects. Even if each of the effects separately produces such a clear-cut interference pattern as that in the Young-type experiment, the net result of summing these patterns would be the disappearance of any interference effect, because in general they are slightly out of phase with each other; that is, the peaks in one of the patterns are slightly displaced compared to those in another. In the example of Schrödinger's linear theater, the entire system in fact consists of the nucleus, the cat and the whole environment surrounding them, although the environment was disregarded for brevity in the preceding section. Thus, the number of degrees of freedom is infinite in effect, and interference will completely disappear.

---

[12] An acronym invented by A. J. Leggett.
[13] This argument is often followed by the statement "Therefore, AND is synonymous with OR for all practical purposes". As emphasized in the ensuing paragraph, however, this sort of *for all practical purposes argument* does not resolve the TAO problem.

The disappearance of interference just argued is called the *decoherence* due to the *environment*.[14] This argument will certainly apply to a majority of situations including a real cat. However, since there is no definite boundary between the microscopic and the macroscopic worlds, possibilities should remain for QIMDS to be detectable with a fairly macroscopic object, so long as quantum mechanics is valid. Even if the number of degrees of freedom is formally infinite, it might not be impossible to reduce the number of those which are harmful to QIMDS; an appropriate control (e.g. cooling to a sufficiently low temperature) over the environment could achieve the desired reduction.

These considerations led Leggett to propose the following program around 1980 (in what follows, "QM ≡ quantum mechanics" and "MR ≡ macrorealism"):

(0) Search both experimentally and theoretically for a macroscopic system which is expected, provided that QM remains valid, to show evidence of QIMDS under an appropriately controlled environment.

(1-0) If the experimental result can be interpreted, on the basis of QM, to show evidence of QIMDS, proceed to the step (2) below.

(1-1) If the experimental result unambiguously denies QIMDS against the quantum-mechanical prediction, then QM may be concluded to be invalid for a system as macroscopic as the one in question. Proceed to modify QM in the light of the negative result.

(2) Scrutinize, without invoking QM, whether or not the experimental result is compatible with MR.

(2-0) If it is, the experiment in question can not decide between QM and MR. Go back to the step (0) and refine the experiment.

(2-1) If it is not, one may conclude that MR is not valid but QM remains to be valid for a system as macroscopic as the one in question. Go back to the step (0) to continue the search for still more macroscopic candidates of QIMDS.

A comment is in order on the step (2). However uncomfortable one might feel with macroscopic superposition predicted by QM, it is illegitimate to reject QM on the basis of one's subjective feeling. A way to quantify this discomfort is to adopt MR, on the basis of which one may derive certain inequalities (Leggett–Garg inequalities[15]) to be satisfied by some measurable quantities (time-correlation functions). Furthermore, it can be shown that there are circumstances where QM

---

[14] The environment here includes many internal degrees of freedom of the macroscopic system in question (e.g. the cat) apart from those which are reserved to distinguish the macroscopically distinct states (e.g. to distinguish whether the cat is alive or dead). Of course, interference between microscopically different states can also be affected and often washed out by environment. The thesis of the above traditional argument is that the decoherence is inevitable and fatal in the case of QIMDS.

[15] They correspond and have the same mathematical structure as Bell's inequalities which are appropriate in the Einstein–Podolsky–Rosen problem, namely, testing QM against the local realism.

violates these inequalities. A sufficiently skilful experiment should be able to reveal this discrepancy, thereby deciding between QM and MR. See Chapter 9 for details.

Hidden in this program is the expectation that quantum mechanics will cease to be valid for a sufficiently macroscopic system.[16] The program itself, however, is independent both of this expectation and of a belief that QM is absolutely valid; it is a down-to-earth research program to enlarge the range of applicability of QM step-by-step from the microscopic world to the more macroscopic one. This program is to be called the *Leggett program*.[17]

1.3 What is meant by "macroscopic"?

1.3.1 Intuitive consideration

We have frequently used the word *macroscopic*. In order to avoid confusion, it is necessary to agree on its meaning as it is used in this book. As a starting point let us consider a Young-type experiment, where a pair of distinct states is involved; they represent a particle passing through either the upper or the lower slit. If the distance between the two slits is macroscopic (say 0.1 mm), one might be inclined to regard the pair of states as macroscopically distinct from each other, even if the system in question is a microscopic particle such as an electron or a neutron. What is macroscopic here, however, is a mere distance; the number of particles involved is only one. By contrast, the word "macroscopic" in this book refers to those situations where a large number of particles are involved, or to be more precise, the number N of the *dynamical degrees of freedom* is large.

The phrase "dynamical degrees of freedom" (hereafter to be abbreviated as degrees of freedom) should also be used with caution. Imagine a tennis ball passing through a wall without being squeezed. There can be no objection to calling this phenomenon a macroscopic tunneling; if the ball is regarded as a collection of atoms, the number of degrees of freedom involved in this phenomenon is comparable to the number of atoms. However, the same phenomenon can also be described by a single degree of freedom, namely, the center of mass. These two descriptions are related to each other by a transformation of variables and are mathematically equivalent. On the basis of this example, it could be argued that the number of degrees of freedom involved, which is not invariant under transformations of variables, cannot quantify the word "macroscopic"; a physical conclusion should not depend on the choice

---

[16] See A. J. Leggett, *The Problems of Physics*, Oxford University Press (1987), Chapter 5, Skeletons in the cupboard.

[17] The program announced by Felix Klein (1849–1925) on the occasion of his appointment to a professorship at the University of Erlangen, well known as Erlangen Programm, was so insightful that it played a long-lasting leading role in the synthesis of geometry. The Leggett program, which is still under development, will be regarded by the late twenty-first century physicists to have played a role in physics comparable to the Klein program in twentieth century mathematics.

of a mathematical description. This objection, which might look reasonable at first sight, may be disposed of as follows. Our intuition which regards the above phenomenon as a macroscopic tunneling does not rely on the number of *collective degrees of freedom* such as the center of mass but on the number of *microscopic degrees of freedom* such as the positions of constituent atoms. Collective degrees of freedom are the elite degrees of freedom which are singled out by rearranging the microscopic ones, all of which our intuition leads us to treat on an equal footing. In accordance to our intuition, we adopt the democratic way of counting the number of degrees of freedom, which in general is of the same order as the number of particles composing the system in question. Thus, the above objection is irrelevant. Of course, the number of constituent particles depends on what we count as fundamental particles; N neutrons may be counted as $3N$ quarks, for instance. However, the difference between N and $3N$ is irrelevant as well; the word "macroscopic" may be quantified only by orders of magnitude.

1.3.2 S-Cattiness

Until the 1970s, the phrase macroscopic quantum phenomena represented collectively superfluidity and superconductivity. For example, the phenomenon of liquid He creeping up along the wall of a glass and flowing out of it is both undoubtedly macroscopic and explicable only in terms of quantum mechanics. In this kind of phenomena, however, microscopic interference at the level of one (or two) particles is enhanced by virtue of cooperation of many particles (or, many pairs of particles), resulting in an effect of macroscopic scale; QIMDS is not involved. Today they are often called *macroscopic quantum phenomena of the first kind* and are distinguished from *macroscopic quantum phenomena of the second kind* in which QIMDS is involved.

Let us elaborate on the difference between the two kinds. Consider, as a typical example of the first kind, superfluid ^{4}He flowing out of a glass. The wavefunction representing its state is conceptually of the form:[18]

$$\prod_{k=1}^{N}\{\psi(\mathbf{r}_k - \mathbf{d}/2) + \psi(\mathbf{r}_k + \mathbf{d}/2)\}, \qquad (1.3.1)$$

where $\mathbf{r}_k$ denotes the position of the k-th of the N atoms composing the liquid ^{4}He, $\mathbf{d}/2$ that of the center of the glass, and $-\mathbf{d}/2$ a position outside the glass (Fig. 1.1). Equation (1.3.1) implies that each of the atoms is in a state of superposition $\psi(\mathbf{r} - \mathbf{d}/2) + \psi(\mathbf{r} + \mathbf{d}/2)$ and that the state of the entire liquid is their product.

---

[18] It should be noted that the following expression is merely schematic. Bose–Einstein condensation by itself is not enough to give rise to superfluidity; interaction among atoms is necessary for a state roughly of the following form to be kept stable.

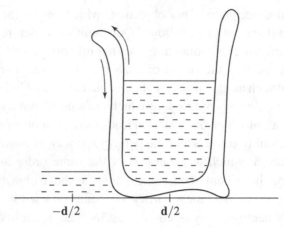

$$-\mathbf{d}/2 \qquad\qquad \mathbf{d}/2$$

Fig. 1.1. Superfluid ^{4}He flowing out of a glass (macroscopic quantum phenomena of the first kind).

It shows that all the atoms are in one and the same "one-particle state". This is nothing but the situation called the *Bose–Einstein condensation*, which realizes the cooperation of many particles as mentioned above.

The representative of macroscopic quantum phenomena of the second kind is, of course, the S-Cat.[19] Unfortunately, however, quantum states representing a real cat are hard to write down. Consider instead a ball passing through double slits. If the ball is assumed to behave as a rigid body, its state is represented conceptually by a wavefunction of the form

$$\prod_{k=1}^{N} \psi_k(\mathbf{r}_k - \mathbf{d}/2) + \prod_{k=1}^{N} \psi_k(\mathbf{r}_k + \mathbf{d}/2), \qquad (1.3.2)$$

where $\mathbf{r}_k$ denotes the position of the k-th of the N atoms composing the ball, and $\pm\mathbf{d}/2$ are the positions of the upper and the lower slit, respectively. In this wavefunction the first and the second terms represent the state in which the center of the ball is located near $\pm\mathbf{d}/2$, respectively (Fig. 1.2).

Each ψ in Eq. (1.3.2) carries the subscript k, which takes account of the fact that each atom occupies a different position within the ball. This is not important, however, in the comparison of (1.3.1) and (1.3.2). What is essential is the difference in the order of the sum (i.e. linear combination) and the product; the sum is the first to be taken in (1.3.1), whereas the product is the first in (1.3.2).

In what follows, a macroscopic quantum phenomenon of the second kind is to be simply called a macroscopic quantum phenomenon and abbreviated as MQP.

---

[19] The present author has not found out why Schrödinger invoked a cat instead of a dog for instance. Perhaps, a cat is more suitable for germinating a sense of strange uneasiness. Also, S-Cat is reminiscent of Carroll's cat in *Alice in Wonderland*.

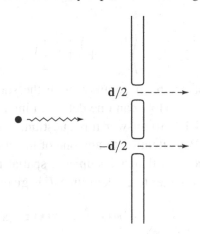

Fig. 1.2. A ball passing through double slits (MQP).

It is desirable to have a measure to quantify the extent to which a given MQP is macroscopic. Such a measure is to be called *S-Cattiness* and denoted by **D**. It may be defined[20] roughly as:

> the maximum number of those democratically-counted degrees of freedom which are involved in an irreducible linear combination,

where an <u>irreducible linear combination</u> is a superposition that can not be factorized into linear combinations involving fewer degrees of freedom. According to this definition, **D** ∼ 1 in the state (1.3.1), **D** ∼ N in the state (1.3.2), and so on. The larger **D** becomes, the closer the given QIMDS will be to the full-fledged S-Cat and the harder to detect experimentally.

1.4 Macroscopic quantum tunneling

1.4.1 Leggett program and macroscopic quantum tunneling

In the previous section, we have briefly described the traditional argument against QIMDS. Let us examine it in more detail. Consider for simplicity a system of N spins $\{\hat{\mathbf{s}}^{(1)}, \hat{\mathbf{s}}^{(2)}, \ldots, \hat{\mathbf{s}}^{(N)}\}$ with each of the spins being of magnitude 1/2, and suppose that it has been prepared in the following state $|\Psi\rangle$ (see Appendix D for the notation concerning spin):

$$|\Psi\rangle := c_+|\Psi_+\rangle + c_-|\Psi_-\rangle, \tag{1.4.1}$$

$$|\Psi_\pm\rangle \equiv \prod_{k=1}^{N}\left|\pm\frac{1}{2}\right\rangle^{(k)}, \quad |c_+|^2 + |c_-|^2 = 1, \tag{1.4.2}$$

---

[20] A more precise mathematical definition has been given by Leggett in Ref. [1] in the bibliography, where the measure was called the *disconnectivity*. Later it was also informally called cattiness.

where

$$\hat{s}_3^{(k)} \left| \pm \frac{1}{2} \right\rangle^{(k)} = \pm \frac{1}{2} \left| \pm \frac{1}{2} \right\rangle^{(k)}. \tag{1.4.3}$$

This state $|\Psi\rangle$ has the structure in common with the state (1.3.2). Accordingly its S-Cattiness is of order N. How can one detect an interference effect predicted by the superposition (1.4.1)? To answer this question, it is helpful to review the Young-type experiment. Pay attention to any one of the counters and suppose that its center is situated at $\mathbf{x}_0$ and that it occupies a spatial region $C(\mathbf{x}_0)$. Then the probability $P(\mathbf{x}_0)$ for the counter to click at time T is given roughly as

$$P(\mathbf{x}_0) = \int_{C(\mathbf{x}_0)} d\mathbf{x} \, |\psi(\mathbf{x})|^2, \quad \psi(\mathbf{x}) \equiv \langle \mathbf{x} | \psi \rangle, \tag{1.4.4}$$

where $|\psi\rangle$, being the state of the particle at the time T, is a superposition of $|\psi_{\pm}\rangle$ which describe the particle having passed through the upper and the lower slit, respectively:

$$|\psi\rangle = |\psi_+\rangle + |\psi_-\rangle. \tag{1.4.5}$$

By use of the *projector*[21] onto the region $C(\mathbf{x}_0)$ defined by

$$\hat{\Pi}_{\mathbf{x}_0} := \int_{C(\mathbf{x}_0)} d\mathbf{x} \, |\mathbf{x}\rangle\langle\mathbf{x}|, \tag{1.4.6}$$

the above probability may be rewritten as

$$\begin{aligned} P(\mathbf{x}_0) &= \langle \psi | \hat{\Pi}_{\mathbf{x}_0} | \psi \rangle \\ &= \langle \psi_+ | \hat{\Pi}_{\mathbf{x}_0} | \psi_+ \rangle + \langle \psi_- | \hat{\Pi}_{\mathbf{x}_0} | \psi_- \rangle + 2\Re \langle \psi_+ | \hat{\Pi}_{\mathbf{x}_0} | \psi_- \rangle. \end{aligned} \tag{1.4.7}$$

This is just a roundabout rephrasing of the elementary result; the last term is what is called the interference term:

$$\langle \psi_+ | \hat{\Pi}_{\mathbf{x}_0} | \psi_- \rangle = \int_{C(\mathbf{x}_0)} d\mathbf{x} \, \psi_+^*(\mathbf{x}) \psi_-(\mathbf{x}), \quad \psi_{\pm}(\mathbf{x}) \equiv \langle \mathbf{x} | \psi_{\pm} \rangle. \tag{1.4.8}$$

The lesson to be learnt from this example is the following:

> There exists an operator (i.e. $\hat{\Pi}_{\mathbf{x}_0}$) such that (a) it corresponds to the experimental procedure of counting a particle, and (b) its off-diagonal element (i.e. (1.4.8)) does not vanish.

It is this condition that renders the Young-type interference effect detectable.

This consideration may be generalized to conclude that the following condition is necessary for the interference between $|\Psi_{\pm}\rangle$ in the superposition (1.4.1) to be detectable:

---

[21] Abbreviation of *projection operator.*

There exists an operator $\hat{O}$ such that (a) it corresponds to the actual measurement, and (b)

$$\langle\Psi_+|\hat{O}|\Psi_-\rangle \neq 0. \qquad (1.4.9)$$

If it were not for an operator with this property, the superposition (1.4.1) would be indistinguishable from the *mixture*[22] represented by the following density operator $\hat{\rho}$:

$$\hat{\rho} = |c_+|^2|\Psi_+\rangle\langle\Psi_+| + |c_-|^2|\Psi_-\rangle\langle\Psi_-|. \qquad (1.4.10)$$

For example, the operator $\hat{O} \equiv \hat{s}_1^{(1)}\hat{s}_1^{(2)}$ does not satisfy the above condition:

$$\langle\Psi_+|\hat{O}|\Psi_-\rangle = \prod_{k=1}^{2}{}^{(k)}\left(\frac{1}{2}\Big|\hat{s}_1^{(k)}\Big|-\frac{1}{2}\right)^{(k)} \prod_{k=3}^{N}{}^{(k)}\left(\frac{1}{2}\Big|-\frac{1}{2}\right)^{(k)} = 0.$$
$$(1.4.11)$$

Hence, no measurement of the physical quantity represented by this operator can reveal the desired interference effect. Any operator of the form of a product of $N-1$ spins or less will not do either. A product involving all of the N spins is needed. However, an experimental realization of such an operator would be impossible for a system with a large N (say, $N \sim 10^{10}$).

It might seem hopeless to dispute this argument. Fortunately, however, any physical system is endowed with a time-evolution operator $\hat{U}(t)$. This operator contains all the powers (i.e. $\hat{H}$, $\hat{H}^2$, $\hat{H}^3$, ...) of the Hamiltonian $\hat{H}$, which in turn contains all the degrees of freedom relevant to the system. Therefore, even if an operator $\hat{O}$ contains only a few degrees of freedom, its time evolution

$$\hat{O}(t) \equiv \hat{U}^\dagger(t)\hat{O}\hat{U}(t) \qquad (1.4.12)$$

can have the property

$$\langle\Psi_+|\hat{O}(t)|\Psi_-\rangle = \langle\Psi_+(t)|\hat{O}|\Psi_-(t)\rangle \neq 0, \quad |\Psi_\pm(t)\rangle \equiv \hat{U}(t)|\Psi_\pm\rangle. \quad (1.4.13)$$

Thus, although an experimenter can not prepare a desired operator, nature can; it is sufficient for an experimenter to measure $\hat{O}$ for the system in the state $|\Psi_\pm(t)\rangle$ instead of the original state $|\Psi_\pm\rangle$. Detection of QIMDS will be possible only if this naturally endowed property is successfully exploited. To start with, of course, $|\Psi_+\rangle$ and $|\Psi_-\rangle$ should be macroscopically distinct from each other. This condition may be guaranteed if a sufficiently large potential barrier separates these two states. In order to exploit (1.4.13), however, $|\Psi_+\rangle$ should be able to evolve in time to $|\Psi_-\rangle$

---

[22] It is often the case that a mixture is called a "mixed state" in contrast to a "pure state" which is synonymous with the phrase "quantum state" as used in this book. Readers are warned not to confuse the word "mixed" in "mixed state" with the word "mixing" used in "mixing angle", "s–p mixing" and so on. In these phrases, "mixing" implies a linear combination; for instance, an "s–p-mixed state" ($\equiv$ "state with s–p mixing") is a "pure state".

and vice versa. Both of these requirements may be met by a situation where the macroscopic system is able to undergo quantum tunneling.

For this reason, macroscopic quantum tunneling, the title of this book, occupies the central position in the Leggett program. To recapitulate, its significance lies in the following features[23] which are obviously interrelated:

- It is genuinely quantum mechanical. Although quantum-mechanical effects are involved in one way or another in any physical phenomena, they may be identified unambiguously in a phenomenon without classical-mechanical analogue. Quantum tunneling is a typical case.
- It guarantees macroscopic distinctness. Two states, which can evolve into each other only via quantum tunneling, are prohibited to do so classically. Hence, an experimenter can unambiguously distinguish them with a macroscopic parameter by preparing them in a nearly classical situation.

Before closing this subsection, we should dispose of an opinion, which insists that it is impossible to detect quantum tunneling in a macroscopic system. The argument goes as follows:[24]

In general the probability of tunneling is proportional to $\exp(-2S/\hbar)$, where S is the action characteristic of the tunneling in question.[25] Let N be the number of degrees of freedom involved in the tunneling, then

$$S/\hbar \sim N \gg 1. \qquad (1.4.14)$$

Thus, the probability of tunneling, which decreases exponentially as N increases, practically vanishes for a macroscopic system. Furthermore, a system may be said to be macroscopic with rigor only in the limit

$$N \to \infty. \qquad (1.4.15)$$

In this limit, the probability of tunneling vanishes rigorously.

This opinion would certainly be right if the word "macroscopic" is defined by the condition (1.4.14). Although this condition is expected to hold in many cases, it does not necessarily agree with what we mean intuitively by "macroscopic". Even if many particles are involved and the two states in question are macroscopically distinct, the potential barrier separating the two states might not necessarily be large. Such a situation as

$$S/\hbar = \mathcal{O}(1) \qquad (1.4.16)$$

---

[23] These are necessary for MQP in general, but quantum tunneling is not. A candidate of MQP without tunneling is a superposition of macroscopically distinct coherent states of light (i.e. electromagnetic field), which, however, is outside the scope of this book.

[24] This is another example of a *for all practical purposes* argument. cf. Footnote 13.

[25] This action is proportional roughly to the square root of the area of the potential barrier.

could be realized; even if S is formally proportional to the large number N, the coefficient of proportionality could be controlled to be small. The situation of interest in this book is not the mathematical limit (1.4.15) but a realistic circumstance where the condition (1.4.16) holds in spite of $N \gg 1$.

1.4.2 Classification of macroscopic quantum tunneling: MQC and MQT

Given a macroscopic system, let R be the set of relevant macroscopic degrees of freedom (i.e. collective degrees of freedom) to describe its quantum tunneling. The number of such degrees of freedom need not be one. For example, in the case of the ball discussed in the preceding section, R is the center-of-mass position which is a three-component vector, that is, R consists of three degrees of freedom. The entire number of degrees of freedom of the given macroscopic system may be rearranged into R and the rest. For simplicity, suppose that R consists of a single degree of freedom and that its quantum-mechanical behavior is governed by the Schrödinger equation of the same form as that for a particle:

$$i\hbar \frac{d}{dt}|\psi(t)\rangle = \hat{H}_S|\psi(t)\rangle, \tag{1.4.17}$$

$$\hat{H}_S := \frac{1}{2M}\hat{P}^2 + U(\hat{R}), \tag{1.4.18}$$

$$[\hat{R}, \hat{P}] = i\hbar, \tag{1.4.19}$$

where M is a positive constant (effective inertial mass), $\hat{P}$ is the momentum operator conjugate to the position operator $\hat{R}$, and $U(\hat{R})$ is an appropriate potential to be specified later. Hereafter, the macroscopic degrees of freedom R is to be called the *macrosystem* as distinguished from the original macroscopic system as a whole; the set of the remaining degrees of freedom is called the *environment* (i.e. the environment for R). Accordingly the above $\hat{H}_S$, which refers to the macrosystem alone, is to be called the *macrosystem Hamiltonian*.[26] The Hamiltonian for the entire system consists of three parts: the macrosystem Hamiltonian, the part referring to the environment alone, and the part specifying the interaction of the macrosystem with the environment. It is, therefore, not obvious whether the fundamental quantum-mechanical description can be reduced to the form (1.4.17)–(1.4.19) which is closed with respect to R. This issue is to be discussed in detail in the fourth and subsequent chapters. For the moment, we assume that the closed form (1.4.17)–(1.4.19) is valid; the aim of this subsection and the following two chapters is to achieve a conceptual understanding of macroscopic quantum tunneling.

Macroscopic quantum tunneling is grossly classified into MQC (macroscopic quantum coherence) and MQT (macroscopic quantum tunneling in the narrow

---

[26] The subscript S stands for the system, namely, not the entire macroscopic system but the macrosystem which is of primary interest.

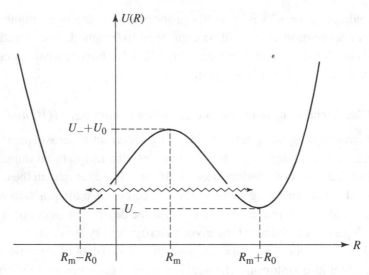

Fig. 1.3. MQC-situation: symmetric double well.

sense). In a phenomenon classified as MQC, a macrosystem R oscillates in time
between macroscopically distinct states. Such an oscillation may be said to be
a temporal counterpart of the Young-type interference pattern that is spatial (see
Section 2.2); QIMDS is involved directly. Thus, MQC is a convenient substitute
for Young-type experiments, which are presumably difficult for a macrosystem. As
mentioned in the preceding subsection, one need not construct a delicate double-slit
apparatus to produce a desired macroscopic superposition, which may be prepared
naturally by the time evolution of the macrosystem itself. A typical potential giving
rise to a MQC-situation is the symmetric double well (Fig. 1.3). The central barrier is
supposed to be so large that the inter-well distance $2R_0$ is a macroscopic quantity and
that the macrosystem initially localized in the left well cannot go over to the right
well classically but can do so only via quantum tunneling. In this way, the <u>left state</u>
(i.e. the state in which the macrosystem is localized in the left well) is guaranteed to
be macroscopically distinct from the <u>right state</u>. The oscillation of the macrosystem
between the two wells is analogous to the classical-mechanical resonance of a pair
of coupled pendulums or tuning forks with a common eigenfrequency. Hence, MQC
may as well be called *macroscopic quantum resonant oscillation*.

   In a phenomenon classified as MQT, on the other hand, a macrosystem R tunnels
only once from a state to another with the latter being macroscopically distinct from
the former. A typical potential giving rise to a MQT-situation is the *bumpy slope*[27]
(Fig. 1.4), where the barrier is supposed to be sufficiently large as in the MQC-
situation. If the macrosystem is initially localized in the well, it cannot escape towards the right of the barrier classically but can do so only via quantum tunneling.

----

[27] A better nickname may be desirable.

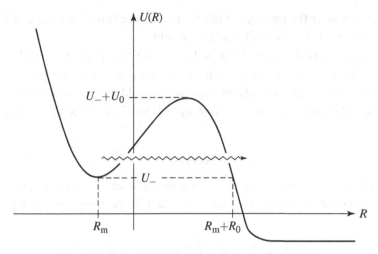

Fig. 1.4. MQT-situation: bumpy slope.

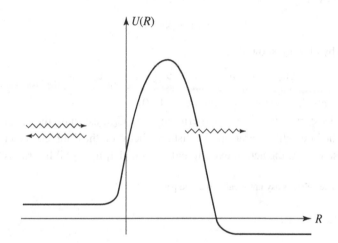

Fig. 1.5. Free-tunneling situation: simple barrier.

This situation is analogous to the α-decay of a nucleus, for instance; the initial state is *metastable*, and the well around $R = R_m$ is called the *metastable well* . Thus, MQT may as well be called *macroscopic quantum decay* of a *metastable state*. As in the case of the MQC-situation, the large barrier guarantees that this metastable state (i.e. the state in which the macrosystem is localized in the metastable well) is macroscopically distinct from the decayed state (i.e. the state in which the macrosystem has escaped to the right of the barrier). Although MQT cannot provide direct evidence for QIMDS (see Chapter 9), its detection would dramatically demonstrate that a macrosystem can exhibit a quantum-mechanical behavior without a classical analogue.

Perhaps the most familiar type of quantum tunneling is the situation depicted in Fig. 1.5, which may be called the <u>free tunneling</u>, since no force acts on the particle either before or after the passage of the barrier.[28] It seems, however, difficult to set up a corresponding situation for a macrosystem.

Incidentally, MQT is often described as a free tunneling followed by several oscillations inside the metastable well. In a situation where this popular account is valid, however, the probability of tunneling would be too small for MQT to be detected. It is therefore advisable to distinguish MQT from free tunneling.

Exercises

Exercise 1.1. Devise a mathematically precise definition of the S-Cattiness (cf. Ref. [1]).

Exercise 1.2. Given a state $|\Psi\rangle$, the density operator $\hat{\rho}_\Psi$ representing it is defined by the property

$$\text{Tr}(\hat{\rho}_\Psi \hat{A}) = \langle\Psi|\hat{A}|\Psi\rangle \quad \text{for an arbitrary operator } \hat{A}, \tag{1.4.20}$$

or equivalently as

$$\hat{\rho}_\Psi := |\Psi\rangle\langle\Psi|. \tag{1.4.21}$$

For $|\Psi\rangle$ given by (1.4.1), show that

$$\hat{\rho}_\Psi = \hat{\rho} + c_+ c_-^* |\Psi_+\rangle\langle\Psi_-| + c_- c_+^* |\Psi_-\rangle\langle\Psi_+|, \tag{1.4.22}$$

where $\hat{\rho}$ represents the mixture as given by (1.4.10).

Exercise 1.3. Design a Young-type experiment with C_{60}; estimate the required orders of magnitude of the inter-slit separation, the distance between the double slits and the array of counters, and so on (cf. the actual experiment in Ref. [10] in the bibliography).

---

[28] Hence, the particle behaves asymptotically as a free particle.

2

Overview of macroscopic quantum tunneling

This chapter summarizes those general aspects of MQC and MQT which are independent of details specific to individual macrosystems.

2.1 Standard form of Hamiltonian and Schrödinger equation: dimensional analysis

The first thing to be done is the dimensional analysis which identifies the dimensionless parameters to quantify the extent to which a given system is expected to behave quantum mechanically. In both MQT- and MQC-situations, the potential has the characteristic length[1] R_0 and the characteristic energy U_0 (see Figs. 1.3 and 1.4), which we adopt as the units of length and energy, respectively. The corresponding characteristic time τ_0 may be estimated as the time needed for a particle of mass M to traverse the distance R_0 at a constant speed with the kinetic energy of the order of U_0:

$$(R_0/\tau_0)^2 M = U_0. \tag{2.1.1}$$

Being determined by the height and the width of the potential barrier, τ_0 may be called the characteristic time of tunneling.[2] We thus adopt

$$\tau_0 := \frac{R_0}{(U_0/M)^{1/2}} \tag{2.1.2}$$

as the unit of time. Likewise, the unit of momentum is to be taken as $P_0 := (MU_0)^{1/2}$. Hereafter we measure all quantities in these units and define the dynamical variables q and p, the potential $V(q)$, the Hamiltonian $\hat{H}_S$ and the time t, all of which are

---

[1] Actual dimension of R depends on the system in question (see Chapter 3). In this chapter, it is supposed to be that of the length. Accordingly, M in Eq. (1.4.18) has the dimension of mass.

[2] It is often called a "tunneling time". At this stage, however, its relationship to tunneling, if any, is Eq. (2.1.1) alone.

dimensionless, as follows:

$$q := (R - R_{\mathrm{m}})/R_0, \quad p := P/P_0, \quad t := t_{\mathrm{conv}}/\tau_0, \qquad (2.1.3)$$

$$V(q) := (U(R_0 q + R_{\mathrm{m}}) - U_-)/U_0, \quad \hat{H}_S := \hat{H}_{S\mathrm{conv}}/U_0, \qquad (2.1.4)$$

where $\hat{H}_{S\mathrm{conv}}$ and t_{conv} are the same as the conventional $\hat{H}_S$ and t in (1.4.17). Accordingly, Eqs. (1.4.17)–(1.4.19) can be rewritten as

$$\mathrm{i}h\frac{\mathrm{d}}{\mathrm{d}t}|\psi(t)\rangle = \hat{H}_S|\psi(t)\rangle, \qquad (2.1.5)$$

$$\hat{H}_S = \frac{1}{2}\hat{p}^2 + V(\hat{q}), \qquad (2.1.6)$$

$$[\hat{q}, \hat{p}] = \left(MU_0 R_0^2\right)^{-1/2}[\hat{R}, \hat{P}] = \mathrm{i}h, \qquad (2.1.7)$$

where

$$h := \frac{\hbar}{U_0 \tau_0} = \frac{\hbar}{P_0 R_0} = \left\{\frac{\hbar^2/MR_0^2}{U_0}\right\}^{1/2}. \qquad (2.1.8)$$

The position representations (q-representation) of Eqs. (2.1.5) and (2.1.6) are given by

$$\mathrm{i}h\frac{\partial}{\partial t}\psi(q, t) = H_S\psi(q, t), \quad \psi(q, t) \equiv \langle q|\psi(t)\rangle, \qquad (2.1.9)$$

$$H_S := -\frac{h^2}{2}\frac{\partial^2}{\partial q^2} + V(q). \qquad (2.1.10)$$

The position representation $\psi(q, t)$ of the state $|\psi(t)\rangle$ is called the *position-represented wavefunction*, which is to be abbreviated as the *wavefunction*. Equations (2.1.5)–(2.1.7) or Eqs. (2.1.9)–(2.1.10) are to be called the underlined standard form[3] of the Schrödinger equation and the Hamiltonian. The dimensionless parameter h represents Planck's constant measured in units of the action $U_0 \tau_0 (= P_0 R_0)$ characterizing the tunneling in question.[4] It quantifies the extent to which the situation of tunneling in question is quantum-mechanical. As shown in Eq. (2.1.8), it is the ratio of U_0 and the zero-point energy of the particle confined to a region whose size is of the order of R_0. The situation where the inequality

$$h \ll 1 \qquad (2.1.11)$$

holds is called the *quasi-classical situation*.

---

[3] Beware of the underlined section; this standard form is a particular dimensionless form resulting from our choice of units. Other choices are possible. For example, a purely quantum-mechanical person would prefer to work with units such that h in Eqs. (2.1.9)–(2.1.10) is replaced by 1. A classical person, on the other hand, is anxious to keep track of h which indicates where quantum mechanics is manifestly involved.

[4] Those readers who customarily use h for $2\pi\hbar$ might feel uneasy with the notation (2.1.8). However, h is not synonymous with $2\pi\hbar$; the symbol h stands for the height of a triangle in plane geometry, the dimensionless Hubble constant in cosmology, and so on. Without following the convention $h \equiv 2\pi\hbar$, we use h in place of a cumbersome notation $\hbar_{\mathrm{dimensionless}}$.

(a)   (b)

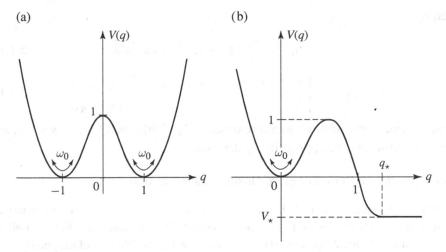

Fig. 2.1. Standardized potentials (a) symmetric double well, and (b) bumpy slope.

All the phenomena treated in this book are supposed to satisfy the above condition. It is then natural to arrange wavefunctions and physical quantities according to their orders of magnitude with respect to h. We use the following notation:

$$\mathcal{O}(1) \equiv \mathcal{O}(h^0), \tag{2.1.12}$$

$$\mathcal{O}(e^{-1/h}) \equiv \mathcal{O}(e^{-B/h}) \quad : B \text{ is a positive constant of } \mathcal{O}(1). \tag{2.1.13}$$

Of course, quantum tunnelings are impossible to detect if h is too small. Its typical envisaged value is about 0.1. In some cases, however, a fairly small value such as 0.01 turns out to be enough for the detection, which may be achieved by observing a large number of events (e.g. the case of quantum nucleation discussed in Section 3.2).

The scale of q-dependence of the <u>standardized potentials</u>, namely $V(q)$ related to $U(R)$ via Eq. (2.1.4), is supposed to be of $\mathcal{O}(1)$ with the curvature at the minima being denoted by ω_0^2 in both MQC- and MQT-situations (Fig. 2.1):

$$V(q) \simeq \begin{cases} \frac{1}{2}\omega_0^2(q \pm 1)^2 & : q \sim \mp 1 \text{ (symmetric double well)}, \\ \frac{1}{2}\omega_0^2 q^2 & : q \sim 0 \text{ (bumpy slope)}, \end{cases} \tag{2.1.14}$$

where ω_0 is a positive number of $\mathcal{O}(1)$. The bumpy slope is supposed to take a negative constant value $V_\star$ for $q > q_\star$ with $q_\star$ being a constant greater than 1.[5] The simplest analytic forms corresponding to Fig. 2.1 (a) and (b) are the fourth-order polynomial V^{SSDW} and the truncated third-order polynomial V^{SBS}, respectively,[6]

[5] This assumption is made merely for simplicity; the potential has only to tend to $V_\star$ sufficiently rapidly as $q \to \infty$.
[6] SSDW $\equiv$ <u>s</u>pecial <u>s</u>ymmetric <u>d</u>ouble <u>w</u>ell, SBS $\equiv$ <u>s</u>pecial <u>b</u>umpy <u>s</u>lope.

given by

$$V^{\text{SSDW}}(q) := \frac{\omega_0^2}{8}(q^2 - 1)^2, \quad \omega_0 \equiv \sqrt{8}, \tag{2.1.15}$$

$$V^{\text{SBS}}(q) := \begin{cases} \frac{1}{2}\omega_0^2 q^2(1 - q), & \omega_0 \equiv \sqrt{27/2} : q < q_\star, \\ V_\star & : q > q_\star. \end{cases} \tag{2.1.16}$$

These special forms turn out to approximate $V(q)$ fairly well for many macrosystems if experiments are appropriately designed.

A few remarks are in order before closing this subsection.

  (i) Even in the quasi-classical situation, it is possible for a system to be in a genuinely quantum-mechanical state (i.e. a state without a classical analogue). For example the system may be localized around the bottom of the metastable well of the bumpy slope; such a state, which is essentially the same as the ground state in the harmonic potential $\omega_0^2 q^2/2$, cannot be described in classical terms.

 (ii) Occasionally readers might encounter such an informal expression as "the system behaves semi-classically if $\hbar$ is small". With $\hbar$ not being dimensionless, however, the unqualified statement " $\hbar$ is small" is meaningless as it stands;[7] it is to be understood as an abbreviation of (2.1.11).

(iii) In spite of the definition (2.1.8), h is not simply proportional to $\hbar$ in general. For instance, in the case of the superconducting quantum interference device (SQUID) (see Section 3.1.1), the inertial mass M represents the electrostatic capacitance of the Josephson junction and implicitly contains $\hbar$. Moreover, the quasi-classical situation (2.1.11) is to be distinguished from a fictitious world with "$\hbar \sim 0$ (i.e. $e^2/\hbar c \gg 1$)". Such a world, which it might be amusing to ponder,[8] allows neither SQUID nor hydrogen atoms to exist and has nothing to do with this book.

2.2 Overview of MQC

Generalizing Fig. 2.1(a), let us consider the tilted double well $V(q; \varepsilon)$ as depicted in Fig. 2.2, where the broken curve shows the original symmetric double well, namely $V(q; 0)$. Let $|-\rangle$ be the state in which the macrosystem is localized in the left well and $|+\rangle$ be that in the right well. The state $|-\rangle$ is supposed to be essentially the same as the ground state of the left well treated by itself, and similarly for $|+\rangle$. Furthermore, they are supposed to be orthonormalized:

$$\langle +|+\rangle = \langle -|-\rangle = 1, \quad \langle +|-\rangle = 0. \tag{2.2.1}$$

The wavefunctions $\psi_\pm(q) \equiv \langle q|\pm\rangle \, (\in \mathbf{R})$ representing them are shown in Fig. 2.3.[9] Confining attention to the subspace spanned by these two states alone, that is,

---

[7] It is as meaningless as "1 cm is long" or "20 K is a low temperature".
[8] See e.g. J. D. Barrow and F. J. Tipler, *The Anthropic Cosmological Principle*, Clarendon Press (1986), Section 4.5.
[9] The phases of $|\pm\rangle$ are to be so chosen that $\psi_\pm(q)$ are positive for almost all q.

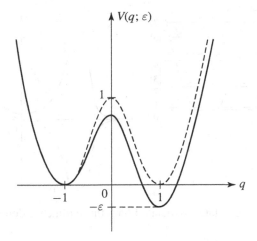

Fig. 2.2. Tilted double well.

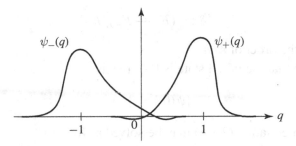

Fig. 2.3. States localized in the left or the right of the tilted double well.

making the two-state approximation,[10] one may express the Hamiltonian $\hat{H}_S$ as

$$\hat{H}_S = E_+|+\rangle\langle+| + E_-|-\rangle\langle-| - \frac{1}{2}h\Delta\left(|+\rangle\langle-| + |-\rangle\langle+|\right), \qquad (2.2.2)$$

where $E_\pm$ are the ground-state energies associated with the right (left) well, respectively, and Δ is a positive constant.[11] If Δ were 0, the left and the right world would be mutually independent; Δ is a measure of the strength of the tunneling connecting the two worlds. Readjusting the origin of energy to be at $(E_+ + E_-)/2$, one may rewrite the above expression as

$$\hat{H}_S = h\hat{S}_3\Delta - h\hat{S}_2\delta, \qquad (2.2.3)$$

$$\hat{S}_3 := -\frac{1}{2}(|+\rangle\langle-| + |-\rangle\langle+|), \qquad (2.2.4)$$

$$\hat{S}_2 := \frac{1}{2}(|+\rangle\langle+| - |-\rangle\langle-|), \qquad (2.2.5)$$

---

[10] In general, a system whose states lie in a two-dimensional Hilbert space is called a two-state system.
[11] To be precise, it is possible to choose the relative phase between $|\pm\rangle$ so that Δ is positive; Fig. 2.3 represents this choice.

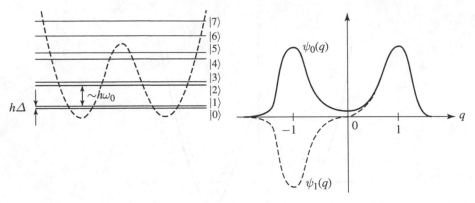

Fig. 2.4. Eigenstates associated with the symmetric double well.

where

$$\delta := (E_- - E_+)/h \qquad (2.2.6)$$

is a quantity of the order of ε/h.

Now, suppose that the initial state is $|-\rangle$, that is,

$$|\psi(0)\rangle = |-\rangle. \qquad (2.2.7)$$

The Schrödinger equation (2.1.5) may be solved as follows.

(i) Case of $\delta = 0$: The ground state $|0\rangle$, the excited state $|1\rangle$ and the associated eigenvalues E_0 and E_1 of $\hat{H}_S$ are found as

$$|1\rangle = (|+\rangle - |-\rangle)/\sqrt{2}, \quad E_1 = +h\Delta/2, \qquad (2.2.8)$$
$$|0\rangle = (|+\rangle + |-\rangle)/\sqrt{2}, \quad E_0 = -h\Delta/2. \qquad (2.2.9)$$

It follows that $E_1 - E_0 = h\Delta$ (Fig. 2.4); Δ is called the *ground-state energy splitting* (or the *tunnel splitting* of the ground-state energy).[12] Hence, one finds that

$$|\psi(t)\rangle = \exp(-i\hat{H}_S t/h)|-\rangle = \frac{1}{\sqrt{2}}\left(e^{it\Delta/2}|0\rangle - e^{-it\Delta/2}|1\rangle\right)$$

$$= \frac{1}{2}\left\{\left(e^{it\Delta/2} + e^{-it\Delta/2}\right)|-\rangle + \left(e^{it\Delta/2} - e^{-it\Delta/2}\right)|+\rangle\right\}. \qquad (2.2.10)$$

Pay attention to the coefficient of $|+\rangle$, for example, in the last line. It is a sum of the coefficients of $|0\rangle$ and $|1\rangle$ in the preceding line. This corresponds to the superposition of the upper and the lower waves in the Young-type interference; the states $|0\rangle$ and $|1\rangle$

---

[12] In the absence of tunneling (i.e. $\Delta = 0$), the ground state is doubly degenerate; the two states $|\pm\rangle$ have a common energy. This degeneracy is lifted by the tunnel effect and the degenerate energy level is split into two levels: hence the word "splitting".

play the role of the double slits. Let $\mathcal{P}_+(t)$ be the probability for the macrosystem to be found in the state $|+\rangle$ at time t, then

$$\mathcal{P}_+(t) = \{\sin(t\Delta/2)\}^2. \tag{2.2.11}$$

(ii) Case of $0 < |\delta| \ll \omega_0$: Although one may begin with diagonalizing $\hat{H}_S$ as in the previous case, the following intuitive method is more illuminating. Let

$$\hat{S}_1 := \frac{1}{i}[\hat{S}_2, \hat{S}_3] = \frac{i}{2}(|+\rangle\langle-| - |-\rangle\langle+|), \tag{2.2.12}$$

then the set $\{\hat{S}_a \mid a = 1, 2, 3\}$ constitutes a spin operator of magnitude $1/2$. Introducing a three-dimensional Euclidean space $\mathbf{E}^3$ endowed with the orthonormal basis

$$\{\mathbf{e}_j \mid j = x, y, z : \mathbf{e}_j \cdot \mathbf{e}_k = \delta_{jk}\}, \tag{2.2.13}$$

one may define vectors $\hat{\mathbf{S}}$ and $\mathbf{B}$ as

$$\hat{\mathbf{S}} := \hat{S}_1\mathbf{e}_x + \hat{S}_2\mathbf{e}_y + \hat{S}_3\mathbf{e}_z, \tag{2.2.14}$$

$$\mathbf{B} := -\mathbf{e}_y\delta + \mathbf{e}_z\Delta. \tag{2.2.15}$$

The Hamiltonian can then be cast into the form

$$\hat{H}_S = h\mathbf{B} \cdot \hat{\mathbf{S}}, \tag{2.2.16}$$

which may be regarded as the energy of a spin[13] located in the spatially-uniform magnetic field $\mathbf{B}$. The expectation value of the spin

$$\mathbf{S}(t) \equiv \langle\hat{\mathbf{S}}\rangle(t) := \langle\psi(t)|\hat{\mathbf{S}}|\psi(t)\rangle \tag{2.2.17}$$

undergoes precession around the $\mathbf{B}$-axis. The initial condition (2.2.7) implies that

$$\mathbf{S}(0) = -\frac{1}{2}\mathbf{e}_y. \tag{2.2.18}$$

Therefore, $\mathbf{S}(t)$ traces out a cone of vertical angle 2ϑ, where ϑ is the angle between $\mathbf{B}$ and $-\mathbf{e}_y$ (Fig. 2.5). It follows that

$$\max_t S_2(t) = -\frac{1}{2}\cos 2\vartheta. \tag{2.2.19}$$

This equation combined with the identity

$$\mathcal{P}_+(t) = \frac{1}{2} + S_2(t) \tag{2.2.20}$$

gives

$$\max_t \mathcal{P}_+(t) = (\sin\vartheta)^2 = \frac{\Delta^2}{\Delta^2 + \delta^2}. \tag{2.2.21}$$

---

[13] To be precise, the energy of the magnetic moment associated with the spin; hereafter, the magnitude of the magnetic moment is to be absorbed in the magnetic field.

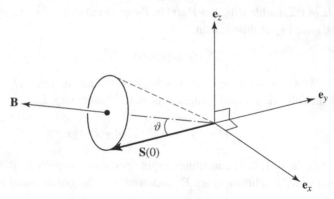

Fig. 2.5. Precession of the fictitious spin.

If a double well prepared in the laboratory is not perfectly symmetric, that is, if $\varepsilon \neq 0$ and the situation is slightly off resonance, the amplitude of the quantum resonant oscillation is suppressed by the factor given on the right-hand side of Eq. (2.2.21). As shown in the next section, typical magnitudes of Δ are of $\mathcal{O}(e^{-1/h})$. Since it is expected to be rather difficult to control ε/h (and hence δ) to take such a small value, one is likely to be in the situation of $\delta \gg \Delta$ and fail to detect the quantum resonant oscillation. This is the first difficulty to be overcome by an experimenter who tries to detect MQC.

So far, we have viewed the quantum resonant oscillation as a precession of the fictitious spin. In terms of the original variable q, this precession should emerge as a periodic motion of the wave packet initially localized in the left well; in the case of $\delta = 0$, for example, the wavefunction representing the state (2.2.10) is given as

$$\psi(q, t) \simeq \cos(t\Delta/2)\, \psi_-(q) + \mathrm{i} \sin(t\Delta/2)\, \psi_+(q). \qquad (2.2.22)$$

The associated probability current $J(q, t)$ does not vanish even under the barrier (i.e. around $q = 0$) and changes its sign with period $2\pi/\Delta$, as it should:

$$J(q, t) := \Re \left\{ \psi^*(q, t) \frac{h}{\mathrm{i}} \frac{\partial}{\partial q} \psi(q, t) \right\} \qquad (2.2.23)$$

$$\simeq \frac{h}{2} \left\{ \psi_-(0)\psi_+'(0) - \psi_+(0)\psi_-'(0) \right\} \sin(t\Delta) : q \sim 0. \qquad (2.2.24)$$

Incidentally, as the potential depicted in Fig. 2.2 is tilted further (i.e. as ε increases), a situation arises in which the ground-state energy associated with the left well coincides with the first-excited-state energy associated with the right well. In this situation, quantum resonant oscillation is possible again. Its amplitude is given, in place of the right-hand side of Eq. (2.2.21), by

$$\frac{\tilde{\Delta}^2}{\tilde{\Delta}^2 + (\delta - \Omega)^2} \quad : \tilde{\Delta} \sim \Delta, \quad \Omega \sim \omega_0. \qquad (2.2.25)$$

This situation is a variant of the MQC-situation discussed above; the result (2.2.25) may be derived by invoking an appropriate two-state system which is different from but analogous to the one used above.

There is another difficulty to be overcome by an experimenter searching for MQC; the quantum resonant oscillation could be destroyed by the very act of detecting it, since the detector necessarily becomes entangled[14] with the macrosystem in question. This is a fundamental difficulty, which is inevitable as long as quantum mechanics is valid. However, the extent of the destruction depends on the experimental setup; it should be possible to devise an ingenious experiment to detect the quantum resonant oscillation while <u>partially destroying it</u> (see Chapter 9).

2.3 Overview of MQT

Given a phenomenon of macroscopic quantum decay, what is the metastable state[15] to be taken as the initial state? To answer this question, it is necessary to know the previous history; that is, how the macrosystem has arrived at the purported metastable state. This is out of the scope of the present discussion,[16] which deals with the quantum decay on the basis of the Schrödinger equation (2.1.5) assuming that an appropriate initial state is given; we have to be content with adopting a physically reasonable quantum state localized in the metastable well as a model for the initial metastable state.

The ground state associated with the metastable well in Fig. 2.1(b) is to be called the *virtual ground state* (see Appendix B for the precise definition). It is nearly the same as the ground state associated with the harmonic potential $\omega_0^2 q^2 / 2$; its wavefunction is a Gaussian function centered at the origin. Hereafter we approximate the initial metastable state with the virtual ground state, which is to be denoted by $|\psi_0\rangle$. The problem of the quantum decay then reduces to solving the Schrödinger equation (2.1.5) with $|\psi(0)\rangle = |\psi_0\rangle$ as the initial condition. The probability $\mathcal{P}(t)$ for the macrosystem to be found in the virtual ground state at time t is to be called

---

[14] In the sense of the entangled state as mentioned in Section 1.1.

[15] The state mentioned in Section 1.4.2. This word refers to a state which can be prepared in the laboratory as opposed to the virtual ground state (see below) which is a theoretical concept.

[16] The situation does not change even if environment is taken into account as well. In general, a discussion on the initial condition itself has necessarily to consider states preceding the initial time, and this consideration in turn has necessarily to refer to states preceding them, and *ad inf.*

the *persistence probability*[17] at time t:

$$\mathcal{P}(t) := \int_{-\infty}^{1} \mathrm{d}q \, |\psi(q, t)|^2. \tag{2.3.1}$$

The upper limit of integration has been chosen to be 1 merely for convenience; any positive number of $\mathcal{O}(1)$ will do. Similarly for the lower limit; $-\infty$ may be replaced by a negative number of order -1.[18] This probability is to be interpreted physically as the probability that the macrosystem has not yet decayed at time t. An explicit calculation gives (see Appendix B)

$$\mathcal{P}(t) \simeq \exp(-\Gamma t), \tag{2.3.2}$$

where Γ is a positive constant of $\mathcal{O}(e^{-1/h})$ (see Eq. (2.4.5) in the next section). This expression satisfies the equation

$$\mathrm{d}\mathcal{P}(t)/\mathrm{d}t \simeq -\Gamma \mathcal{P}(t) < 0, \tag{2.3.3}$$

which appears to describe a classical stochastic process. On the basis of this result, Γ is called the *decay rate* (and, equivalently, Γ^{-1} is called the *lifetime*) of the virtual ground state. The state $|\psi(t)\rangle$, called the *decaying state* for brevity, describes how the virtual ground state decays. Let us formally denote with $|\text{out}\rangle$ the state in which the macrosystem is found outside of the metastable well (i.e. to the right of the barrier in Fig. 2.1(b)). Then, the decaying state may be written schematically[19] as

$$|\psi(t)\rangle = e^{-\Gamma t/2}|\psi_0\rangle + (1 - e^{-\Gamma t})^{1/2}|\text{out}\rangle, \tag{2.3.4}$$

where $|\text{out}\rangle$ is assumed to be normalized (i.e. $\langle\text{out}|\text{out}\rangle = 1$). The right-hand side is the macroscopic superposition in the case of MQT.

A remark is in order. The notion of the "decay rate" (i.e. the probability of decay "per unit time") is alien to quantum mechanics. The latter predicts a *probability for an event to be found at a moment of time* only.[20] We depict the details of $\mathcal{P}(t)$ schematically in Fig. 2.6, where $\bar{\epsilon}$ denotes the indeterminacy of energy in the initial state $|\psi_0\rangle$. Since $|\psi_0\rangle$ is the virtual ground state, $\bar{\epsilon}$ must be much smaller than the separation among the energy levels associated with the harmonic potential which approximates the metastable well:

$$\bar{\epsilon} \ll h\omega_0. \tag{2.3.5}$$

---

[17] Often called "survival probability".

[18] Regardless of specific choices, $\mathcal{P}(0)$ is very close to 1.

[19] It should be noted that this is only schematic. See Appendix B for the precise expression.

[20] Readers may have learned that Fermi's golden rule gives a "transition rate" in general. Fermi's golden rule as applied to quantum decay, however, is equivalent to the computation of $(1 - \mathcal{P}(t))/t$ in the temporal domain (2.3.6), where $\mathcal{P}(t)$ is a probability at a moment of time as defined above.

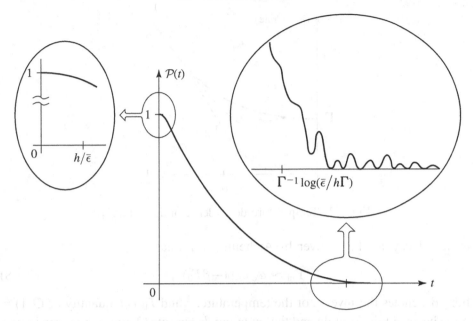

Fig. 2.6. Persistence probability.

Since a quasi-classical situation is under consideration, there exists a temporal domain such that

$$h/\bar{\epsilon} \ll t \ll \Gamma^{-1} \log(\bar{\epsilon}/h\Gamma). \tag{2.3.6}$$

Equation (2.3.2) gives an approximate formula valid only in this domain. Outside this domain, $\mathcal{P}(t)$ deviates from the form (2.3.2); it is not even a monotonically decreasing function of t. If the decay phenomenon in question were a classical stochastic process, $\mathcal{P}(t)$ could represent a "probability distribution for the time t at which the decay occurs". Such an interpretation, however, is impossible in the case of quantum decay. Indeed, the long-time tail of $\mathcal{P}(t)$ decreases slowly, typically[21] as $1/t^2$. If one were to insist on the above interpretation and compute the average value of t, one would not obtain the lifetime Γ^{-1}:

$$\int_0^\infty dt \, t \, \mathcal{P}(t) \bigg/ \int_0^\infty dt \, \mathcal{P}(t) = \infty. \tag{2.3.7}$$

So far we have been considering the limiting situation of the absolute zero temperature. On the other hand, if the temperature is high enough, the system initially in the metastable state may climb over the free-energy barrier with the help of thermal fluctuation. This mechanism of decay is to be called the _classical decay via thermal fluctuation_. Let F_0 be the height of the free-energy barrier, then the

---

[21] Detailed shape of the tail is sensitive to the choice of the initial state. This is a familiar circumstance in general decay phenomena not necessarily involving tunnel effect. For example, the tail may decrease as t^{-3}.

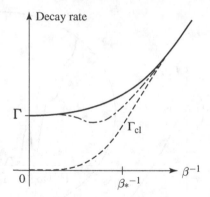

Fig. 2.7. Temperature-dependence of decay rate.

classical decay rate Γ_{cl} is given by Arrhenius' formula:

$$\Gamma_{cl} \sim \omega_0 \exp(-\beta F_0),\qquad\qquad(2.3.8)$$

where β denotes the inverse of the temperature,[22] and F_0 is a quantity of $\mathcal{O}(1)$.[23] On the basis of this formula and the *quantum decay rate* (2.4.5) to be given below, the decay rate as a function of temperature is expected to be roughly of the form of the solid curve in Fig. 2.7.[24] The constant β_*^{-1} in the figure denotes the temperature at which the classical decay rate as extrapolated to the low-temperature region (the broken curve) coincides with the quantum decay rate:

$$\beta_*^{-1} \sim h/B.\qquad\qquad(2.3.9)$$

The main mechanism of decay changes from thermal fluctuation to quantum fluctuation (i.e. tunneling) around the temperature β_*^{-1}, which is therefore called the *crossover temperature*.

An experimental detection of the behavior depicted in Fig. 2.7 would be evidence of MQT. However, the theory of quantum decay at finite temperatures,[25] with which an experimental result is to be compared, is far from being well-established. Since our main interest is in quantum tunneling, we focus our attention on the low-temperature region ($\beta^{-1} \ll \beta_*^{-1}$), where it is legitimate to put $\beta^{-1} = 0$.

---

[22] For brevity, the conventional temperature multiplied by the Boltzmann constant k_B is to be called the temperature; $\beta := 1/(\text{temperature})$. Although the Boltzmann constant was of historical importance, it is a mere conversion factor between units (say, between Kelvin and Joule). With the following re-definition, k_B disappears completely from thermodynamics and statistical mechanics; $(\text{temperature}) := (\text{temperature})_{\text{conv}} k_B$, $(\text{entropy}) := (\text{entropy})_{\text{conv}}/k_B$.

[23] Energy is measured throughout in units of U_0 defined in Fig. 1.4. It is quite a subtle problem to determine the precise value of F_0.

[24] To be precise, Γ in this figure is the rate of quantum decay with environmental effects taken into account (see Chapter 7). The amount of influence of environment could depend on temperature in such a way as to lead to the dot-dashed curve in the figure.

[25] "Finite temperature", which is jargon used frequently, actually means "finite inverse temperature", namely, non-zero temperature.

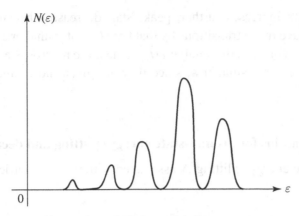

Fig. 2.8. Expected experimental result (schematic) for macroscopic resonant tunneling.

There is a class of phenomena called *macroscopic resonant tunneling*[26] which lies between MQC and macroscopic quantum decay. In a phenomenon of this class, the expected resonant oscillation of amplitude (2.2.25) is not observed due perhaps to influences of environment and/or the measurement apparatus itself, but instead a one-way transition from the left well to the right well is detected. This is called "resonant" tunneling because it occurs only when the ground-state energy associated with the left well coincides nearly with an excited-state energy associated with the right well. These phenomena are in the same status as macroscopic quantum decay in the sense that it cannot be a direct evidence for QIMDS (cf. Chapter 9). Hence it is appropriate to classify them as a subclass of MQT. A typical experiment of macroscopic resonant tunneling will proceed as follows. First, an ensemble of macrosystems[27] is prepared. At the initial time 0, the double well is tilted leftwards ($\varepsilon = \varepsilon(0) < 0$) so that all the macrosystems are found in the left-well ground state $|-\rangle$. Then, ε is increased gradually:[28]

$$0 < \dot{\varepsilon}(t)/|\varepsilon(0)| \ll \omega_0 . \tag{2.3.10}$$

As ε is swept in this way, the double well becomes symmetric ($\varepsilon(t) = 0$), slightly tilted rightwards, and then further tilted rightwards ($\varepsilon(t) > 0$). During this process, the experimenter monitors whether or not macrosystems make transition to the right well. Let $N(\varepsilon)$ be the number of macrosystems which do so at the time when $\varepsilon(t) = \varepsilon$, then the result expected typically is the one depicted schematically in Fig. 2.8. As ε starts increasing in the positive value, both the height and the width of the barrier decrease thereby facilitating tunneling, and as a result Lorentzian

---

[26] *Macroscopic resonant quantum tunneling*, to be precise.

[27] A large number of macrosystems of the same property.

[28] To put it formally, *adiabatically*. If the *adiabatic condition* (2.3.10) is not satisfied, the macrosystem may jump up to one of the left-well excited states before tunneling to the right well. Of course one could intentionally violate the condition to investigate the effects of this jumping.

peaks grow. As ε increases further, peaks start decreasing because most of the macrosystems have made transitions by that time (i.e. at a smaller ε). Details of the figure may depend on $\dot{\varepsilon}(t)$; the smaller $\dot{\varepsilon}(t)$ is, the more macrosystems are expected to make transitions at a smaller ε, since they are given more time to tunnel at a given ε.

2.4 Formulas for ground-state energy splitting and decay rate

The ground-state energy splitting Δ associated with the symmetric double well is found as

$$\Delta = \tilde{A}\exp(-I/h), \tag{2.4.1}$$

$$\tilde{A} := \frac{\omega_0}{\pi}\left\{\frac{4\pi\omega_0}{h}\exp(2\tilde{\xi})\right\}^{1/2}, \tag{2.4.2}$$

$$I := \int_{-1}^{1} dq\sqrt{2V(q)}, \tag{2.4.3}$$

$$\tilde{\xi} := \int_{0}^{1} dq\left\{\frac{\omega_0}{\sqrt{2V(q)}} - \frac{1}{1-q}\right\}. \tag{2.4.4}$$

The quantum decay rate Γ of the virtual ground state associated with the bumpy slope is found to be given by a formally analogous formula:

$$\Gamma = A\exp(-B/h), \tag{2.4.5}$$

$$A := \frac{\omega_0}{2\pi}\left\{\frac{4\pi\omega_0}{h}\exp(2\zeta)\right\}^{1/2}, \tag{2.4.6}$$

$$B := 2\int_{0}^{1} dq\sqrt{2V(q)}, \tag{2.4.7}$$

$$\zeta := \int_{0}^{1} dq\left\{\frac{\omega_0}{\sqrt{2V(q)}} - \frac{1}{q}\right\}. \tag{2.4.8}$$

The essential factors in these formulas can easily be obtained by use of the naive WKB method. In general, certain conditions need be met for the WKB method to be applicable; the kinetic energy of the particle must be sufficiently high (that is, the wavelength associated with the particle must be much shorter than the spatial scale of variation of the potential). Also, the so-called connection formula can be used only in one direction (either to the right or to the left). What we call the naive WKB method ignores these conditions and rules; it applies WKB method to the ground state and liberally uses the connection formulas to both directions.

The computation of the ground-state energy splitting is easy conceptually; it is sufficient to deal with bound-state energies only. The result is

$$\Delta^{\text{naiveWKB}} = \frac{\omega_0}{\pi} \exp\{-\tilde{S}(h\omega_0/2)/h\}. \tag{2.4.9}$$

The exponent is the sub-barrier action for the case where the energy E coincides with the left (or right)-well ground-state energy $h\omega_0/2$:

$$\tilde{S}(E) := \int_{-q[E]}^{q[E]} dq \sqrt{2\{V(q) - E\}}, \tag{2.4.10}$$

where $q[E]$ is the positive root of $V(q) = E$. The sub-barrier action may be expanded in terms of E as[29]

$$\tilde{S}(E) = I - \frac{E}{\omega_0} \log\left(\frac{2\omega_0^2 \exp(1 + 2\tilde{\zeta})}{E}\right) + \mathcal{O}(E^{3/2}). \tag{2.4.11}$$

Combining this with Eq. (2.4.9), one gets Eq. (2.4.1). To be precise, one obtains the extra factor

$$(e/\pi)^{1/2} = 0.930 \cdots \tag{2.4.12}$$

in the pre-exponential factor $\bar{A}$.[30] The correct formula (2.4.1) can be obtained with the proper WKB method which is free from the deficiencies of the naive one.

Some preparatory argument is needed to compute the decay rate of the virtual ground state. It follows from the theory summarized in Appendix B that one has only to determine energy-eigenfunction $\mathcal{U}_E$ as a function of the energy E. If one computed $\mathcal{U}_E$ with the naive WKB method, one would find

$$\Gamma^{\text{naiveWKB}} = \frac{\omega_0}{2\pi} \exp\{-2S(h\omega_0/2)/h\}. \tag{2.4.13}$$

The exponent is twice the sub-barrier action for the case where the energy E coincides with the ground-state energy $h\omega_0/2$ of the metastable well:

$$S(E) := \int_{q_-[E]}^{q_+[E]} dq \sqrt{2\{V(q) - E\}}, \tag{2.4.14}$$

where $q_\pm[E]$ are the roots of $V(q) = E$ such that $0 < q_-[E] < q_+[E]$. The expansion of $2S(E)$ with respect to E takes the same form as that of $\tilde{S}(E)$; it is necessary only to substitute $\{B, \zeta\}$ for $\{I, \tilde{\zeta}\}$ in Eq. (2.4.11). Thus, one obtains the same extra factor (2.4.12) in the pre-exponential factor A as in the case of the ground-state energy splitting. The correct formula (2.4.5) can be obtained by applying the proper WKB method to compute a complex eigenvalue associated with

---

[29] For details of the calculation, see, e.g. Appendix B of Ref. [20].
[30] It is rather surprising that the naive WKB method gives the correct result (2.4.1) apart from this numerical factor of $\mathcal{O}(1)$.

the *quasi-stationary wave* (see Appendix B). In actual comparisons of experiments with theory, however, it is rarely the case that a factor of $\mathcal{O}(1)$ matters; experiments are rarely so precise. Hence, the result of the naive WKB method is enough for practical purposes.

The numerical constants I, $\tilde{\zeta}$, B and ζ may be computed explicitly for the potentials (2.1.15) and (2.1.16) as

$$I^{\text{SSDW}} = \frac{2}{3}\omega_0 = \frac{4}{3}\sqrt{2} = 1.88\cdots, \quad \tilde{\zeta}^{\text{SSDW}} = \log 2, \tag{2.4.15}$$

$$B^{\text{SBS}} = \frac{8}{15}\omega_0 = \frac{4}{5}\sqrt{6} = 1.95\cdots, \quad \zeta^{\text{SBS}} = \log 4. \tag{2.4.16}$$

2.5 Quantum decay and irreversible processes

The quantum decay rate Γ describes phenomena quite different from those described by the ground-state energy splitting Δ in spite of the formal similarity between their formulas.

So long as environmental effects are neglected, the theoretical problem of MQC reduces to one of finding discrete eigenvalues of a relevant Hamiltonian. The corresponding energy-eigenfunctions[31] represent bound states. This problem is conceptually simple.

By contrast, the energy spectrum of a Hamiltonian governing a MQT phenomenon is continuous and the associated energy-eigenfunctions $\mathcal{U}_E(q)$ are not normalizable. The decaying state $|\psi(t)\rangle$ is not an energy-eigenfunction; if it were, the absolute value of the corresponding wavefunction $\psi(q, t)$ would not change in time. This wavefunction is a superposition of $\mathcal{U}_E(q)$ over the energy-range of the order of $\bar{\epsilon}$:

$$\psi(q, t) = \int_{E_0-\bar{\epsilon}}^{E_0+\bar{\epsilon}} dE \, c(E) e^{-iEt/h} \mathcal{U}_E(q). \tag{2.5.1}$$

At $t = 0$, this is a wavepacket $\psi_0(q)$ localized around the origin ($q = 0$); the coefficients $c(E)$ of the superposition are so chosen that the interference among the superposed waves $\mathcal{U}_E(q)$ is <u>maximally constructive</u> around the origin. At $t \neq 0$, each wave $\mathcal{U}_E(q)$ acquires the phase factor $\exp(-iEt/h)$, which modifies the pattern of interference. Around the origin, the interference becomes less constructive than at $t = 0$ and the absolute value of the wavefunction decreases in the initial temporal domain (i.e. for small t). In general, the persistence probability $\mathcal{P}(t)$ behaves initially as

$$\mathcal{P}(t) \simeq 1 - (t/\tau_Q)^2 \quad : t \ll \tau_Q \sim h/\bar{\epsilon}. \tag{2.5.2}$$

---

[31] Eigenfunctions associated with a Hamiltonian are conventionally called energy-eigenfunctions.

This initial temporal domain is called the *quantum Zeno regime* (see Section 4. 5). As time goes on, the interference can never become maximally constructive again but can occasionally become more constructive than at a preceding time; see the damped oscillation in the long-time domain ($t \gg \Gamma^{-1}$) in Fig. 2.6. This oscillation, which would be difficult to understand if the decay is regarded as a stochastic process, reflects quite a natural behavior for a wave; in general, there is no reason for the degree of constructiveness of an interference to decrease monotonically in time. The monotonic decrease in the quantum Zeno regime is a result of the initial condition that the interference was maximally constructive at $t = 0$. Furthermore, as emphasized in Section 2. 3, the exponential decrease (2.3.2) holds only in the temporal domain (2.3.6).

Each of the energy-eigenfunctions involved in the above superposition is itself a superposition of a wave reflected and another transmitted, respectively, by the potential barrier. The wavefunction $\psi(q, t)$ as a whole, however, does not behave as a well-localized wavepacket which is reflected back and forth within the metastable well and finally passes through the barrier; a wavepacket localized so well would require a correspondingly large $\bar{\epsilon}$ in Eq. (2.5.1) in violation of the condition (2.3.5).

Although $\mathcal{P}(t)$ can oscillate occasionally, it exhibits an overall decrease in time, which is a kind of irreversibility. Where does this come from? The answer is simple. Consider, instead of (2.5.1), the following function of time:

$$\phi_N(t) \equiv \sum_{n=0}^{N} c_n e^{-i\omega_n t} = e^{-i\omega_0 t} \sum_{n=0}^{N} c_n e^{-i\Omega_{n0} t} \quad : \Omega_{n0} \equiv \omega_n - \omega_0 > 0. \quad (2.5.3)$$

For $N = 1$, $|\phi_1(t)|$ recovers the initial value at $t = 2\pi/\Omega_{10}$. For $N \geq 2$, however, the situation is different. If there exists a constant τ_P and a set of integers $\{M_1, M_2, \cdots, M_N\}$ such that

$$\Omega_{n0}\,\tau_P = 2\pi M_n, \quad (2.5.4)$$

then $|\phi_N(t)|$ recovers the initial value at $t = \tau_P$. For this to be the case, each frequency Ω_{n0} must be an integral multiple of a common fundamental frequency. This circumstance occurs only for such a special case as a wavepacket for a harmonic oscillator. If at least one of the ratios among the Ω_{n0}'s is an irrational number, $|\phi_N(t)|$ never recovers the initial value. However, if the condition (2.5.4) holds approximately, $|\phi_N(\tau_P)|$ can be fairly close to $|\phi_N(0)|$. This phenomenon, which is essentially the same as a beat of sound, is called the *Poincaré cycle* and τ_P is called the *Poincaré period*.[32] In short, if one waits for a sufficiently long time, $|\phi_N(t)|$

---

[32] More formally, *Poincaré recurrence time*.

returns to a value as close to $|\phi_N(0)|$ as one desires. This may be stated formally as follows:

Poincaré recurrence theorem:[33] Given a positive ϵ, there exists $\tau_P^{(\epsilon)}$ such that

$$\max_t \left| \left| \phi_N\left(t + \tau_P^{(\epsilon)}\right)\right| - |\phi_N(t)| \right| < \epsilon, \tag{2.5.5}$$

where $\tau_P^{(\epsilon)}$ may be called the Poincaré recurrence time for the accuracy ϵ. The integral in Eq. (2.5.1), which may be regarded as a limit of a sum, contains as it were, a continuously-infinite set of frequencies. Hence, the Poincaré period of $\psi(0, t)$ is ∞, and so is that of $\mathcal{P}(t)$. This is the origin of the above-mentioned irreversibility.

Although we have mentioned irreversibility, the phenomenon just discussed is in fact reversible, since the Schrödinger equation is form-invariant under time reversal; for a given T, the function

$$\bar{\psi}(q, t) := \psi^*(q, T - t) \tag{2.5.6}$$

obeys the same Schrödinger equation as the one obeyed by $\psi(q, t)$, and satisfies the following conditions:

$$\bar{\psi}(q, 0) = \psi^*(q, T), \tag{2.5.7}$$
$$\bar{\psi}(q, T) = \psi^*(q, 0). \tag{2.5.8}$$

Thus, $|\bar{\psi}(q, t)|$ is a time-reversed version of $|\psi(q, t)|$. In reality, however, a decayed state is never observed to return to the initial metastable state. This is because it is extremely hard to realize the initial condition (2.5.7); in other words, it is practically impossible to reproduce precisely the coefficients $c^*(E)e^{+iET/\hbar}$ occurring in the superposition on the right-hand side. Therefore, although the phenomenon in question is described by a mathematically reversible equation, it is irreversible in practice. The nature of this *arrow of time* is the same as that of a wave on the surface of a pond.[34]

2.6 Euclidean action: bounce and instanton

There is an interesting interpretation for the constant B appearing in the exponent of the formula (2.4.5).[35] Imagine a fictitious particle[36] which moves under the

---

[33] This theorem was originally formulated for an "area-conserving dynamical system", whose essence is extracted in the text.

[34] It is a familiar scene that a circular wave emanates from a stone dropped on the surface of a pond, but no circular wave is ever seen to emerge from the edge and converge to the center to let a stone jump out of the surface. For a detailed account, see, e.g. P. C. W. Davies, *The Physics of Time Asymmetry*, University of California Press (1977), Chapters 5–6.

[35] This is a purely theoretical interpretation, which is totally irrelevant to analyses of experimental data. Nevertheless, experimentalists will be better off with the knowledge of jargon used in this interpretation so as not to be left astray by theoreticians' pedantry.

[36] The word "fictitious" is not meant to imply that the interpretation under discussion is void or fruitless. Readers are encouraged to enjoy an open-minded imagination.

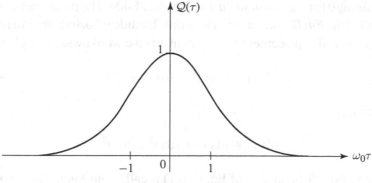

Fig. 2.9. Bounce.

potential $-V(q)$, and let τ be the fictitious time to describe the motion. Consider in particular the case that the energy is zero, and denote with $\mathcal{Q}(\tau)$ the position of the zero-energy fictitious particle at the fictitious time τ, then

$$\frac{1}{2}[\dot{\mathcal{Q}}(\tau)]^2 - V(\mathcal{Q}(\tau)) = 0, \tag{2.6.1}$$

where the overdot $\cdot$ denotes differentiation with respect to τ. Since the choice of the origin of time is at our disposal, we suppose without loss of generality that

$$\mathcal{Q}(0) = 1. \tag{2.6.2}$$

This trajectory starts from $q = 0$ at $\tau = -\infty$, bounces back at $q = 1$ at $\tau = 0$, and returns to $q = 0$ at $\tau = +\infty$ (Fig. 2.9). Hence, $\mathcal{Q}$ is called the *bounce*. To be more explicit, it is defined by the equation

$$|\tau| = \int_{\mathcal{Q}(\tau)}^{1} \frac{dq}{\{2V(q)\}^{1/2}}. \tag{2.6.3}$$

$\mathcal{Q}(\tau)$ is even with respect to τ. Hence

$$\dot{\mathcal{Q}}(0) = 0. \tag{2.6.4}$$

It is helpful to recall the law of conservation of energy for an ordinary particle which moves under the potential $V(q)$:

$$\frac{1}{2}\left(\frac{dq(t)}{dt}\right)^2 + V(q(t)) = \text{constant}. \tag{2.6.5}$$

Equation (2.6.1) may be obtained from this by making the formal replacement

$$t = i\tau, \quad q(i\tau) = \mathcal{Q}(\tau), \tag{2.6.6}$$

and substituting 0 for the constant on the right-hand side. The parameter τ is conventionally called the *Euclidean time*.[37] The word 'Euclidean' originates in a relativistic context; the formal replacement (2.6.6) converts the Minkowski line element

$$-(\mathrm{d}t)^2 + (\mathrm{d}x)^2 + (\mathrm{d}y)^2 + (\mathrm{d}z)^2 \qquad (2.6.7)$$

to the Euclidean one

$$(\mathrm{d}\tau)^2 + (\mathrm{d}x)^2 + (\mathrm{d}y)^2 + (\mathrm{d}z)^2. \qquad (2.6.8)$$

Accordingly, the left-hand side of Eq. (2.6.1) is called the *Euclidean energy*.

The bounce, which increases monotonically for $\tau < 0$, establishes a one-to-one correspondence between $q \in (0, 1)$ and $\tau \in (-\infty, 0)$ via

$$q = \mathcal{Q}(\tau). \qquad (2.6.9)$$

By use of this correspondence, the constant B may be rewritten as

$$B = 2\int_{-\infty}^{0} \mathrm{d}\tau\, \dot{\mathcal{Q}}(\tau)\dot{\mathcal{Q}}(\tau) = \int_{-\infty}^{\infty} \mathrm{d}\tau\, [\dot{\mathcal{Q}}(\tau)]^2. \qquad (2.6.10)$$

Noting the *law of conservation of Euclidean energy* (2.6.1), one finds that

$$B = S[\mathcal{Q}], \qquad (2.6.11)$$

where

$$S[q] := \int_{-\infty}^{\infty} \mathrm{d}\tau \left\{ \frac{1}{2}[\dot{q}(\tau)]^2 + V(q(\tau)) \right\}. \qquad (2.6.12)$$

The integrand on the right-hand side may be interpreted as the Lagrangian governing the fictitious particle, for $V = -(-V)$. Accordingly, $S[q]$ is called the *Euclidean action functional*.[38] Equation (2.6.11) shows that B is the value of this functional evaluated at the bounce. Hence B is called the *bounce action*.

Let $\mathcal{F}_\infty$ be the set (or the space) of functions which are smooth and vanish at $\tau = \pm\infty$:

$$\mathcal{F}_\infty := \{q \mid q(\pm\infty) = 0\}. \qquad (2.6.13)$$

---

[37] Often called the "imaginary time" as well. Remember, however, that $\tau \in \mathbf{R}$.
[38] The argument of a functional is to be embraced by the square bracket [ ]; S is a function of the variable q, which is to be distinguished from $q(\tau)$, namely the value of the function q at time τ (see Appendix C for details).

Adopt $\mathcal{F}_\infty$ as the domain on which $S[q]$ is to be defined, then the bounce is a non-trivial[39] *stationary point* of $S[q]$:

$$\left.\frac{\delta S[q]}{\delta q(\tau)}\right|_{q=\mathcal{Q}} = 0, \tag{2.6.14}$$

$$\mathcal{Q}(0) = 1, \quad \dot{\mathcal{Q}}(0) = 0, \tag{2.6.15}$$

where the symbol $|_{q=\mathcal{Q}}$ stipulates that $\mathcal{Q}$ is to be substituted for q after the differentiation. In the special case of $V = V^{\text{SBS}}$, the bounce is found explicitly as

$$\mathcal{Q}^{\text{SBS}}(\tau) = \left(\cosh\frac{\omega_0\tau}{2}\right)^{-2}. \tag{2.6.16}$$

The constant I may be interpreted analogously. It suffices to modify the above argument for B as follows; replace $\{B, \mathcal{Q}\}$ with $\{I, \mathcal{Q}_I\}$, replace Eqs. (2.6.2) and (2.6.15) with

$$\mathcal{Q}_I(0) = 0, \quad \dot{\mathcal{Q}}_I(0) > 0, \tag{2.6.17}$$

replace Eq. (2.6.3) with

$$\tau = \int_0^{\mathcal{Q}_I(\tau)} \frac{dq}{\{2V(q)\}^{1/2}}, \tag{2.6.18}$$

and adopt

$$\tilde{\mathcal{F}}_\infty := \{q \mid |q(\pm\infty)| = 1\} \tag{2.6.19}$$

in place of $\mathcal{F}_\infty$ as the domain of definition of $S[q]$. We depict the trajectory $\mathcal{Q}_I(\tau)$ schematically in Fig. 2.10. The "action density" (i.e. Euclidean Lagrangian)

$$\frac{1}{2}[\dot{\mathcal{Q}}_I(\tau)]^2 + V(\mathcal{Q}_I(\tau)) \tag{2.6.20}$$

for $\mathcal{Q}_I$ is localized in the interval of width of the order of ω_0^{-1} on the τ-axis; in the time scale much longer than ω_0^{-1}, it has a positive value at the "instant" of $\tau \sim 0$ and vanishes otherwise. Hence, $\mathcal{Q}_I$ is called the *instanton*.[40] In the special case of $V = V^{\text{SSDW}}$, the instanton is found explicitly as

$$\mathcal{Q}_I^{\text{SSDW}}(\tau) = \tanh\frac{\omega_0\tau}{2}. \tag{2.6.21}$$

Since the action density is localized on the Euclidean-time axis in the case of the bounce $\mathcal{Q}$ as well, $\mathcal{Q}_I$ and $\mathcal{Q}$ are collectively called *instantons* in some literature.

---

[39] The word "non-trivial" means that $\mathcal{Q} \neq 0$ (i.e. $\mathcal{Q}(\tau)$ is not identically zero). The condition $\mathcal{Q}(0) = 1 \neq 0$ guarantees that $\mathcal{Q}$ is non-trivial.

[40] For details, see, e.g. R. Rajaraman: *Solitons and Instantons* (North-Holland Publishing Co., 1982).

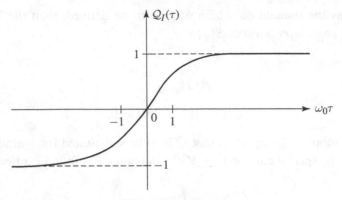

Fig. 2.10. Instanton.

Certainly, they are quite similar as far as the properties displayed in this section are concerned. However, as has been emphasized, they occur in quite different physical contexts, namely the energy splitting and the quantum decay, respectively. Ignoring this distinction, one is liable to be led to conceptual confusion. Therefore, it is advisable to make a clear distinction between the bounce Q and the instanton Q_I.

Exercises

Exercise 2.1. Suppose that $V(q;\varepsilon)$ is a fourth order polynomial in q and satisfies the following conditions:

$$V(-1;\varepsilon) = 0, \quad V(1;\varepsilon) = -\varepsilon, \quad V'(\pm1;\varepsilon) = 0, \quad V(q;0) = V^{\mathrm{SSDW}}(q).$$

Show that $V(q;\varepsilon)$ is equal to the special tilted double well $V^{\mathrm{STDW}}(q;\varepsilon)$ defined as

$$V^{\mathrm{STDW}}(q;\varepsilon) := V^{\mathrm{SSDW}}(q) - 4\varepsilon\left(\frac{q+1}{2}\right)^3 + 3\varepsilon\left(\frac{q+1}{2}\right)^4. \qquad (2.6.22)$$

Exercise 2.2.
(1) Let $V_{\pm}^{\mathrm{HO}}(q;\varepsilon) =: \frac{1}{2}\omega_{\pm}^2(q \mp 1)^2$ be the harmonic approximant of $V^{\mathrm{STDW}}(q;\varepsilon)$ in the neighborhood of $q = \pm 1$, respectively, and $|\pm'\rangle$ be the ground state associated with $V_{\pm}^{\mathrm{HO}}$, respectively. These two states span a two-dimensional Hilbert space. For this space, construct an orthonormal basis $|\pm\rangle$ such that $\||+\rangle - |+'\rangle\|^2 + \||-\rangle - |-'\rangle\|^2$ is minimized.
(2) Use the above basis to derive the two-state-approximated Hamiltonian (2.2.3). In particular, consider the quasi-classical situation and express δ and Δ in terms of h and ε for the case of $0 < \varepsilon \ll 1$.

Exercise 2.3.
(1) Find the eigenstates and eigenvalues associated with the Hamiltonian (2.2.3).
(2) Use the above result to compute $|\psi(t)\rangle$ and $\mathcal{P}_+(t)$.

(3) Suppose that the initial state is $|-'\rangle$, which is one of the states $|\pm'\rangle$ defined in Exercise 2.2 above. Compute the probability $\mathcal{P}_{+'}(t)$ for the system to be found in the state $|+'\rangle$ at time t.

Exercise 2.4. Compute and display the probability current (2.2.23) in detail.

Exercise 2.5. Use the naive WKB method to derive Eqs. (2.4.1) and (2.4.5) (both multiplied by $(e/\pi)^{1/2}$).

Exercise 2.6. In what circumstances is it valid to view a MQT-phenomenon as a "free tunneling through the barrier followed by repeated oscillations within the metastable well?"

Exercise 2.7. Estimate, perhaps with a numerical computation, the Poincaré period $\tau_P^{(\epsilon)}$ of the function $g_N(\tau) := N^{-1} \sum_{n=1}^{N} \cos(\sqrt{n}\tau)$ for various ϵ's and N's.

Exercise 2.8. For a free particle, usual Gaussian wavepackets spread monotonically as time passes. However, this need not be the case. Construct a Gaussian wavepacket that starts out contracting and then spreads later.

Exercise 2.9. Derive Eqs. (2.6.16) and (2.6.21). For generic bumpy slopes and symmetric double wells, check the schematic forms depicted in Figs. 2.9 and 2.10, respectively, and analytically determine their asymptotic behavior for $\omega_0 \tau \gg 1$.

3

Some candidate systems for macroscopic quantum tunneling

This chapter is an introduction to typical systems in which MQP are expected to be detectable. Explanations are meant to be self-contained and comprehensible without prior knowledge, but are necessarily somewhat sketchy and imprecise. Unsatisfied readers may wish to consult appropriate literature quoted in the text.[1]

3.1 SQUID

SQUID[2] is a superconducting ring with the superconductivity destroyed deliberately at a place. The ring shown in Fig. 3.1, for example, is made of superconducting niobium with a membrane of aluminum oxide inserted at the hatched portion. This portion is not superconducting. (To be precise, superconductivity leaks somewhat into this portion out of the neighboring niobium.) In Fig. 3.1, the aluminum-oxide membrane mediates the contact between the upper and the lower parts of the superconductor. This part of the figure, which is called the *Josephson junction*, is much smaller than the rest of the ring as exemplified in the figure. Let us begin by explaining about this junction.

3.1.1 Josephson junction and equivalent circuit for SQUID

The junction possesses three functions. First, it allows the supercurrent $I_J(t)$ called the *Josephson current* to flow through it. In general, a superconducting state is

---

[1] A reference will contain several other references, which in turn contain yet others, *ad infinitum*. Thus, a sober student faces an exponential increase in the number of references to be consulted (the difficulty of divergence). At some stage, therefore, one should pretend that one has understood enough material and proceed further. What is hard is to decide precisely at which stage one should do so. Perhaps, all one can do to deepen understanding is to return occasionally to the problem but view it through a surface that is different from the previous one as though one were migrating along an infinite-dimensional Klein's bottle.

[2] To be precise, the rf-SQUID. SQUID is the acronym for *superconducting quantum interference device*. (This device does not possess ten legs, though.) The "interference" here refers to a phenomenon belonging to the macroscopic quantum phenomena of the first kind (see Section 1.3). For details on SQUID and superconductivity in general, see, e.g. M. Tinkham, *Introduction to Superconductivity*, McGraw-Hill, New York (1975); T. Tsuneto, *Superconductivity and Superfluidity*, Cambridge University Press (1998).

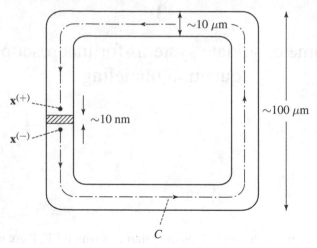

Fig. 3.1. Typical SQUID.

characterized by the *Ginzburg–Landau order-parameter field* $\boldsymbol{\Psi}_{\mathrm{GL}}(\mathbf{x}, t)$. In the case of the superconducting niobium in Fig. 3.1, the order-parameter field is a single complex number:

$$\boldsymbol{\Psi}_{\mathrm{GL}}(\mathbf{x}, t) \equiv |\boldsymbol{\Psi}_{\mathrm{GL}}(\mathbf{x}, t)| \exp\{i\theta(\mathbf{x}, t)\}. \tag{3.1.1}$$

Thus, the order-parameter is endowed with a phase $\theta(\mathbf{x}, t)$. To a good approximation, $\theta(\mathbf{x}, t)$ is independent of $\mathbf{x}$ within each cross-section perpendicular to the central line C (the dot-dashed curve in Fig. 3.1) of the ring. Suppose that C is oriented anti-clockwise, and let $\mathbf{x}^{(-)}$ be its starting point (the lower edge of the junction) and $\mathbf{x}^{(+)}$ be its ending point (the upper edge of the junction). Then, the *phase difference* $\theta(t)$ may be defined as

$$\theta(t) := \theta(\mathbf{x}^{(-)}, t) - \theta(\mathbf{x}^{(+)}, t), \tag{3.1.2}$$

which is the difference between the phase at the upper edge and that at the lower edge of the junction. In terms of this phase difference, the Josephson current is expressed as

$$I_{\mathrm{J}}(t) = -I_{\mathrm{c}} \sin\theta(t). \tag{3.1.3}$$

The convention for the sign of the electric current is such that a positive electric current flows from the upper to the lower edge through the junction. The coefficient I_{c} is a constant which depends on the details of the junction (i.e. the kind of superconducting material, the kind of material inserted at the junction, the width and shape of the junction, and so on). The above equation, which relates the supercurrent flowing through the junction to the phase difference, is called *Josephson's current–phase relation*. The Josephson current flows without voltage (i.e. electric potential

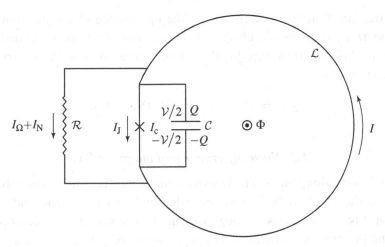

Fig. 3.2. Equivalent circuit for SQUID.

difference) between the upper and the lower edges of the junction. Its maximum value is I_c as is clear from the above relation. Therefore, if a current greater than I_c is forced to flow through the junction, the current cannot entirely be supercurrent and a voltage arises.[3]

Second, the junction functions as a capacitor (i.e. condenser) as well; electric charge can accumulate at both of its edges. We denote with $\mathcal{C}$ the effective electric capacitance of the junction and with $Q(t)$ the charge accumulated at the upper edge (hence the charge at the lower edge is given by $-Q(t)$).

Third, the junction may function as a resistance as well; if a voltage $\mathcal{V}(t)$ arises between the edges, a current $I_\Omega(t)$ proportional to the voltage flows through the junction. Let $\mathcal{R}$ be the effective resistance of the junction, then

$$Q(t) = \mathcal{C}\mathcal{V}(t), \quad I_\Omega(t) = \mathcal{V}(t)/\mathcal{R}. \tag{3.1.4}$$

The current $I_\Omega(t)$ is not a supercurrent but an ordinary ohmic current accompanied by Joule's heat and hence by energy dissipation. It then follows from the *fluctuation–dissipation theorem* that there should also be a fluctuating current flowing through the junction. This fluctuating current (or the noise current), called the Nyquist current (or the Nyquist noise), is to be denoted by $I_N(t)$.

The SQUID ring has a self-inductance, to be denoted by $\mathcal{L}$, determined by its geometrical shape. What has been mentioned so far may be summarized in the form of an equivalent circuit representing the SQUID. In Fig. 3.2, the origin of the electric potential has been so adjusted that the latter is equal to $\mathcal{V}(t)/2$ at the upper edge (hence, $-\mathcal{V}(t)/2$ at the lower edge), and the total current $I(t)$ flowing along the ring is shown as well as the currents $I_J(t)$, $I_\Omega(t)$ and $I_N(t)$; the arrows

---

[3] The subscript c in I_c signifies that the latter is the critical current in this sense.

indicate the direction of the currents. At the upper edge of the junction, during the period $(t, t+dt)$, electric charge of the amount $I(t)dt$ flows in and that of the amount $\{I_J(t) + I_\Omega(t) + I_N(t)\}dt$ flows out. Hence, the law of conservation of electric charge gives

$$\dot{Q}(t) = I(t) - I_\Omega(t) - I_J(t) - I_N(t). \tag{3.1.5}$$

3.1.2 Phase difference and magnetic flux

A current flowing along the ring induces a *magnetic flux* (hereafter to be abbreviated as *flux*) threading through the ring. Suppose that an external magnetic field is applied and let $\Phi(t)$ be the total flux[4] through the ring and $\mathbf{A}(\mathbf{x}, t)$ be the vector potential to describe the total magnetic field. The supercurrent density $\mathbf{J}_s(\mathbf{x}, t)$ at a point $\mathbf{x}$ in the superconductor is expressed in terms of the order-parameter field as[5]

$$\mathbf{J}_s(\mathbf{x}, t) = \frac{(-2e)}{m^*} \Re \left[\Psi_{GL}^*(\mathbf{x}, t) \left\{ \frac{\hbar}{i} \nabla - (-2e)\mathbf{A}(\mathbf{x}, t) \right\} \Psi_{GL}(\mathbf{x}, t) \right]. \tag{3.1.6}$$

This expression has the same form as that for the quantum-mechanical probability current (2.2.23); the symbol corresponding to $\frac{\hbar}{i}\nabla$ in the latter is also replaced by one corresponding to $\frac{\hbar}{i}\nabla - (-e)\mathbf{A}(\mathbf{x}, t)$ if a magnetic field is present. Hence it is *gauge-invariant*; let the right-hand side of Eq. (3.1.6) be denoted by $\mathcal{J}_s\{\Psi_{GL}, \mathbf{A}\}$, then, the *gauge transformation*

$$\tilde{\Psi}_{GL}(\mathbf{x}, t) = \Psi_{GL}(\mathbf{x}, t) \exp\{-i\chi(\mathbf{x})\}, \tag{3.1.7}$$

$$\tilde{\mathbf{A}}(\mathbf{x}, t) = \mathbf{A}(\mathbf{x}, t) + \frac{\hbar}{2e} \nabla \chi(\mathbf{x}), \tag{3.1.8}$$

where $\chi(\mathbf{x})$ is an arbitrary smooth function, leaves $\mathcal{J}_s$ unchanged:

$$\mathcal{J}_s\{\tilde{\Psi}_{GL}, \tilde{\mathbf{A}}\} = \mathcal{J}_s\{\Psi_{GL}, \mathbf{A}\}. \tag{3.1.9}$$

The fact that Cooper pairs rather than individual electrons are responsible for superconductivity is reflected in the electric charge $(-2e)$ occurring in Eq. (3.1.6). The positive constant m^* is called the effective mass of the Cooper pair. In spite of the above-mentioned similarity, however, it is to be noted that $\Psi_{GL}(\mathbf{x}, t)$ is not a quantum-mechanical wavefunction;[6] Eq. (3.1.6) does not represent a probability current but an actual electric current. By use of Eq. (3.1.1), the expression (3.1.6)

---

[4] The sign of the flux is taken to be positive when it points <u>upwards</u> as indicated in Fig. 3.2.
[5] The electric charge of an electron is denoted by $-e$.
[6] Although the Ginzburg–Landau order-parameter field is often called a "macroscopic wavefunction", this terminology is quite misleading. Readers are advised to refrain from using it.

may be rewritten as

$$\mathbf{J}_s(\mathbf{x}, t) = -\frac{2e}{m^*} |\Psi_{GL}(\mathbf{x}, t)|^2 \{\hbar \nabla \theta(\mathbf{x}, t) + 2e\mathbf{A}(\mathbf{x}, t)\}. \qquad (3.1.10)$$

Let us recall the *Meissner effect*: a magnetic field can penetrate into a superconductor only up to a distance of the order of *London's penetration depth* λ_L from its surface. Therefore, if the thickness d of the ring is much greater than λ_L, no magnetic field can reach the central line C of the ring. The supercurrent along C also vanishes since a current is to be accompanied by a magnetic field. Furthermore, at temperatures much lower than the critical temperature for superconductivity, the order parameter is developed so well that $|\Psi_{GL}(\mathbf{x}, t)| \neq 0$ at least along C. Hence, the quantity embraced by the curly bracket $\{\ \}$ on the right-hand side of Eq. (3.1.10) should vanish along C. Thus, integrating this quantity along C, one obtains

$$\int_C d\mathbf{x} \cdot \mathbf{A}(\mathbf{x}, t) = -\frac{\hbar}{2e} \int_C d\mathbf{x} \cdot \nabla \theta(\mathbf{x}, t). \qquad (3.1.11)$$

Since the contour C of integration is so oriented that it starts from $\mathbf{x}^{(-)}$ and ends at $\mathbf{x}^{(+)}$, the integral on the right-hand side gives $-\theta(t)$. On the left-hand side, it is legitimate to augment C with the infinitesimal contour through the junction starting from $\mathbf{x}^{(+)}$ and ending at $\mathbf{x}^{(-)}$, thereby replacing C with a closed contour encircling the ring.[7] Hence, the left-hand side gives the total flux $\Phi(t)$ through the ring. One thus finds

$$\theta(t) = 2\pi \Phi(t)/\Phi_0, \qquad (3.1.12)$$

where

$$\Phi_0 := 2\pi\hbar/2e \qquad (3.1.13)$$

is called the *flux quantum* (or *fluxon*). The result (3.1.12) shows that the phase difference θ is not a mere angle but represents a physical quantity, namely the flux; θ and $\theta + 2\pi$ correspond to different physical situations.[8]

Combining Eq. (3.1.12) with Faraday's law of induction

$$\dot{\Phi}(t) = -\mathcal{V}(t), \qquad (3.1.14)$$

one finds

$$\dot{\theta}(t) = -\frac{2\pi}{\Phi_0} \mathcal{V}(t) = -2e\mathcal{V}(t)/\hbar, \qquad (3.1.15)$$

which is called *Josephson's acceleration equation*.

---

[7] Without loss of generality, the vector potential $\mathbf{A}(\mathbf{x}, t)$ can be supposed to be a smooth function of $\mathbf{x}$. Hence, the added infinitesimal contour gives a negligible contribution to the integral.

[8] By contrast, the phase difference in the so-called CBJJ (current-biased Josephson junction) is a mere angle; θ and $\theta + 2\pi$ correspond to the same physical situation.

3.1.3 Equation of motion for magnetic flux

Let Φ_{ex} be that portion of the total flux which is due to the external magnetic field (supposed to be time-independent), then[9]

$$\Phi(t) = \Phi_{ex} + \mathcal{L}I(t). \tag{3.1.16}$$

Combining this equation with Eq. (3.1.5) and expressing all the variables in terms of $\Phi(t)$ with the aid of Eqs. (3.1.3), (3.1.4), (3.1.12) and (3.1.14), one obtains

$$\mathcal{C}\ddot{\Phi}(t) + \frac{1}{\mathcal{R}}\dot{\Phi}(t) + I_c \sin\left(2\pi\frac{\Phi(t)}{\Phi_0}\right) + \frac{1}{\mathcal{L}}(\Phi(t) - \Phi_{ex}) = I_N(t). \tag{3.1.17}$$

This equation has the form of the equation of motion for a particle under friction with the right-hand side interpretable as a fluctuating force. Thus, this is a kind of Langevin equation (see Section 5.6).

In the idealized case of vanishing fluctuation ($I_N = 0$) and vanishing dissipation ($\mathcal{R} = \infty$),[10] the above equation reduces to

$$\mathcal{C}\ddot{\Phi}(t) + I_c \sin\left(2\pi\frac{\Phi(t)}{\Phi_0}\right) + \frac{1}{\mathcal{L}}(\Phi(t) - \Phi_{ex}) = 0. \tag{3.1.18}$$

This equation, which has the form of Newton's equation of motion, may be derived via the principle of least action from the Lagrangian

$$L(\Phi, \dot{\Phi}) := \frac{1}{2}\mathcal{C}\dot{\Phi}^2 - U(\Phi), \tag{3.1.19}$$

where

$$U(\Phi) := \frac{1}{2\mathcal{L}}(\Phi - \Phi_{ex})^2 - \frac{I_c\Phi_0}{2\pi}\cos\left(2\pi\frac{\Phi}{\Phi_0}\right). \tag{3.1.20}$$

Hence, the momentum conjugate to Φ is given by

$$P := \partial L(\Phi, \dot{\Phi})/\partial\dot{\Phi} = \mathcal{C}\dot{\Phi}, \tag{3.1.21}$$

and the Hamiltonian H_Φ corresponding to (3.1.19) is found as

$$H_\Phi := \{P\dot{\Phi} - L(\Phi, \dot{\Phi})\}_{\dot{\Phi}=P/\mathcal{C}} = \frac{1}{2\mathcal{C}}P^2 + U(\Phi). \tag{3.1.22}$$

Of course, the right-hand side of Eq. (3.1.19) may be multiplied by an arbitrary constant, which does not affect Eq. (3.1.18). However, Eq. (3.1.19) as it stands has

---

[9] Equation (3.1.16) holds to a good approximation, since the width of the junction is much smaller than the circumference of the ring.

[10] It might sound strange that an infinite resistance corresponds to vanishing dissipation. Nevertheless, this statement is correct: if $\mathcal{R} = \infty$, ohmic current does not flow and hence Joule's heat is not generated. To be more quantitative, ohmic current $\propto \mathcal{R}^{-1}$ and Joule's heat $\propto \mathcal{R} \times$ (ohmic current)$^2 \propto \mathcal{R}^{-1}$.

the virtue of allowing the following physical interpretation:

$$\frac{1}{2}\mathcal{C}\dot{\Phi}^2 = \frac{1}{2}\mathcal{C}\mathcal{V}^2 = \frac{Q^2}{2\mathcal{C}}$$

= the electric energy due to the charge accumulated at the junction, (3.1.23)

$$\frac{1}{2\mathcal{L}}(\Phi - \Phi_{\text{ex}})^2 = \frac{1}{2}\mathcal{L}I^2$$

= the magnetic energy due to the current flowing along the ring. (3.1.24)

The second term on the right-hand side of Eq. (3.1.20) is characteristic of a Josephson junction and is called the *Josephson-coupling energy*; the microscopic theory of superconductivity indeed shows that the energy due to the presence of the junction is precisely of this form.

The fact that electric and magnetic energies play the role of kinetic and potential energies, respectively, is not peculiar to a SQUID, but is already familiar in the canonical theory of electrodynamics *in vacuo*.[11] Equation (3.1.21) shows that the momentum canonically conjugate to the flux is nothing but the electric charge,[12] and correspondingly Eq. (3.1.22) shows that the effective inertial mass[13] governing the motion of the flux is the electric capacitance. We somewhat abuse language to call H_Φ and $U(\Phi)$ the <u>flux Hamiltonian</u> and the <u>flux potential</u>, respectively.

3.1.4 Naive quantum theory

The theory presented so far is a classical phenomenology.[14] We now make a bold assumption that the flux Φ obeys quantum mechanics as follows:

$$i\hbar \frac{\mathrm{d}}{\mathrm{d}t}|\psi(t)\rangle = \hat{H}_\Phi |\psi(t)\rangle, \qquad (3.1.25)$$

---

[11] The energy density $\{(\mathbf{E}(\mathbf{x}, t))^2 + (\mathbf{B}(\mathbf{x}, t))^2\}/2$ of the electromagnetic field *in vacuo* can be expressed in terms of the vector potential as $\{(\dot{\mathbf{A}}(\mathbf{x}, t))^2 + (\nabla \times \mathbf{A}(\mathbf{x}, t))^2\}/2$ with the speed of light chosen as the unit of speed. The latter expression is manifestly of the form of the sum of kinetic and potential energies (and, moreover, that for a harmonic oscillator).

[12] It would not be entirely off the point to regard this charge–flux relation as a kind of the so-called "number–phase relation" (i.e. the purported relationship that the phase is canonically conjugate to the particle number); the flux is linked with the phase difference and the charge is proportional to the difference in the number of electrons at the opposite edges of the capacitor. In the "number–phase relation", however, the particle number is non-negative definite and the phase is a pure angle, whereas in the charge–flux relation the electron-number difference can be negative and the phase difference is not a pure angle. In view of this important distinction, it is advisable to distinguish the charge–flux relation from the "number–phase relation". In any case, the latter is not a precisely formulated concept.

[13] Of course, its actual dimension is not that of a mass.

[14] The existence of superconductivity and the justification of its description in terms of a complex order-parameter field do not follow from a classical theory. Once they are guaranteed by quantum mechanics, however, one may regard the order-parameter field as a given classical field and develop a phenomenological <u>classical theory</u>, which is the one presented in Section 3.1.3. This circumstance is essentially the same as for the classical gas theory, which deals with a collection of atoms with classical mechanics; the existence of atoms can be guaranteed only by quantum mechanics.

$$\hat{H}_\Phi := \frac{1}{2C} \hat{P}^2 + U(\hat{\Phi}), \qquad (3.1.26)$$

$$[\hat{\Phi}, \hat{P}] = i\hbar. \qquad (3.1.27)$$

Various symbols in the above are to be used in the same sense as in the quantum mechanics for a particle; $|\psi(t)\rangle$ stands for the quantum-mechanical state for the flux at time t, and its Φ-represented wavefunction $\psi(\Phi, t)$ is defined with the help of the ket $|\Phi\rangle$ satisfying

$$\hat{\Phi}|\Phi\rangle = \Phi|\Phi\rangle \qquad (3.1.28)$$

as

$$\psi(\Phi, t) := \langle\Phi|\psi(t)\rangle, \qquad (3.1.29)$$

which gives the probability amplitude for the flux to be found with the value Φ at time t. It should be admitted that the above assumption is merely a plausible one which lacks a solid foundation at the stage of the phenomenological theory under discussion.[15]

In terms of the dimensionless quantities

$$\theta \equiv 2\pi\Phi/\Phi_0, \quad \theta_{ex} \equiv 2\pi\Phi_{ex}/\Phi_0, \qquad (3.1.30)$$

the flux potential (3.1.20) may be rewritten as

$$U(\Phi) = \frac{I_c\Phi_0}{2\pi}\tilde{U}(\theta), \qquad (3.1.31)$$

$$\tilde{U}(\theta) := \frac{\gamma}{2}(\theta - \theta_{ex})^2 - \cos\theta, \qquad (3.1.32)$$

$$\gamma \equiv \Phi_0/2\pi\mathcal{L}I_c = I_{\Phi_0}/2\pi I_c, \quad I_{\Phi_0} \equiv \Phi_0/\mathcal{L}, \qquad (3.1.33)$$

where I_{Φ_0} is the strength of the current which induces the flux precisely of the amount Φ_0 through the ring.

If the constant γ and the external magnetic field (or equivalently θ_{ex}) are appropriately adjusted,[16] the following typical situations may be realized:

---

[15] It would require a considerable amount of theoretical preparation to justify this assumption on the basis of a microscopic theory. Since such a task is beyond the scope of this book, a concerned reader is referred to Ref. [6] in the bibliography.

[16] If the Josephson junction in Fig. 3.1 is replaced by a dc-SQUID, the equivalent circuit is still given by Fig. 3.2 with I_c now depending on flux Φ'_{ex} applied through the dc-SQUID. This property may be used to vary I_c continuously.

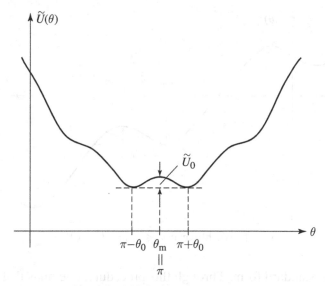

Fig. 3.3. MQC-situation : $\theta_{\text{ex}} = \pi$.

(a) $\theta_{\text{ex}} = \pi$:

$$\tilde{U}(\theta) = \frac{\gamma}{2}(\theta - \pi)^2 + \cos(\theta - \pi). \tag{3.1.34}$$

This is an even function of $\theta - \pi$. In particular, in the case of $\gamma < 1$, the neighborhood of $\theta = \pi$ constitutes a symmetric double well giving rise to a MQC-situation (Fig. 3.3).

(b) $\gamma \ll 1$ and $b \equiv \gamma \theta_{\text{ex}} = \mathcal{O}(1)$: For $\theta = \mathcal{O}(1)$, the first term of (3.1.32) may be approximated as

$$\frac{\gamma}{2}(\theta - \theta_{\text{ex}})^2 \simeq \frac{\gamma}{2}\theta_{\text{ex}}^2 - b\theta. \tag{3.1.35}$$

The first term $\gamma \theta_{\text{ex}}^2/2$ on the right-hand side is a constant which merely shifts the origin of energy and does not affect the motion of the flux. Dropping this term, one finds

$$\tilde{U}(\theta) \simeq -\cos\theta - b\theta, \tag{3.1.36}$$

which is called the *washboard*. In particular, in the case of $b \lesssim 1$, the neighborhood of $\theta = \pi/2$ in Fig. 3.4 constitutes a bumpy slope giving rise to a MQT-situation.

The dimensionless variable θ has been introduced with Φ_0 adopted as the unit of flux. For the purpose of considering the MQC- and the MQT-situations depicted in Figs. 3.3 and 3.4, respectively, the more convenient unit of flux is

$$\delta\Phi := (\Phi_0/2\pi)\theta_0, \tag{3.1.37}$$

which is the <u>distance in the space of flux</u> representing the width θ_0 of the potential barrier. This quantity therefore corresponds to R_0 in Section 2.1. Likewise, C corresponds to M there. In accordance with Section 2.1, one may convert the flux

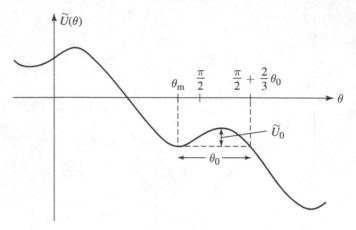

Fig. 3.4. Washboard: $\gamma \ll 1$ and $b = \mathcal{O}(1)$.

potential to the standard form. Through this procedure, one finds that

$$q = (\theta - \theta_\mathrm{m})/\theta_0, \tag{3.1.38}$$

$$\tau_0 = (\mathcal{C}/U_0)^{1/2}\delta\Phi, \tag{3.1.39}$$

and accordingly that

$$h \equiv \frac{\hbar}{\tau_0 U_0} = \frac{2e}{\theta_0\sqrt{\mathcal{C}U_0}} = \frac{h_0}{\theta_0\sqrt{\tilde{U}_0}}, \quad h_0 \equiv 2\left(\frac{e^2/\mathcal{C}}{I_\mathrm{c}\Phi_0/2\pi}\right)^{1/2}. \tag{3.1.40}$$

In this fashion, h can be expressed in the form which apparently does not contain $\hbar$ (cf. end of Section 2.1). Let N be the number of atoms (or electrons) constituting the Josephson junction, then one may estimate that $\mathcal{C} = \mathcal{O}(N)$ and $I_\mathrm{c} = \mathcal{O}(N)$, thereby arriving at a formal estimate $h_0 = \mathcal{O}(N^{-1})$. In the case of a small junction such as the one in Fig. 3.1, however, it turns out that the value of h_0 is large enough for h to be of the order of 0.1; this may be realized by adjusting the constant γ as discussed below. Thus, tunneling is expected to be detectable since the constants I and B in Eqs. (2.4.1) and (2.4.5), respectively, are of order 1.

Suppose that MQC has been detected for a SQUID of the size shown in Fig. 3.1. The associated S-Cattiness $\mathbf{D}$ introduced in Section 1.3 is expected to be of the order of the number of electrons which contribute to the current I_{Φ_0} corresponding to the flux quantum Φ_0, hence $\mathbf{D} \sim 10^{15}$. In spite of this large number of electrons, the height U_0 of the potential barrier controlling the tunneling in question is of the order of 10 K (as measured in units of Kelvin), which is far smaller than the binding energy (of the order of 10^5 K) of an electron in hydrogen atom. This is an important point which exemplifies the possibility (1.4.16).

Let us explicitly compute the constants θ_0 and $\tilde{U}_0$ to obtain h for typical cases:

(a') MQC-situation: $\gamma \lesssim 1$

Expansion of Eq. (3.1.34) with respect to $\vartheta \equiv \theta - \pi$ gives

$$\tilde{U}(\theta) = 1 - \frac{1}{2}(1 - \gamma)\vartheta^2 + \frac{1}{4!}\vartheta^4 - \frac{1}{6!}\vartheta^6 + \cdots$$

$$= 1 + \frac{\theta_0^4}{4!}(-2q^2 + q^4) - \frac{\theta_0^6}{6!}q^6 + \cdots, \tag{3.1.41}$$

$$q \equiv \vartheta/\theta_0, \quad \theta_0 \simeq \{6(1 - \gamma)\}^{1/2}. \tag{3.1.42}$$

Hence, focusing on the region $|q| \lesssim 1$ and neglecting the terms of $\mathcal{O}((1 - \gamma)^3)$, one finds the special symmetric double well with

$$\tilde{U}_0 \simeq \frac{3}{2}(1 - \gamma)^2, \quad h \simeq \frac{h_0}{3(1 - \gamma)^{3/2}}. \tag{3.1.43}$$

(b') MQT-situation: $b \lesssim 1$

Expansion of Eq. (3.1.36) with respect to $\vartheta \equiv \theta - \pi/2$ gives

$$\tilde{U}(\theta) \simeq -\frac{\pi}{2}b + (1 - b)\vartheta - \frac{1}{3!}\vartheta^3 + \frac{1}{5!}\vartheta^5 + \cdots. \tag{3.1.44}$$

Hence, neglecting $\mathcal{O}((1 - b)^{5/2})$, one finds the special bumpy slope with

$$\theta_m \simeq \frac{\pi}{2} - \frac{\theta_0}{3}, \quad \theta_0 \simeq \{18(1 - b)\}^{1/2}, \tag{3.1.45}$$

$$\tilde{U}_0 \simeq \frac{2}{3}\{2(1 - b)\}^{3/2}, \quad h \simeq \frac{1}{\sqrt{6}}\frac{h_0}{\{2(1 - b)\}^{5/4}}. \tag{3.1.46}$$

It is somewhat <u>unexpected</u>, in view of the naive nature of the underlying theory, that some of the results of this section have been confirmed quantitatively by experiments. MQT-situations were set up to measure decay rates; results corresponding to Fig. 2.7 were obtained. Also, experiments of resonant tunneling (see Section 2.3) were performed with an external magnetic field controlled in such a way that the value of θ_{ex} was swept around π; results corresponding to Fig. 2.8 were obtained and the measured inter-peak distances were in precise agreement with the naive-quantum-theoretical prediction inferred from the numerically computed energy levels associated with the left and the right well, respectively, of the flux potential (3.1.20). (The width of each of the peaks was found to be much larger than that expected from Eq. (2.2.25). A detailed explanation is yet to come, but some of the intrinsic causes have been envisaged, such as the lifetime effect (i.e. the effect due to the flux making transition to the right-well ground state immediately after tunneling from the left well to one of the right-well excited states) and the finite-temperature effect.)

3.1.5 Friction coefficient

Some supplementary remarks are in order on Eq. (3.1.17). Adopting τ_0 as the unit of time, one may rewrite this equation in the dimensionless form for $\theta(t)$ as

$$\ddot{\theta}(t) + \eta\dot{\theta}(t) + \frac{(\Omega_0\tau_0)^2}{\gamma}\sin\theta(t) + (\Omega_0\tau_0)^2(\theta(t) - \theta_{\text{ex}}) = r_{\text{N}}(t), \quad (3.1.47)$$

$$\Omega_0 \equiv (\mathcal{L}\mathcal{C})^{-1/2}, \quad r_{\text{N}}(t) \equiv \frac{2\pi}{\Phi_0}\frac{\tau_0^2}{C}I_{\text{N}}(t), \quad (3.1.48)$$

$$\eta := \frac{\tau_0}{\mathcal{R}C} = \frac{(\delta\Phi)^2}{\mathcal{R}\hbar}h = \frac{\theta_0^2}{8\pi}\frac{\mathcal{R}_{\text{K}}}{\mathcal{R}}h, \quad (3.1.49)$$

where

$$\mathcal{R}_{\text{K}} := 2\pi\hbar/e^2 \simeq 2.58 \times 10^4\,\Omega \quad (3.1.50)$$

is called the *von Klitzing constant*. Equation (3.1.47) shows that the dimensionless constant η quantifies the friction. Accordingly, η is to be called the *friction coefficient*, which together with the fluctuating force on the right-hand side summarizes environmental effects on the classical dynamics of the SQUID.

In general, influence of the environment reduces the ground-state energy splitting and the decay rate below the values expected from Eqs. (2.4.1) and (2.4.5) (see Chapters 6 and 7). Environmental effects on the tunneling of SQUID can be quantitatively expressed in terms of η as in the classical case. The advantage with a SQUID is that an experimenter can control the value of $\mathcal{R}$ (hence η) almost at will. The low-temperature limiting value (i.e. the constant value in the low-temperature region in Fig. 2.7) of the experimentally measured decay rates agrees well with the *Caldeira–Leggett formula*, namely the quantum decay rate with environmental effects taken into account (see Chapter 7), and moreover, tends to the value (2.4.5) as $\mathcal{R}$ is increased.

The SQUID presented in this section is an over-simplified model[17] of a real SQUID. The origin of $\mathcal{R}^{-1}$ is not quite clear. Some of the candidates are[18]

(1) quasi-particles in the superconductor (their effect, however, is expected to vanish as the temperature is lowered), and
(2) metallic foils used to shield electromagnetic radiation.[19]

Note added in the English edition: In 2000, only 3 years after the Japanese edition was published, the ground-state energy splitting as well as the corresponding

---

[17] Called the *resistively shunted junction model* as to the treatment of the resistance.
[18] In the case of CBJJ, noises from external circuits are also candidates.
[19] In the resonant-tunneling experiment mentioned in Section 3.1.4, for example, noises (i.e. noise currents) may arise in the AuPd film used to shield the radiation of frequency higher than 1 GHz, this value being chosen in consideration of $\omega_0/2\pi \sim 40\,\text{GHz}$.

quantity relevant to excited states was detected by microwave spectroscopy experiments. See Refs. [12] and [13] in the bibliography for excited states and ground state, respectively. These results may be regarded as a fairly convincing evidence for a macroscopic superposition. In order to compare quantum mechanics with macrorealism, however, it is necessary to detect a time evolution corresponding to Eq. (2.2.11); see Chapter 9 for details. Such a time-domain experiment remains to be a challenge for experimentalists.

3.2 Liquid ^{3}He –^{4}He mixture

3.2.1 Phase separation in liquid ^{3}He – ^{4}He mixture

There are two kinds of stable isotopes of He, namely ^{3}He and ^{4}He. Each of them constitutes a substance which does not freeze but remains a liquid (at pressures below about 25 atm) down to the absolute-zero temperature. At temperatures below 0.8 K, it is not possible to mix them in arbitrary proportions. Below 0.1 K, in particular, the concentration of ^{3}He soluble in liquid ^{4}He is less than about 0.06 (i.e. 6%). Beyond this saturation concentration, the liquid mixture separates into a ^{3}He-rich phase of almost pure ^{3}He and a ^{3}He-dilute phase in which the concentration of ^{3}He is about 6%. This phenomenon is an example of phase separation.[20] In a recent experiment, a supersaturated liquid mixture, in which the concentration of ^{3}He was higher than the saturation concentration by a tiny amount Δx, was prepared and then Δx was gradually increased.[21] As a result, it was found that the phase separation took place when Δx reached a critical value Δx_c. Moreover, this phenomenon was confirmed to occur below 0.1 K down to 0.4 mK, and the value of Δx_c was about 2×10^{-3} independently of temperature below 10 mK. In view of the fact that the liquid ^{3}He –^{4}He mixture is an ideal system almost completely free from impurities, it is likely that the above result reflects intrinsic properties of the system.[22]

---

[20] In a laboratory on the Earth, the ^{3}He-rich phase floats on top of the ^{3}He-dilute phase, since a ^{3}He atom is less massive than a ^{4}He atom.

[21] The crux of the experiment was the precise control of Δx. This was achieved by an ingenious use of the superfluidity of liquid ^{4}He at temperatures in question; superfluid ^{4}He was pumped into or out of the liquid mixture through a superleak, that is, a porous material through which the superfluid component of ^{4}He can pass but ^{3}He can not. Since the entropy of the superfluid component is zero and hence is not accompanied by heat, this method, if carried out sufficiently slowly, enables the concentration of ^{3}He to be altered while keeping the liquid mixture in thermal equilibrium with the heat bath. For details about superfluid ^{4}He, see, e.g. Tsuneto: ibid.

[22] As explained below, the essence of this phenomenon is nucleation. In general, the vast majority of everyday nucleation is triggered by impurities such as a dust or by the wall of a container. In the case of liquid ^{3}He –^{4}He mixture, the wall is likely to be covered preferentially by ^{4}He. (He atoms are attracted to the wall by the van der Waals force. Since ^{4}He is more massive and hence its zero-point vibration is less than ^{3}He, the former is adsorbed to the wall more easily than the latter.) Hence, the wall is unlikely to trigger the nucleation of ^{3}He. Nucleation to be considered in the text is the <u>intrinsic nucleation</u>, which is caused by intrinsic properties of the system without the aid of external agents such as a dust or a wall. (Intrinsic nucleation is customarily called

Let us begin by discussing the mechanism of phase separation of the saturated liquid mixture at temperature β^{-1}. Although each of the atoms in the liquid performs thermal motion randomly, there is a certain probability that a large number of ^{3}He atoms gather together at a site by chance with the site also determined probabilistically. Such a thermal fluctuation creates a region of stable state (i.e. ^{3}He-rich phase) in the metastable state (i.e. the spatially-homogeneous supersaturated phase). Suppose that the region is spherical and call it a *bubble*.[23] Suppose that the free energy per unit volume of the metastable state is higher by the amount of ε_b than that of the stable state (ε_b is proportional to the excess concentration Δx).[24] On formation of a bubble of radius R, the free energy decreases by the amount of $(4\pi R^3/3)\varepsilon_b$ inside the bubble,[25] and at the same time increases by the amount of $4\pi R^2 \sigma_b$ at the interface (i.e. the surface of the bubble), where σ_b is the interfacial free energy per unit area ("surface tension"). Hence the free energy of the system increases by the amount of $F(R)$ given by

$$F(R) := 4\pi R^2 \sigma_b - (4\pi R^3/3)\varepsilon_b. \tag{3.2.1}$$

This function has the same form as the bumpy slope depicted in Fig. 1.4 (with $R_m = 0$). The peak position R_c and the peak height F_0 are determined by ε_b and σ_b. Even if a bubble of radius less than R_c emerges, the system will roll downhill to the left and return to the metastable state ($R = 0$). However, if a bubble of radius greater than R_c emerges, the system will roll downhill to the right; that is, the bubble will grow. Thus, R_c is the radius of the *critical bubble*; once a critical bubble is formed by thermal fluctuation, there can emerge a bubble that can actually grow thereby triggering a phase separation. Such an event is called *thermal nucleation* (or *classical nucleation* in contrast to the quantum nucleation discussed below),[26] and the frequency $\Gamma_{\text{cl nuc}}$ of its ocurrence is given by

$$\Gamma_{\text{cl nuc}} = N_{\text{nuc}}\, \nu\, \exp(-\beta F_0). \tag{3.2.2}$$

"homogeneous nucleation". But the latter terminology, which envisages nucleation taking place uniformly in space in the absence of spatially inhomogeneous agents (e.g. a wall), is not entirely appropriate; intrinsic nucleation is usually homogeneous in space, but not necessarily vice versa.) It goes without saying that the interpretation of the experimental data has to be completely changed if there are overlooked sources of extrinsic nucleation.

23 We use this terminology (rather than, say, "droplet") in view of the fact that the density of liquid ^{3}He is lower than that of the mixture.

24 The subscript b stands for "bubble".

25 The free energy inside the bubble is proportional to the volume of the bubble because the force acting between He atoms is of short range. In such a case, the bubble is most likely to be spherical since the surface area (hence the surface energy as well) for a given volume is minimized for a sphere. This is the reason why we are considering a spherical bubble from the outset. In case a long-range force should be involved, the free energy inside the bubble would depend not only on the volume but also on the shape of the bubble. For a charged fluid, for instance, the bubble is likely to be cigar-shaped so as to reduce the effect of Coulomb repulsion.

26 Accordingly, the critical bubble is formally called the *critical nucleus*.

Here, the constant N_{nuc} is the number of candidate sites at which nucleation may take place and is estimated to be of the order of the number ($\sim 10^{23}$) of ^{3}He atoms in the liquid mixture. Another constant ν is the so-called attempt frequency, namely the frequency at which a group of atoms attempt to form a bubble at each nucleation site. Since attempting to form a bubble is nothing but causing a local fluctuation of concentration, ν may be roughly estimated as

$$\nu^{-1} \sim \frac{\text{mean inter-atomic distance}}{\text{speed of sound}} \sim \frac{2\,\text{Å}}{50\,\text{m/s}} \sim 10^{-11}\,\text{s}. \qquad (3.2.3)$$

The physical meaning of $\Gamma_{cl\,nuc}$ is such that if the time of the order of $\Gamma_{cl\,nuc}^{-1}$ elapses after the system is prepared in the metastable state, it is quite probable that a critical bubble is formed. In this sense, $\Gamma_{cl\,nuc}^{-1}$ is called the lifetime of the metastable state with respect to the thermal nucleation.

The peak height F_0, which depends on the excess concentration Δx through ε_b, is estimated to be about 10 K if the experimentally found critical value ($\sim 2 \times 10^{-3}$) in the low-temperature limit is substituted for Δx. It follows, for $\beta^{-1} < 0.01$ K for instance, that

$$\Gamma_{cl\,nuc}^{-1} > 10^{-34} \times \exp(1000)\,\text{s} \sim 10^{390}\,\text{year}. \qquad (3.2.4)$$

This value far exceeds even the age of the present universe. Hence, at least below 10 mK, it is impossible for thermal nucleation to occur in a laboratory.[27] As mentioned above, however, many events of phase separation have been detected in experiments performed at temperatures much lower than 10 mK.

A conceivable mechanism to trigger the phase separation at such low temperatures is a quantum fluctuation rather than a thermal one. If the constants σ_b and ε_b on the right-hand side of Eq. (3.2.1) are replaced by their respective zero-temperature counterparts, the free energy $F(R)$ may be interpreted as the potential energy $U(R)$; a state of the system (i.e. the liquid mixture) may be represented by the bubble radius R, and the time evolution of the system may be viewed as the motion of a particle (whose position is R) under the potential $U(R)$. In other words, the bubble radius R plays the role of the macrosystem R in the sense of Section 1.4.2. Since the kinetic energy of the particle (namely, the energy of the metastable state of the liquid mixture) is less than U_0, the particle is not allowed to go over the potential barrier in classical mechanics. Quantum mechanics, however, may allow the particle to tunnel through the barrier; a bubble of radius of the order of R_0 (see Fig. 1.4) may be formed via the tunnel effect (this is called *quantum nucleation*), and then grow as in the thermal case.[28]

---

[27] This conclusion remains intact even if the estimated values for N_{nuc} and ν were off by several orders of magnitude and/or that for F_0 were off by the factor of 10.

[28] In the case of quantum fluctuation, R_c loses its significance and is taken over by R_0.

3.2.2 Rayleigh–Plesset model

In order to describe a tunneling phenomenon, it is not enough to know the potential. The knowledge of the kinetic energy associated with the expanding and/or contracting motion of the bubble is necessary as well. Rayleigh derived the latter more than 80 years ago.[29] According to him,[30] the relevant motion of a bubble in d-dimensional space is governed by the following Lagrangian:

$$L(R, \dot{R}) = \frac{1}{2} M(R) \dot{R}^2 - U(R), \qquad (3.2.5)$$

$$M(R) := R^m \mu_b, \qquad (3.2.6)$$

$$U(R) := \mathcal{S}_{d-1} R^{d-1} \sigma_b - \mathcal{B}_d R^d \varepsilon_b, \qquad (3.2.7)$$

where μ_b, σ_b and ε_b are positive constants, and $M(R)$ is the effective inertial mass associated with the motion in question[31] and is to be called the *effective bubble mass* for brevity. The form (3.2.6) has been assumed for simplicity; the exponent m (> 0) is to be called the *mass exponent*. In Eq. (3.2.7), the constants $\mathcal{S}_{d-1}$ and $\mathcal{B}_d$ denote the area of the $(d-1)$-dimensional unit sphere and the volume of the d-dimensional unit ball, respectively:

$$\mathcal{S}_{d-1} = d\mathcal{B}_d = \frac{2\pi^{d/2}}{\Gamma(d/2)}. \qquad (3.2.8)$$

Although this model (to be called the *Rayleigh–Plesset model*) was not conceived particularly with nucleation phenomena in mind, it is frequently used in dealing with quantum nucleation. As shown shortly, it is expected that $m = d = 3$ for a bubble in a fluid (the original Rayleigh–Plesset model).

Let us derive the formula (3.2.6) for the effective bubble mass. Consider for simplicity a single-component incompressible fluid in d-dimensional space. Suppose that there is a spherical bubble of radius $R(t)$ in the fluid at time t, and that the fluid is in the stable phase with mass density $\rho_<$ inside the bubble and in the metastable phase with mass density $\rho_>$ outside, where the subscripts $<$ and $>$ stand for inside and outside, respectively (Fig. 3.5). Suppose also that the mass–density field is spherically symmetric with respect to the center of the bubble, and denote the field with $\rho(r, t)$, then

$$\rho(r, t) = \rho_> \, \theta(r - R(t)) + \rho_< \, \theta(R(t) - r), \qquad (3.2.9)$$

---

[29] "... reading O. Reynolds's description of the sounds emitted by water in a kettle as it comes to the boil, and their explanation as due to the partial or complete collapse of bubbles as they rise through cooler water", Rayleigh investigated the dynamics of expansion and contraction of a bubble: Rayleigh, On the Pressure Developed in a Liquid During the Collapse of a Spherical Cavity, *Phil. Mag.* **xxxiv** (July–December 1917), 94.

[30] The surface-energy term in the potential was considered later by M. S. Plesset, The Dynamics of Cavitation Bubbles, *J. Appl. Mech.* **16** (1949), 277. In the text we work with spatial dimension d for generality.

[31] We leave out translation motion of the bubble.

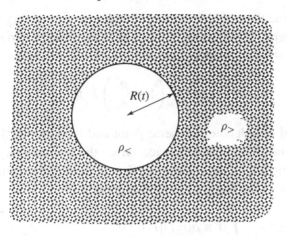

Fig. 3.5. Bubble.

where r denotes the distance from the center of the bubble and θ is the step function.[32] On the other hand, it follows from the law of conservation of mass that

$$\frac{\partial \rho(r, t)}{\partial t} + \frac{1}{r^{d-1}} \frac{\partial}{\partial r} \{r^{d-1} \rho(r, t) v(r, t)\} = 0, \qquad (3.2.10)$$

where $v(r, t)$ denotes the radial component of the velocity field of the fluid. These two equations together lead to

$$-(\rho_> - \rho_<) \dot{R}(t) \delta(r - R(t)) + \frac{1}{r^{d-1}} \frac{\partial}{\partial r} \{r^{d-1} \rho(r, t) v(r, t)\} = 0. \quad (3.2.11)$$

Except at the surface of the bubble, the first term on the left-hand side vanishes and the density is constant. Hence

$$v(r, t) \propto r^{-(d-1)} : r \neq R(t). \qquad (3.2.12)$$

This equation, combined with the requirement that the velocity field should be finite at the center of the bubble, gives

$$v(r, t) = 0 : r < R(t). \qquad (3.2.13)$$

Thus, the velocity field may be expressed in the form

$$v(r, t) = v_R(t) \left(\frac{R(t)}{r}\right)^{d-1} \theta(r - R(t)). \qquad (3.2.14)$$

[32] The step function θ is defined as

$$\theta(x) := \begin{cases} 1 & : \ x > 0, \\ 0 & : \ x < 0. \end{cases}$$

One may determine the coefficient $v_R(t)$ by extracting remaining information from Eq. (3.2.11); integrating both sides of the equation with respect to r over the interval $(R(t) - 0, \ R(t) + 0)$, and making use of Eq. (3.2.14), one finds

$$v_R(t) = \left(1 - \frac{\rho_<}{\rho_>}\right) \dot{R}(t). \tag{3.2.15}$$

Now, let $\mathbf{x}$ and $d^d\mathbf{x}$ denote a generic point and the volume element around it, respectively, of the d-dimensional space. Since the kinetic energy L_K associated with the expanding and/or contracting motion of the bubble should originate from the kinetic energy of the fluid, L_K can be expressed as

$$
\begin{aligned}
L_K &= \frac{1}{2} \int d^d\mathbf{x} \, \rho(r, t)\{v(r, t)\}^2 \\
&= \frac{1}{2}\{\dot{R}(t)\}^2 \mathcal{S}_{d-1}\rho_{\text{eff}} \int_{R(t)}^{\infty} dr \, r^{d-1} \left(\frac{R(t)}{r}\right)^{2(d-1)} \\
&= \frac{1}{2} \rho_{\text{eff}} \frac{\mathcal{S}_{d-1}}{d-2}\{R(t)\}^d \{\dot{R}(t)\}^2,
\end{aligned}
\tag{3.2.16}
$$

$$\rho_{\text{eff}} := \left(1 - \frac{\rho_<}{\rho_>}\right)^2 \rho_>. \tag{3.2.17}$$

Comparison of this result with Eqs. (3.2.5)–(3.2.6) gives[33]

$$m = d, \quad \mu_{\text{b}} = \frac{\mathcal{S}_{d-1}}{d-2} \rho_{\text{eff}}. \tag{3.2.18}$$

3.2.3 Naive quantum theory

Let us cast the equations into the standard form in the spirit of Section 2. 1. In view of the fact that the effective mass now depends on R, we modify the definition of the unit of time τ_0 as follows:

$$M(R_0)\left(\frac{2R_0}{(m+2)\tau_0}\right)^2 = U_0. \tag{3.2.19}$$

---

[33] The following result is valid only for $d \geq 3$; the integral in Eq. (3.2.16) diverges for $d \leq 2$. This divergence is caused by the long tail of the velocity field, which in turn is caused by the assumption of the incompressibility (hence the infinite speed of sound) of the fluid. If the two-dimensional case should be considered, the idealization of incompressible fluid is not allowed. Also, for systems other than a simple fluid, the effective bubble mass is a complicated function of R in general. For example, in the case of crystallization where a crystal is nucleated out of a fluid (i.e. a solution containing the elements of the crystal as a solute) at the surface of a "seed crystal" put in the fluid, it has been suggested that $M(R)$ is proportional to $R \log R$ if the nucleating crystal is supposed to be a disk of radius R; see M. Uwaha, *J. Low Temp. Phys.* **52** (1983), 15.

Definitions of R_0 and U_0 are the same as those given in Fig. 1.4 (with $R_m = 0$ and $U_- = 0$), that is,

$$R_0 = \frac{\mathcal{S}_{d-1}\sigma_b}{\mathcal{B}_d\varepsilon_b}, \tag{3.2.20}$$

$$U_0 = U\left(\frac{d-1}{d}R_0\right) = \frac{(d-1)^{d-1}}{d^d}\frac{(\mathcal{S}_{d-1}\sigma_b)^d}{(\mathcal{B}_d\varepsilon_b)^{d-1}}, \tag{3.2.21}$$

$$\tau_0 = \frac{2}{m+2}\left\{\frac{d^d}{(d-1)^{d-1}}\frac{(\mathcal{S}_{d-1}\sigma_b)^{m-d+2}}{(\mathcal{B}_d\varepsilon_b)^{m-d+3}}\mu_b\right\}^{1/2}. \tag{3.2.22}$$

Adopting these constants as fundamental units, we introduce dimensionless variables as

$$t := t_{conv}/\tau_0, \quad q := (R/R_0)^{(m+2)/2}, \tag{3.2.23}$$

$$L(q, \dot{q}) := L_{conv}(R, \dot{R})/U_0 = \frac{1}{2}\dot{q}^2 - V(q), \tag{3.2.24}$$

where the overdot $\dot{}$ on the right-hand side denotes, of course, the differentiation with respect to the dimensionless time t, and

$$V(q) := U(R)/U_0 = \frac{d^d}{(d-1)^{d-1}}((R/R_0)^{d-1} - (R/R_0)^d) \tag{3.2.25}$$

$$= \frac{d^d}{(d-1)^{d-1}}\left(q^{2(d-1)/(m+2)} - q^{2d/(m+2)}\right). \tag{3.2.26}$$

The classical Hamiltonian corresponding to the Lagrangian (3.2.24) may be naively cast into the (q-representation of) quantum-mechanical Hamiltonian H of the form of (2.1.10) with

$$h \equiv \frac{\hbar}{U_0\tau_0} = \frac{m+2}{2}\left\{\frac{d^d}{(d-1)^{d-1}}\frac{(\mathcal{B}_d\varepsilon_b)^{m+d+1}}{(\mathcal{S}_{d-1}\sigma_b)^{m+d+2}}\frac{1}{\mu_b}\right\}^{1/2}\hbar. \tag{3.2.27}$$

In the case of $m \neq 0$, we are confronted with a unfamiliar problem of the quantum mechanics of a particle whose mass depends on its position: our intuition is likely to fail. In terms of the variable q defined by Eq. (3.2.23), however, the problem is converted into one of a constant mass and is easier to deal with. This is the reason why we work with q instead of R/R_0, Of course, the price is paid by $V(q)$ (hereafter to be called the _bubble potential_) whose shape is markedly different from that of Fig. 1.4. The difference is particularly conspicuous near the origin, where the bubble potential is so steep that its derivative diverges (Fig. 3.6).

Although the variable q is restricted to be positive, the Hamiltonian H is Hermitian if either the Dirichlet or the Neumann boundary condition is imposed on the wavefunction at $q = 0$. One may solve the Schrödinger equation (2.1.9) under either of the boundary conditions with the naive WKB method to obtain the decay

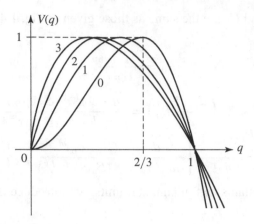

Fig. 3.6. Bubble potential (the case of $d = 3$ with the number denoting m).

rate of the virtual ground state. The result in the case of $d = 3$ is found as

$$\Gamma \sim h^{-m/(m+4)} \exp\left[-B/h + 2bh^{-m/(m+4)} + \mathcal{O}(h^\xi)\right] : m > 0, \qquad (3.2.28)$$

$$\xi \equiv \min\left\{0, \ -\frac{m-2}{m+4}\right\}. \qquad (3.2.29)$$

The coefficient b of the second term in the exponent is a positive number of $\mathcal{O}(1)$. For a given h, the value of the decay rate predicted by the above formula is far greater than that predicted by Eq. (2.4.5) for the case of $m = 0$, and the difference increases as h decreases. The reason is simple; since the bubble potential for $m \neq 0$ is steep near the origin, the associated virtual-ground-state energy E_0 (namely, the zero-point energy in the <u>wedge-shaped metastable well</u>) is much higher than that for the case of $m = 0$, and the system is far easier to tunnel. A dimensional analysis by use of the indeterminacy relation (or, more formally, a calculation on the basis of the variational principle) gives

$$E_0 \sim \min_q \left\{\frac{h^2}{q^2} + q^{4/(m+2)}\right\} \sim h^{4/(m+4)}, \qquad (3.2.30)$$

$$S(E_0) = S(0) + \mathcal{O}\left(h^{4/(m+4)}\right), \qquad (3.2.31)$$

where $S(E)$ is the sub-barrier action defined by Eq. (2.4.14). The last term in the above equation gives the second term in the exponent of (3.2.28).

Since (3.2.28) is to be interpreted as the rate of formation of a single bubble, the quantum nucleation rate $\Gamma_{\text{qu nuc}}$ to be compared with $\Gamma_{\text{cl nuc}}$ is the right-hand side of (3.2.28) multiplied by N_{nuc} and τ_0^{-1}, where the latter factor is needed to recover the ordinary unit:

$$\Gamma_{\text{qu nuc}} \sim N_{\text{nuc}} \, \tau_0^{-1} h^{-m/(m+4)} \exp\left[-B/h + 2bh^{-m/(m+4)}\right]. \qquad (3.2.32)$$

This is to be called the *modified Lifshitz–Kagan formula of quantum nucleation rate*.[34]

Assume that this formula is applicable to the ^3He –^4He liquid mixture. Equation (3.2.27) with $m = 3$ gives

$$h \propto \varepsilon_b^{7/2} \propto (\Delta x)^{7/2}. \qquad (3.2.33)$$

Since h^{-1} appears in the exponent, the quantum nucleation rate increases sharply as Δx is increased even slightly. Hence it is expected that nucleation (and the consequent phase separation) will take place as soon as Δx has reached a certain critical value. This theoretical expectation is in accord with the experimental result that the critical excess concentration, Δx_c, is independent of temperature below 10 mK.[35] Comparison of the experimental data with the theory leads to the estimate $h \sim 0.015$ and $R_0 \sim 20$ Å. The emergence of a bubble of radius 20 Å of the stable phase (with the ^3He-concentration of almost 100%) in the homogeneous supersaturated liquid mixture (with the ^3He-concentration of about 6%) must involve He atoms within the radial distance of 50 Å (namely, about 10^4 atoms) at least. In this sense, the homogeneous metastable state is macroscopically distinct from the state with a bubble. Thus, provided that the possibility of an extrinsic nucleation can be ruled out in the experiment, it may be tentatively concluded that the observed phase separation is due to *macroscopic quantum nucleation* and that it is a kind of MQT with the S-Cattiness estimated[36] as **D** $\sim 10^4$. Incidentally, it should be noted that

$$\Gamma_{qu\ nuc}^{(m=3)} / \Gamma_{qu\ nuc}^{(m=0)} > 10^{20} : h < 0.1. \qquad (3.2.34)$$

Therefore, the second term in the exponent of Eq. (3.2.32) is crucial in an analysis of experimental data.

In general, a change of the state of matter from a metastable state to a stable one is called a first-order phase transition. It is thought to be ubiquitous, encompassing not only such familiar phenomena as the emergence of a droplet in a supercooled vapor but also the purported emergence of a region of true vacuum in the homogeneous false vacuum at an early stage of the universe. Being a kind of MQT, macroscopic quantum nucleation occupies a significant role in the Leggett program as mentioned in Chapter 1. Apart from this, it has a universal significance as a mechanism to trigger first-order phase transitions at very low temperatures.

---

[34] In the classic work (Ref.[19]) by Lifshitz and Kagan, who pioneered a quantum-mechanical treatment of the first-order phase transition, the formula without the second term in the exponent was derived; the effect of the zero-point energy (3.2.30) was left out of consideration. This is the reason for the adjective *modified*.

[35] To be rigorous, the value of Δx_c should vary from one experimental run to another. Its fluctuation, however, is expected to be quite small; if the numerical estimate quoted below is correct, $\Gamma_{qu\ nuc}$ increases by the factor of 10 if Δx is increased only by the amount of 10 ppm ($= 10^{-5}$) around the value of 2×10^{-3}.

[36] Although this value is much smaller than that for a SQUID, the phenomenon in question may be said to be "semi-macroscopic" if not macroscopic.

Experiments in the search for quantum nucleation are under way with liquid ^{4}He, too. As liquid ^{4}He is depressurized and reaches a certain critical pressure that is low enough below the equilibrium pressure, a cavity (a bubble of vapor) will emerge suddenly in the otherwise homogeneous liquid. This expected phenomenon is called *cavitation*, and is being analyzed with the help of the theory introduced above.

It should be mentioned, however, that this theory has various problems. First, the validity of the model itself is somewhat dubious. Although the Rayleigh–Plesset model may certainly be appropriate to an everyday-life-scale bubble such as the one in the boiling water in a kettle, it is not warranted to be applicable to a bubble of radius 20 Å or so, not to speak of a nearly zero-radius bubble (i.e. a bubble just being born out of the homogeneous liquid). Second, the validity of the naive quantum theory (2.1.9)–(2.1.10) is not quite obvious. Can one go over to the quantum theory via the Lagrangian (3.2.24) treating q as if it were a position variable for a particle? Furthermore, even if this is legitimate, the Lagrangian (3.2.24) does not uniquely lead to the Hamiltonian (2.1.10). To adopt the latter corresponds to adopting a particular *operator ordering* between $\hat{R}$ and its conjugate momentum $\hat{P}$ in the kinetic-energy term as is seen from

$$-\frac{h^2}{2}\frac{\partial^2}{\partial q^2} \propto -\frac{\hbar^2}{2}\{M(R)\}^{-1/2}\frac{\partial}{\partial R}\{M(R)\}^{-1/2}\frac{\partial}{\partial R}. \qquad (3.2.35)$$

To change the operator ordering is equivalent to augmenting the potential by a term of $\mathcal{O}(h^2)$.[37] Third, there is no guiding principle to decide the boundary condition to be imposed on the wavefunction at $q = 0$. Fourth, the environmental effects, which have been neglected, can be important.

All of these problems could be settled only by starting from a more microscopic theory (e.g. quantum field theory) and precisely sorting out relevant macroscopic degrees of freedom.

3.2.4 Field theory of nucleation

One would naturally hesitate to "quantize" a model derived on the basis of fluid dynamics. Nucleation, which is thought to be a phenomenon of emergence of a bubble in the otherwise spatially-homogeneous metastable state, requires a field-theoretic consideration. It is true that fluid dynamics is a kind (or, rather a prototype) of field theory, but it is a classical one. A convincing description of nucleation should be based on quantum field theory. In this subsection, we start from a field-theoretic model, which should be amenable to a quantum-mechanical treatment, to investigate whether and how it may be reduced to the Rayleigh–Plesset model within classical-mechanical treatment.

[37] This correction term does not contribute to the lowest-order WKB approximation under consideration. It constitutes, however, an important problem in deciding the nature of the quantum mechanics obeyed by the system; in particular, it could dictate the boundary condition mentioned in the following line.

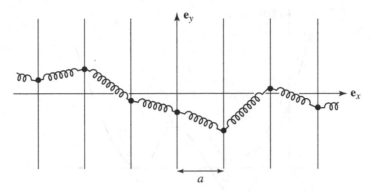

Fig. 3.7. Chain of beads on a broken abacus.

Imagine a broken abacus (Fig. 3.7). The rods, each of which supports a bead, are aligned parallel to the e_y-axis with the mutual separation a. Each bead can slide smoothly along the supporting rod. Each pair of neighboring beads is connected by a spring of which the spring constant is $\kappa^{(a)}$ and the natural length is idealized to be 0. Let u_l be the position (along the e_y-axis) of the l-th bead, which is supposed to feel the potential $U^{(a)}(u_l)$. Suppose also that the mass of each bead is $M_{\text{bead}}^{(a)}$. Then, the dynamics of the chain of beads is governed by the Lagrangian $L_{\text{chain}}^{(a)}$ defined as

$$L_{\text{chain}}^{(a)} := \sum_{l=-\infty}^{\infty} \left\{ \frac{M_{\text{bead}}^{(a)}}{2} \dot{u}_l^2 - \frac{\kappa^{(a)}}{2} (u_l - u_{l-1})^2 - U^{(a)}(u_l) \right\}. \qquad (3.2.36)$$

Introducing a positive constant $\mathcal{K}$ of dimension of [action] $\times$ [mass]$^{-1/2}$ and the re-scaled variable

$$\phi_l := \sqrt{\kappa^{(a)}a}\, u_l/\mathcal{K}, \qquad (3.2.37)$$

we rewrite the above Lagrangian as

$$L_{\text{chain}}^{(a)} = \sum_{l=-\infty}^{\infty} a \left\{ \frac{\mathcal{K}^2}{2c^2} \dot{\phi}_l^2 - \frac{\mathcal{K}^2}{2} \left(\frac{\phi_l - \phi_{l-1}}{a} \right)^2 - \frac{1}{a} U^{(a)}(\mathcal{K}\phi_l/\sqrt{\kappa^{(a)}a}) \right\},$$

$$(3.2.38)$$

where

$$c \equiv \sqrt{\kappa^{(a)}a^2/M_{\text{bead}}^{(a)}} \qquad (3.2.39)$$

is a constant of dimension of speed. Now we introduce a real field[38] $\phi(x, t)$ such that

$$\phi(la, t) = \phi_l(t), \qquad (3.2.40)$$

---

[38] A field whose value is a real number.

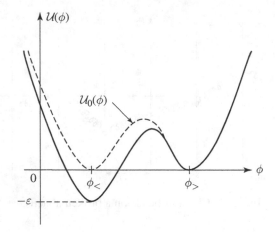

Fig. 3.8. Field potential: tilted double well.

and take the *continuum limit* (i.e. $a \downarrow 0$) to convert the chain of beads into a string:[39]

$$L_{\text{string}} := \lim_{a \downarrow 0} L_{\text{chain}}^{(a)}$$

$$= \int dx \left[\frac{\mathcal{K}^2}{2} \left\{ \left(\frac{\partial \phi(x,t)}{c \, \partial t} \right)^2 - \left(\frac{\partial \phi(x,t)}{\partial x} \right)^2 \right\} - \mathcal{U}(\phi(x,t)) \right], \quad (3.2.41)$$

$$\mathcal{U}(\phi) \equiv \lim_{a \downarrow 0} \frac{1}{a} U^{(a)}(\mathcal{K}\phi / \sqrt{\kappa^{(a)}a}). \quad (3.2.42)$$

Generalizing the above consideration, one may arrive at the *non-linear Klein–Gordon field* governed by the following Lagrangian density $\mathcal{L}$:[40]

$$\mathcal{L} := \frac{\mathcal{K}^2}{2} \left\{ \left(\frac{\partial \phi}{c \, \partial t} \right)^2 - (\nabla \phi)^2 \right\} - \mathcal{U}(\phi), \quad (3.2.43)$$

where $\phi \equiv \phi(\mathbf{x}, t)$ is a real scalar field, $\mathbf{x}$ and ∇ denote a generic point and gradient in the d-dimensional space, and c is a positive constant of dimension of speed. If the *field potential* $\mathcal{U}$ were a quadratic polynomial in ϕ, the above field would be nothing but the Klein–Gordon field. A field potential appropriate to describe nucleation is a slightly tilted double well depicted in Fig. 3.8. This figure is similar to Fig. 2.2, but the abscissa does not represent a particle position but the field.

---

[39] The constants $M_{\text{bead}}^{(a)}$ and $\kappa^{(a)}$ as well as the function $U^{(a)}(u)$ are supposed to depend on the inter-rod separation a in such a way as to ensure the existence of the continuum limit (3.2.41).

[40] The constant $\mathcal{K}$ has been introduced, for the convenience of the application to be made in the latter half of this subsection, so that ϕ has the dimension of $[\text{length}]^{-d/2}$. In conventional literature, this constant is absorbed into the field: $\phi_{\text{conv}} = \mathcal{K}\phi$.

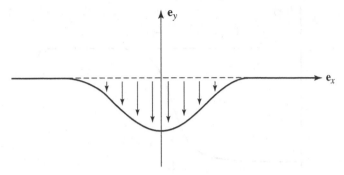

Fig. 3.9. Progressive motion of a dislocation (or process of nucleation).

The motion of a *dislocation* in a crystal may be described by this model with $d = 1$.[41] Roughly speaking, a dislocation is a string lying mostly along the e_x-axis in Fig. 3.7 and moves towards the direction of $-e_y$. It is energetically impossible for the entire linear string $\{\phi(x, t) = \phi_> \mid \forall x\}$ to move over the barrier to the adjacent line $\{\phi(x, t) = \phi_< \mid \forall x\}$ at a stroke. Rather, only a portion of the string moves over the barrier first to produce a situation depicted in Fig. 3.9, and then the portion widens itself sideways thereby allowing the traversal of the entire string over the barrier. If the field ϕ describing the local position of a dislocation is re-interpreted as a field describing some sort of phase of matter (and, accordingly, e_y is re-interpreted as a direction in the space of phases of matter), the progressive motion of a dislocation may be re-interpreted as a process of nucleation in the one-dimensional space.

The field ϕ, which can assume both positive and negative values, may also be regarded as a continuum approximant[42] of a component (along the anisotropy axis) of local spin for the system of spins located at lattice sites of a uniaxial magnet. Suppose that an external magnetic field has been applied along the easy axis of magnetization so that all the spins have been aligned in that direction. If the direction of the magnetic field is suddenly reversed, the system is prepared in the metastable state where all the spins are aligned anti-parallel to the magnetic field (i.e. $\{\phi(\mathbf{x}, t) = \phi_> \mid \forall \mathbf{x}\}$). After some time has elapsed, there will emerge a region in which the spins have been flipped to align parallel to the magnetic field (i.e. $\phi(\mathbf{x}, t) = \phi_<$).[43] Such a region is to be called a *bubble*. We assume, for simplicity, that the bubble is spherical with the center chosen to be the origin of the space.

---

[41] With the proviso that the field potential is not a tilted double well but a washboard reflecting the periodicity of the crystal.

[42] Here, the continuum approximation is to be effected with respect not only to spatial positions but also to the spin density.

[43] If one were concerned with a phase transition of the vacuum, the field ϕ would describe "phases of the vacuum"; $\phi_>$ and $\phi_<$ would correspond to the *false vacuum* and the *true vacuum*, respectively.

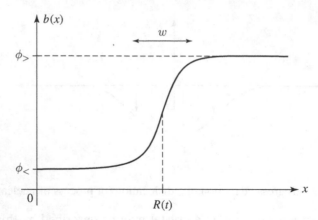

Fig. 3.10. Profile of the interface (surface of the bubble).

The interface (i.e. the surface of the bubble), however, is not a geometrical sphere but possesses a <u>profile</u>; as $r \, (\equiv |\mathbf{x}|)$ increases, the field $\phi(\mathbf{x}, t)$ increases continuously from $\phi_<$ to $\phi_>$. We define the *bubble radius*[44] $R(t)$ as the radius of the sphere on which $\phi(\mathbf{x}, t) = (\phi_< + \phi_>)/2$. In the case of a large bubble for which the profile is clear-cut, the interface may be approximated locally to be planar. If the magnitude ε of the tilt of the field potential is small, the profile will be such as to minimize

$$\int d^d\mathbf{x} \left\{ \frac{\mathcal{K}^2}{2}(\nabla\phi)^2 + \mathcal{U}_0(\phi) \right\}, \tag{3.2.44}$$

which is the *potential energy of the field* under the symmetric double well $\mathcal{U}_0$ depicted by the broken curve in Fig. 3.8; the locations of minima of $\mathcal{U}_0$ are chosen to coincide with those of $\mathcal{U}$.

Hence the field describing the situation with a single bubble of radius $R(t)$ centered at the origin may be expressed as

$$\phi(\mathbf{x}, t) \simeq b(r - R(t)), \tag{3.2.45}$$

where $b(x)$ is the solution to

$$\mathcal{K}^2 b''(x) = \mathcal{U}_0'(b(x)), \quad b(-\infty) = \phi_<, \quad b(\infty) = \phi_>, \tag{3.2.46}$$

with the prime $'$ on b and $\mathcal{U}_0$ denoting x- and ϕ-differentiation, respectively:

$$b'(x) \equiv \frac{db(x)}{dx}, \quad \mathcal{U}_0'(\phi) \equiv \frac{d\mathcal{U}_0(\phi)}{d\phi}. \tag{3.2.47}$$

The solution $b(x)$ is depicted schematically in Fig. 3.10. In general, this sort of configuration of a field is called a *kink*. The width of the interfacial portion is to be denoted by w.

---

[44] The definition of a bubble radius is not unique.

If Eq. (3.2.45) is assumed, the infinitely many degrees of freedom associated with the field may be reduced to the single degree of freedom $R(t)$. Let L be the effective Lagrangian to govern $R(t)$, then

$$L = \int d^d x \, \mathcal{L} \Big|_{\phi=b(r-R(t))} = L_K + L_S + L_B, \tag{3.2.48}$$

$$L_K \equiv \int d^d x \frac{K^2}{2c^2} \{\partial b(r-R(t))/\partial t\}^2$$

$$\simeq \frac{1}{2} \{R(t)\}^{d-1} \{\dot{R}(t)\}^2 \mu_b, \tag{3.2.49}$$

$$\mu_b \equiv \frac{K^2}{c^2} \mathcal{S}_{d-1} \int_{-\infty}^{\infty} dx \{b'(x)\}^2, \tag{3.2.50}$$

$$L_S \equiv -\int d^d x \left\{ \frac{K^2}{2} \{\nabla b(r-R(t))\}^2 + \mathcal{U}_0(b(r-R(t))) \right\}$$

$$\simeq -\{R(t)\}^{d-1} \mathcal{S}_{d-1} \sigma_b, \tag{3.2.51}$$

$$\sigma_b \equiv \int_{-\infty}^{\infty} dx \left\{ \frac{K^2}{2} \{b'(x)\}^2 + \mathcal{U}_0(b(x)) \right\}, \tag{3.2.52}$$

$$L_B \equiv -\int d^d x \{\mathcal{U}(b(r-R(t))) - \mathcal{U}_0(b(r-R(t)))\} = \{R(t)\}^d \mathcal{B}_d \varepsilon_b, \tag{3.2.53}$$

$$\varepsilon_b \equiv -\frac{1}{\{R(t)\}^d \mathcal{B}_d} \int_{r<R(t)} d^d x \{\mathcal{U}(b(r-R(t))) - \mathcal{U}_0(b(r-R(t)))\}$$

$$\simeq \varepsilon, \tag{3.2.54}$$

where ε is the magnitude of the tilt of the field potential (Fig. 3.8), and the approximate inequality $\simeq$ stipulates those equalities which are valid only under the *thin-wall approximation* applicable to the case of $R(t) \gg w$. Incidentally, it follows from the defining equation (3.2.46) that

$$\frac{K^2}{2} \{b'(x)\}^2 = \mathcal{U}_0(b(x)), \tag{3.2.55}$$

and accordingly that

$$\sigma_b = c^2 \mu_b / \mathcal{S}_{d-1} = K^2 \int_{-\infty}^{\infty} dx \{b'(x)\}^2 = K \int_{\phi_<}^{\phi_>} d\phi \sqrt{2\mathcal{U}_0(\phi)}. \tag{3.2.56}$$

The above-obtained pieces L_K, L_S and L_B correspond to the kinetic energy, the interfacial energy and the volume energy, respectively, in the Lagrangian (3.2.5).

Thus, it has been shown that the mass exponent m is equal to $d-1$ in the case of the non-linear Klein–Gordon field: The Rayleigh–Plesset model with $m = d$, which is supposed to describe a bubble in a fluid, does not follow from the non-linear Klein–Gordon field. The reason should be fairly clear from the calculations presented so far; the non-linear Klein–Gordon field does not possess a conservation law corresponding to the law of conservation of mass for a fluid, and as a consequence only the field near the interface contributes to the effective mass.

We therefore turn our attention to the *generalized non-linear Schrödinger field* [45] governed by the following Lagrangian density:

$$\mathcal{L} := \frac{i\mathcal{A}}{2}\frac{\partial\psi}{\partial t}\psi^* + \text{c.c.} - \frac{\mathcal{A}^2}{2m_A}|\nabla\psi|^2 - \mathcal{U}(|\psi|), \qquad (3.2.57)$$

where $\psi \equiv \psi(\mathbf{x}, t)$ is a complex scalar field, $\mathcal{U}$ is the same tilted double well as used in Eq. (3.2.43), and $\mathcal{A}$ and m_A are positive constants of dimension of action and mass, respectively.[46] Let $I[\psi]$ be the action functional associated with the Lagrangian density (3.2.57):

$$I[\psi] := \int dt \int d^d\mathbf{x}\mathcal{L}. \qquad (3.2.58)$$

It turns out to be convenient to express ψ in the polar form, namely in terms of the amplitude and the phase:

$$\psi(\mathbf{x}, t) = \phi(\mathbf{x}, t)\exp\{i\theta(\mathbf{x}, t)\}, \qquad (3.2.59)$$

where $\phi (\geq 0)$ and θ are real fields. Substituting the above expression for ψ in Eq. (3.2.57), one finds

$$\mathcal{L} = -\mathcal{A}\phi^2\frac{\partial\theta}{\partial t} - \frac{m_A}{2}\phi^2\mathbf{v}^2 - \frac{\mathcal{A}^2}{2m_A}(\nabla\phi)^2 - \mathcal{U}(\phi), \qquad (3.2.60)$$

where

$$\mathbf{v} := (\mathcal{A}/m_A)\nabla\theta \qquad (3.2.61)$$

has the dimension of speed. The last equation shows that the phase field $\theta(\mathbf{x}, t)$ plays the role of "velocity potential".

We begin by finding a stationary point of the action functional with respect to θ. Suppose that the system is placed in a container such that the normal component (i.e. the component perpendicular to the container wall) of the velocity field vanishes at the wall. Then the stationary point is determined by

$$0 = \frac{m_A}{\mathcal{A}}\frac{\delta}{\delta\theta}I[\psi] = \frac{\partial\rho}{\partial t} + \nabla\cdot(\rho\mathbf{v}), \qquad (3.2.62)$$

$$\rho := m_A\phi^2. \qquad (3.2.63)$$

---

[45] The phrase 'non-linear Schrödinger field' is conventionally reserved for the case of $\mathcal{U}(|\psi|) = |\psi|^4$. This is the reason for the adjective <u>generalized</u>. Readers, who are unfamiliar with this Lagrangian density and wonder what is the rationale for this apparently bizarre form, may wish to read the footnote following Eq. (3.2.85).

[46] The quantum theory of the free field (i.e. the case of $\mathcal{U} = 0$ in the above Lagrangian density) reveals that m_A is the mass of a particle associated with the field. If the field is interpreted to describe liquid ^{4}He, say, then the particle in question is nothing but a ^{4}He atom. This motivates the subscript A, which stands for 'atom'.

This equation has precisely the form of the law of conservation of mass for a fluid. We are thus led to interpret $\rho(\mathbf{x}, t)$ defined by Eq. (3.2.63) as the mass density of the fluid. Let us eliminate θ from $I[\psi]$ by use of the above equation and denote the result with $\tilde{I}[\phi]$, then

$$\tilde{I}[\phi] = \int dt \int d^d\mathbf{x}\, \tilde{\mathcal{L}} - \mathcal{A} \int d^d\mathbf{x}\, [\phi^2\theta]_{t=-\infty}^{t=\infty}, \qquad (3.2.64)$$

$$\tilde{\mathcal{L}} := \frac{1}{2}\rho\mathbf{v}^2 - \frac{\mathcal{A}^2}{2m_A}(\nabla\phi)^2 - \mathcal{U}(\phi). \qquad (3.2.65)$$

Perhaps, this last form looks more familiar than (3.2.57). Of course, $\mathbf{v}$ is a functional of ρ in accordance with Eq. (3.2.62). Hence, $\tilde{I}[\phi]$ is a functional of ϕ alone. For simplicity, we drop the second term on the right-hand side of Eq. (3.2.64).[47]

So far, a bubble has not been introduced yet. Consider now a situation where there is a single spherical bubble centered at the origin. One may solve Eq. (3.2.62) for $\mathbf{v}$ under the condition that $\mathbf{v}$ should be regular at the bubble center ($r = 0$) to find

$$\mathbf{v} = -\frac{\mathbf{e}_r}{r^{d-1}\rho(r)} \int_0^r dr'\, r'^{d-1} \frac{\partial\rho(r')}{\partial t}, \qquad (3.2.66)$$

$$\mathbf{e}_r \equiv \mathbf{x}/r, \qquad \rho(r) \equiv \rho(r\mathbf{e}_r, t). \qquad (3.2.67)$$

Therefore, the kinetic energy part $\tilde{L}_K$ of the Lagrangian turns out to be a non-local functional of ϕ:

$$\tilde{L}_K := \int d^d\mathbf{x}\, \frac{1}{2}\rho\mathbf{v}^2$$

$$= \frac{1}{2} \int d^d\mathbf{x}' \int d^d\mathbf{x}'' \frac{\partial\rho(r')}{\partial t} \mathcal{M}(r', r'') \frac{\partial\rho(r'')}{\partial t}, \qquad (3.2.68)$$

$$\mathcal{M}(r', r'') \equiv \frac{1}{\mathcal{S}_{d-1}} \int_{\max(r',r'')}^\infty \frac{dr}{r^{d-1}\rho(r)}, \qquad (3.2.69)$$

where the kernel $\mathcal{M}(r', r'')$ embodies the non-locality.[48]

We may reduce the number of degrees of freedom with the same procedure as in the case of the non-linear Klein–Gordon field. To be explicit, we now define $b(x)$ by Eq. (3.2.46) with $\mathcal{K}^2$ replaced by $\mathcal{A}^2/m_A$, and assume Eq. (3.2.45) to obtain

$$\partial\rho(r)/\partial t \simeq -\dot{R}(t)\, \partial\rho(r)/\partial r. \qquad (3.2.70)$$

It then follows from Eq. (3.2.66) that

$$\mathbf{v} \simeq \left(1 - \frac{\rho(0)}{\rho(r)}\right)\left(\frac{R(t)}{r}\right)^{d-1} \dot{R}(t)\, \mathbf{e}_r. \qquad (3.2.71)$$

---

[47] In quantum theory, this term may not be negligible; it could affect the phase of a quantum state.
[48] If $\mathcal{M}(r, r')$ were proportional to $\delta(r - r')$, the above Lagrangian would be "local".

Combining Eqs. (3.2.45) and (3.2.71) with (3.2.65), we find

$$\tilde{I}[\phi] \simeq \int dt\,(L_K + L_S + L_B),\qquad(3.2.72)$$

where

$$L_K \equiv \int d^d\mathbf{x}\,\frac{1}{2}\rho\mathbf{v}^2\bigg|_{\phi=b(r-R(t))} \simeq \frac{1}{2}\{R(t)\}^d\{\dot{R}(t)\}^2\mu_b,\qquad(3.2.73)$$

$$\mu_b \equiv \mathcal{S}_{d-1}\int_0^\infty \frac{dy}{y^{d-1}}\rho(R(t)y)\left\{1 - \frac{\rho(0)}{\rho(R(t)y)}\right\}^2$$

$$\simeq \frac{\mathcal{S}_{d-1}}{d-2}\left(1 - \frac{\rho_<}{\rho_>}\right)^2\rho_>,\qquad(3.2.74)$$

$$\rho_\alpha \equiv m_A\phi_\alpha^2 : \alpha \text{ stands for either } > \text{ or } <.\qquad(3.2.75)$$

The forms of L_S and L_B are the same as in the case of the non-linear Klein–Gordon field with the proviso that the constant $\mathcal{K}$ is replaced by $\mathcal{A}/\sqrt{m_A}$. Thus, the Rayleigh–Plesset model with $m = d$ has been derived from the generalized non-linear Schrödinger field under the thin-wall approximation.

The actual liquid He, which is a system of many atoms interacting via a two-body potential of the Lennard–Jones type,[49] is quite hard to deal with directly. Fortunately, in low-energy phenomena where quantum nucleation may be a possibility, He atoms are likely to move in such a way as to evade head-on collisions. Such a situation is expected to be describable by a *low-energy effective theory* as given by the Lagrangian density (3.2.57). Deriving the equation of motion for the field by applying the principle of least action to the action functional $I[\psi]$, and linearizing the equation around $|\psi(\mathbf{x}, t)| = \phi_>$, one may find normal modes of infinitesimal fluctuations of the field in the metastable state; let ω_k be the frequency of the normal mode of wave number k, then[50]

$$\omega_k = \{1 + (\lambda_0 k)^2\}^{1/2}c_0 k,\qquad(3.2.76)$$

$$c_0 \equiv \{\mathcal{U}''(\phi_>)/4m_A\}^{1/2},\quad \lambda_0 \equiv \frac{\mathcal{A}}{2m_A c_0}.\qquad(3.2.77)$$

This frequency is proportional to k in the long-wavelength region ($k^{-1} \gg \lambda_0$). The corresponding normal mode represents a coupled oscillation of the mass density ρ and the phase θ. Therefore, the constant c_0, which is determined by the curvature of the field potential, may be interpreted as the speed of sound in the metastable

---

[49] The inter-atomic potential energy (i.e. the interaction energy) of a pair of He atoms separated by distance r is roughly proportional to $(a/r)^{12} - (a/r)^6$, where a may be called the atomic radius. The region $r < a$ inside the steep potential wall is called the <u>hard core</u>. Thus, apart from the weak van der Waals attractive force in the region $r \gtrsim a$, an He atom may be regarded as a hard sphere of radius a.

[50] In the quantum-mechanical treatment of the normal modes, $\hbar\omega_k$ is identified as the excitation energy of a *Bogoliubov quasiparticle*.

state. The corresponding Compton wavelength λ_0, being a characteristic length for the system in question, determines the interfacial width w of the kink:

$$w \sim \lambda_0. \tag{3.2.78}$$

If the field potential $\mathcal{U}$ is a gentle function such as the one depicted in Fig. 3.8, its local maximum is of the order of $\phi_0^2 \mathcal{U}''(\phi_>)$ with $\phi_0 \equiv \phi_> - \phi_<$. It then follows from Eq. (3.2.56) that

$$\sigma_b = \frac{\overline{\sigma}}{2} \frac{\mathcal{A}}{\sqrt{m_A}} \{\phi_0^2 \mathcal{U}''(\phi_>)\}^{1/2} \phi_0 = \overline{\sigma} \frac{\mathcal{A}\rho_0 c_0}{m_A}, \tag{3.2.79}$$

$$\rho_0 \equiv 4m_A \phi_0^2,$$

where $\overline{\sigma}$ is a numerical factor of order unity; the value of σ_b is thus estimated to be roughly equal to the energy density $\rho_0 c_0^2$ multiplied by the Compton wavelength λ_0. In this manner, various constants contained in the Lagrangian density (3.2.57) may be related to measurable properties of the actual liquid, with the proviso that the field ψ should be a Bose field, to which the method of the present section is applicable. The liquid ^4He is thought to be describable with a Bose field. To deal with the ^3He–^4He liquid mixture, on the other hand, it is necessary first to eliminate the Fermi field representing ^3He atoms.

In order to discuss quantum nucleation, all of the procedures mentioned above should be carried out quantum-mechanically right from the stage of field theory. In the case of the non-linear Klein–Gordon field, one begins with introducing the momentum Π conjugate to ϕ:

$$\Pi := \frac{\partial \mathcal{L}}{\partial(\partial\phi/\partial t)} = \frac{\mathcal{K}^2}{c^2} \partial\phi/\partial t. \tag{3.2.80}$$

One then constructs the field Hamiltonian

$$H := \int d^d\mathbf{x} \{\Pi \partial\phi/\partial t - \mathcal{L}\}\Big|_{\partial\phi/\partial t = c^2\Pi/\mathcal{K}^2}$$

$$= \int d^d\mathbf{x} \left\{ \frac{c^2}{2\mathcal{K}^2}(\Pi(\mathbf{x}))^2 + \frac{1}{2}(\nabla\phi(\mathbf{x}))^2 + \mathcal{U}(\phi(\mathbf{x})) \right\}, \tag{3.2.81}$$

and finally replaces $\phi(\mathbf{x})$ and $\Pi(\mathbf{x})$ by the *field operators* $\hat{\phi}(\mathbf{x})$ and $\hat{\Pi}(\mathbf{x})$, respectively, which are supposed to obey the canonical commutation relation

$$[\hat{\phi}(\mathbf{x}), \ \hat{\Pi}(\mathbf{x}')] = i\hbar\delta(\mathbf{x} - \mathbf{x}'). \tag{3.2.82}$$

Likewise, in the case of the generalized non-linear Schrödinger field, the momentum conjugate to ψ is found as

$$\frac{\partial \mathcal{L}}{\partial (\partial \psi / \partial t)} = i \mathcal{A} \psi^*. \tag{3.2.83}$$

Consequently, the field Hamiltonian takes the form

$$
\begin{aligned}
H &:= \int d^d \mathbf{x} \{ i \mathcal{A} \psi^* \partial \psi / \partial t - \mathcal{L} \} \\
&= \int d^d \mathbf{x} \left\{ \frac{\mathcal{A}^2}{2m_A} \nabla \psi^*(\mathbf{x}) \nabla \psi(\mathbf{x}) + \mathcal{U}\left(|\psi(\mathbf{x})| \right) \right\},
\end{aligned} \tag{3.2.84}
$$

in which $\psi(\mathbf{x})$ and $\psi^*(\mathbf{x})$ should be replaced by the field operators $\hat{\bar{\psi}}(\mathbf{x})$ and $\hat{\bar{\psi}}^\dagger(\mathbf{x})$, respectively, which are supposed to obey the canonical commutation relation

$$\left[\hat{\bar{\psi}}(\mathbf{x}),\ \hat{\bar{\psi}}^\dagger(\mathbf{x}') \right] = \frac{\hbar}{\mathcal{A}} \delta(\mathbf{x} - \mathbf{x}'). \tag{3.2.85}$$

Conventionally, the re-scaled field $\hat{\psi}(\mathbf{x}) := (\mathcal{A}/\hbar)^{1/2} \hat{\bar{\psi}}(\mathbf{x})$ is used so that[51]

$$[\hat{\psi}(\mathbf{x}),\ \hat{\psi}^\dagger(\mathbf{x}')] = \delta(\mathbf{x} - \mathbf{x}'). \tag{3.2.86}$$

A full-fledged quantum field theory of nucleation, of which a preliminary attempt has been made in the case of the non-linear Klein–Gordon field, is yet to be constructed.

3.3 Single-domain magnet

A usual magnet (ferromagnet) is composed of many domains. By reducing the size of a magnet, however, one may prepare a magnet with a single domain. Figure 3.11 shows the size of a typical single-domain nickel magnet actually prepared in a laboratory. In a domain, all microscopic spins are aligned in one and the same direction and precess in one and the same way under a magnetic field. Hence, in a single-domain ferromagnet, motion of individual microscopic spins may be represented by the motion of the total spin (i.e. the sum of all the microscopic spins). This situation is analogous to that for a rigid ball where translation motions of individual atoms can be represented by the motion of the center of mass.

A typical model for a single-domain ferromagnet is described by the following Hamiltonian $H(\hat{\mathbf{S}})$ with $\hat{\mathbf{S}}$ being the spin operator[52] of magnitude S:

$$H(\hat{\mathbf{S}}) := \frac{K_\parallel}{S^2} \hat{S}_z^2 - \frac{K_\perp}{S^2} \hat{S}_y^2 - \frac{\mathcal{B}'}{S} \hat{S}_x - \frac{\mathcal{B}}{S} \hat{S}_y - \frac{\mathcal{B}_\parallel}{S} \hat{S}_z, \tag{3.3.1}$$

---

[51] Perhaps the readers are more familiar with the Hamiltonian (3.2.84) than the Lagrangian (3.2.57). Indeed, the quantum mechanics for a many-boson system can be cast into the form of Eqs. (3.2.84) and (3.2.86), where the field operators act on a Fock space. It is then possible to define the Heisenberg–picture operator $\hat{\psi}(\mathbf{x}, t)$ and derive the Heisenberg equation of motion obeyed by the latter. If one searches for a Lagrangian which gives the corresponding equation of motion at the classical level, one finds the Lagrangian (3.2.57).

[52] The spin operator here is dimensionless; see Appendix D for relevant notations.

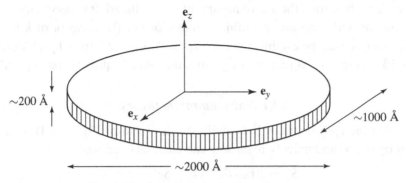

Fig. 3.11. A typical single-domain magnet.

where, $K_\parallel$ and $K_\perp$ are positive constants representing energies due to anisotropy,[53] and $\mathcal{B}_\parallel$, $\mathcal{B}$ and $\mathcal{B}'$ represent components of an externally applied magnetic field.[54] If the magnet in question consists of N microscopic spins each of magnitude s, then

$$S = Ns, \tag{3.3.2}$$

which is of the order of 10^6 for a magnet of the size depicted in Fig. 3.11.

Consider a classical vector $\mathbf{S}$ of length S, and let θ and ϕ be the azimuth angles for its direction:

$$S_z = S\cos\theta, \quad S_x = S\sin\theta\cos\phi, \quad S_y = S\sin\theta\sin\phi. \tag{3.3.3}$$

We take the range of the angle ϕ as

$$-\pi \le \phi \le \pi, \tag{3.3.4}$$

where $\phi = -\pi$ is to be identified with $\phi = \pi$. Let $H(\mathbf{S})$ be the classical version of (3.3.1) with $\hat{\mathbf{S}}$ replaced by $\mathbf{S}$. In the absence of a magnetic field, for instance, $H(\mathbf{S})$ has minima at $\mathbf{S}=S\mathbf{e}_y$ and at $\mathbf{S} = -S\mathbf{e}_y$. Hence, it might be tempting to argue as follows: "for $\mathbf{S}$ to change its direction continuously from $-\mathbf{e}_y$ to $+\mathbf{e}_y$, for instance, it has to pass a direction in which the energy $H(\mathbf{S})$ is higher than in the $\pm\mathbf{e}_y$ directions. This implies a possibility of *quantum tunneling for spin*." Is this argument acceptable? Recall that for a particle a quantum tunneling is the passage through a potential barrier that is higher than the kinetic energy. Accordingly, it makes sense to talk about a quantum tunneling only if the Hamiltonian (3.3.1) can be cast into a sum of kinetic and potential energies as in the case of the SQUID in the previous section. However, the energy $H(\mathbf{S})$ is the total energy, which should

---

[53] The first and the second terms of the Hamiltonian are called the longitudinal anisotropy energy and the transverse anisotropy energy, respectively. Among various origins conceivable for the anisotropy, the principal one tends to be the shape of the magnet; the direction of the demagnetizing field, which is generated by the alignment of microscopic spins, is determined by the shape of the magnet.

[54] The last three terms of the Hamiltonian are Zeeman energies, where the magnetic moment associated with the spin has been absorbed into $\mathcal{B}_\parallel$, $\mathcal{B}$ and $\mathcal{B}'$.

be conserved. Therefore, the above argument is misguided. It is necessary to return to the fundamental level and examine whether or not the concept of kinetic and potential energies can be established for a spin. This is not a trivial problem; a spin behaves like a top and, as readers are aware, the latter is quite unlike a particle.

3.3.1 Naive quantum theory

Let us write the spin operator $\hat{\mathbf{S}}$ formally in the form of Eq. (3.3.3) with (θ, ϕ) replaced by $(\hat{\theta}, \hat{\phi})$ and naively regarding the latter as operators:

$$\hat{S}_z = S\cos\hat{\theta}, \quad \hat{S}_+ = Se^{i\hat{\phi}}\sin\hat{\theta}. \tag{3.3.5}$$

Putting this into the commutation relation

$$[\hat{S}_z, \hat{S}_+] = \hat{S}_+, \tag{3.3.6}$$

one finds

$$\left[\cos\hat{\theta}, e^{i\hat{\phi}}\right] = S^{-1}e^{i\hat{\phi}}. \tag{3.3.7}$$

Assume that one can construct a <u>ϕ-representation</u> for the spin operator. Then, for the above relation to hold, it is sufficient that

the ϕ-representation for $\cos\hat{\theta}$ is given by $\dfrac{S^{-1}}{i}\dfrac{\partial}{\partial\phi}$. $\qquad$ (3.3.8)

Taking this for granted, one obtains

$$[\hat{\phi}, \hbar\hat{S}_z] = i\hbar, \tag{3.3.9}$$

which implies that $\hbar\hat{S}_z$ is the momentum conjugate to $\hat{\phi}$. Furthermore, let us make the following assumption:

Assumption $\mathcal{A}0$: The spin lies almost in the xy-plane always:[55]

$$|\theta - \pi/2| \ll 1. \tag{3.3.10}$$

Under this assumption, the ϕ-representation for $H(\hat{\mathbf{S}})$, to be denoted by H_S, is given as [56]

$$H_S \simeq -\frac{S^{-2}}{2(2K_\parallel)^{-1}}\frac{\partial^2}{\partial\phi^2} - \mathcal{B}_\parallel\frac{S^{-1}}{i}\frac{\partial}{\partial\phi} + U_S(\phi)$$

$$= \frac{1}{2(2K_\parallel)^{-1}}\left(\frac{S^{-1}}{i}\frac{\partial}{\partial\phi} - \frac{\mathcal{B}_\parallel}{2K_\parallel}\right)^2 + U_S(\phi), \tag{3.3.11}$$

$$U_S(\phi) := K_\perp(\cos\phi)^2 - \mathcal{B}'\cos\phi - \mathcal{B}\sin\phi. \tag{3.3.12}$$

---

[55] A more precise statement is the following: it is possible to impose an initial condition such that the inequality (3.3.10) holds at an arbitrary time.

[56] We suppress additive constants $-K_\perp$ and $-\mathcal{B}_\parallel^2/4K_\parallel$ in the first and the second lines, respectively, of Eq. (3.3.11).

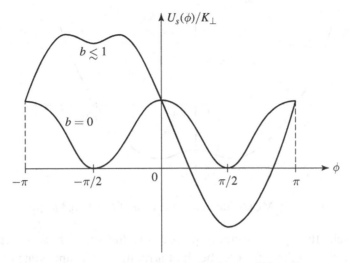

Fig. 3.12. Naive spin potential (case of $b' = 0$).

Thus, in the case of $\mathcal{B}_{\parallel} = 0$, the problem of a spin has been reduced to that of a (neutral) particle under the potential $U_S(\phi)$. The mass of the particle is inversely proportional to the longitudinal anisotropy energy and the role of the Planck constant is played by S^{-1}. Of course, this is merely a suggestive interpretation inferred from the particular form $S^{-2}/(2K_{\parallel})^{-1}$; there is no logical necessity at this stage to write $2K_{\parallel}/S^2$ in that form. In the case of $\mathcal{B}_{\parallel} \neq 0$, the first term of Eq. (3.3.11) is of the same form as the kinetic energy of a charged particle moving along a ring threaded through by a magnetic flux. Hence, physical phenomena governed by the above Hamiltonian are periodic in $\mathcal{B}_{\parallel}$ with period $2K_{\parallel}/S$. (See Section 3.3.4 for details.) In the rest of this section, we suppose that $\mathcal{B}_{\parallel} = 0$. Again, we abuse language to call $U_S(\phi)$ the *naive spin potential*. Typical cases are shown in Figs. 3.12 and 3.13 with the notation

$$a \equiv K_{\perp}/K_{\parallel}, \quad b \equiv \mathcal{B}/2K_{\perp}, \quad b' \equiv \mathcal{B}'/2K_{\perp}. \tag{3.3.13}$$

Consider, for example, the case of $\mathcal{B}' = 0$. As the external magnetic field $\mathcal{B}$ is varied, there occurs a qualitative change in the shape of $U_S(\phi)$ at $b = 1$. Hence, $2K_{\perp}$ is a kind of critical magnetic field. One may prepare the following typical situations by adjusting the value of $\mathcal{B}$:

(i) $b = 0$, $b' = 0$: periodic well (MQC-situation)
(ii) $b \lesssim 1$, $b' = 0$: double bumpy slope (MQT-situation)

These spin potentials are similar in shape to the flux potentials in the case of the SQUID. However, the nature of the variables is different; although the flux can take arbitrary real values, the azimuth angle is constrained by the condition (3.3.4).

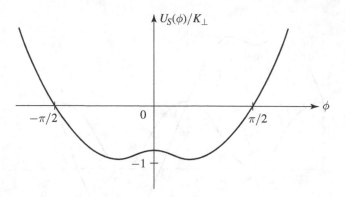

Fig. 3.13. MQC-situation in the case of $b = 0$ and $b' \lesssim 1$.

If, for example, the spin is initially prepared in the state of $\phi \sim -\pi/2$, then it may tunnel either to the left or to the right arriving at the same state of $\phi \sim \pi/2$. The two amplitudes are expected to interfere with each other. Let us consider the MQC-situations in detail.

(i′) MQC-situation 1: $b = b' = 0$
   Following the procedure of Section 2.1, one may put equations into the standard form to find

$$h = \frac{2}{\pi S}\sqrt{\frac{2}{a}}. \tag{3.3.14}$$

It is seen that the situation is quasi-classical in the sense of Eq. (2.1.11) if $\sqrt{a}S \gg 1$. The instanton action I defined by Eq. (2.4.3) may be calculated with the result

$$\frac{1}{2}I = \sqrt{8}/\pi \ (= 0.9003 \cdots \sim 1). \tag{3.3.15}$$

Accordingly, the ground-state energy splitting is found as

$$\Delta = 2\tilde{A}\exp\left(-2\sqrt{a}S\right), \tag{3.3.16}$$

with the coefficient $\tilde{A}$ being given by Eq. (2.4.2). The foremost factor $2\,(= 1 + 1)$ on the right-hand side embodies the above-mentioned interference effect. (To be rigorous, this conclusion is correct only if S is an integer; in the case of a half-integral S, the corresponding factor is $0\,(= 1 - 1)$. In the presence of $\mathcal{B}_{\parallel}$, the factor depends periodically on $\mathcal{B}_{\parallel}$. These details are explained in Sections 3.3.4 and 3.3.5.) Since the value of h^{-1}, which is proportional to S (hence, to the size of the magnet), is of the order of 10^6 for a typical magnet shown in Fig. 3.11 for which $a \sim 1$, at first sight it seems impossible to detect MQC. However, as pointed out shortly, one can reduce the value of h^{-1} by adjusting b' so that $b' \lesssim 1$; this may be accomplished by applying a magnetic field in the x-direction as well,
(i″) MQC-situation 2: $b = 0$, $b' \lesssim 1$

In the region of $\phi \sim 0$, the spin potential may be approximated as

$$U_S(\phi)/K_\perp \simeq 1 - 2b' - (1 - b')\phi^2 + \frac{1}{3}\left(1 - \frac{b'}{4}\right)\phi^4. \qquad (3.3.17)$$

This is essentially the same as the MQC-situation (a') for the SQUID with γ replaced by b', hence

$$h \propto (1 - b')^{-3/2}. \qquad (3.3.18)$$

The instanton action I is given by I^{SSDW} of Eq. (2.4.15). Interference effects are absent in this case.

As to the MQT-situation, apart from the fact that there are two possible directions for tunneling, qualitative features are the same as those for SQUID. In particular, h^{-1} tends to 0 as b tends to the critical value 1:

$$h \propto (1 - b)^{-3/2}. \qquad (3.3.19)$$

Results of the naive quantum theory are supported by experiments; MQT (including resonant tunneling) has been detected in the sense that both kinds of experimental data corresponding to Fig. 2.7 and Fig. 2.8 have been obtained. In particular, as $\mathcal{B}$ is varied, the magnetization increases sharply[57] at specific values of $\mathcal{B}$, which correspond to the peaks in Fig. 2.8 and are in good agreement with those predicted by the naive quantum theory.[58] There is also a report to claim that MQC has been detected in a certain kind of magnet.[59] However, this claim is controversial. It seems to be too early to make a definitive statement that MQC has indeed been detected.

3.3.2 Longitude basis and longitude-represented wavefunction

Many of the readers must feel uneasy about the quantum theory so far developed; it is extremely naive (and sloppy). Let us now provide the naive quantum theory with a sound theoretical basis. For this purpose, we need some preparation.

Readers should of course be quite familiar with the standard quantum mechanics of a spin. The purpose of this section is not to review it, but to examine whether or not "quantum tunneling of a spin" can be a rational concept, and, if it can be, to understand it as intuitively as possible. We therefore try to reformulate the quantum mechanics for a spin in a form as close to the wavemechanics for a

---

[57] In other words, the hysteresis curve for the magnetization exhibits jumps.

[58] In this experiment (Ref. [17] in the bibliography), the magnet is not so large as the one depicted in Fig. 3.11; it is a molecule with $S = 10$.

[59] An antiferromagnetic protein (called ferritin since it stores iron), which was removed from the spleen of a horse in the early 1990s but can now be synthesized artificially. In order to discuss this magnet, the theory in this section has to be so modified as to be applicable to an antiferromagnet. It turns out that the quantity corresponding to S is of $\mathcal{O}(10^3) \sim \mathcal{O}(10^4)$.

particle as possible. Recall that the latter is obtained by constructing the position-representation of states and operators with the aid of the eigenket $|\mathbf{r}\rangle$ of the particle-position operator $\hat{\mathbf{r}}$. Note also that the three-dimensional Euclidean space $\mathbf{E}^3$, of which the particle-position $\mathbf{r}$ is an element, constitutes the *configuration space* for the particle.

A familiar basis for the Hilbert space $\mathcal{H}_S$ to describe a spin of magnitude S is the standard basis

$$\{|m\rangle \mid m = -S, -S+1, \ldots, S : \hat{S}_z|m\rangle = m|m\rangle\}, \qquad (3.3.20)$$

which being discrete is not suitable for the above-mentioned purpose. Another basis of frequent use consists of so-called *spin coherent states*. Let $\mathbf{n}$ be a real unit vector, then the spin coherent state $|\mathbf{n}\rangle$ is defined by the following equation:

$$\mathbf{n} \cdot \hat{\mathbf{S}}|\mathbf{n}\rangle = S|\mathbf{n}\rangle, \quad \langle \mathbf{n}|\mathbf{n}\rangle = 1. \qquad (3.3.21)$$

It goes without saying that the maximum eigenvalue of $\mathbf{n} \cdot \hat{\mathbf{S}}$ is S. Since $|\mathbf{n}\rangle$ is the eigenstate associated with the maximum eigenvalue, it may be said pictorially that

in the state $|\mathbf{n}\rangle$, the spin points towards the $\mathbf{n}$-direction. $\qquad (3.3.22)$

The set $\{|\mathbf{n}\rangle \mid \mathbf{n} \in S^2\}$,[60] constituting a basis, may be used to construct a representation. However, the sphere S^2 is not a configuration space but the *phase space* for a spin (see Appendix D). Hence, this representation is not suitable either for our purpose. Then, what is the configuration space for a spin? The naive quantum theory might suggest the following idea: "only the x- and y-components of $\mathbf{n}$ belong to the configuration space. Accordingly, the latter may be identified with the circle S^1". Unfortunately, this idea does not work since the set $\{|\mathbf{n}\rangle \mid \mathbf{n} \in S^1\}$ does not constitute a basis.

If the naive quantum theory makes any sense at all, the state $|m\rangle$ for the spin should correspond to the momentum eigenket $|\mathbf{p}\rangle$ for a particle. The latter is related to the position eigenket $|\mathbf{r}\rangle$ as

$$|\mathbf{r}\rangle = \int \frac{d\mathbf{p}}{(2\pi\hbar)^{3/2}} \, e^{-i\mathbf{p}\cdot\mathbf{r}/\hbar} \, |\mathbf{p}\rangle. \qquad (3.3.23)$$

This relationship suggests the introduction of the following state:[61]

$$|\phi\rangle_s := \sum_m e^{-im\phi}|m\rangle, \quad \sum_m \equiv \sum_{m=-S}^{S}. \qquad (3.3.24)$$

---

[60] S^d stands for the d-dimensional unit sphere.

[61] This state is utilized in an attempt to construct a "phase operator" for a harmonic oscillator (or a photon) as well. See, e.g. D. F. Walls and G. J. Milburn, *Quantum Optics*, Springer (1994), Section 2. 8.

It can be shown that

$$\int_{-\pi}^{\pi} \frac{d\phi}{2\pi} |\phi\rangle_s \,_s\langle\phi| = \sum_m |m\rangle\langle m| = 1, \tag{3.3.25}$$

that is, the closure holds for the set

$$\{|\phi\rangle_s \mid -\pi < \phi < \pi\}. \tag{3.3.26}$$

This set is to be called the _longitude basis_. Of course, the dimension of $\mathcal{H}_S$ being $2S + 1$, this basis is not a linearly independent set and its elements are not mutually orthogonal:

$$_s\langle\phi'|\phi\rangle_s = \sum_m e^{i(\phi'-\phi)m} = \frac{\sin\left\{\left(S + \frac{1}{2}\right)(\phi' - \phi)\right\}}{\sin\left\{\frac{1}{2}(\phi' - \phi)\right\}}. \tag{3.3.27}$$

In other words, the basis (3.3.26) is an over-complete basis.

The closure (3.3.25) allows an arbitrary $|\psi\rangle \,(\in \mathcal{H}_S)$ to be expanded in terms of the longitude basis:

$$|\psi\rangle = \left\{\int_{-\pi}^{\pi} \frac{d\phi}{2\pi} |\phi\rangle_s \,_s\langle\phi|\right\} |\psi\rangle = \int_{-\pi}^{\pi} \frac{d\phi}{2\pi} \psi_s(\phi)|\phi\rangle_s, \tag{3.3.28}$$

$$\psi_s(\phi) := \,_s\langle\phi|\psi\rangle = \sum_m e^{im\phi}\langle m|\psi\rangle. \tag{3.3.29}$$

Suppose that $|\psi\rangle$ is normalized to unity, then it follows from the closure that

$$\int_{-\pi}^{\pi} \frac{d\phi}{2\pi} |\psi_s(\phi)|^2 = \langle\psi|\psi\rangle = 1. \tag{3.3.30}$$

The function $\psi_s(\phi)$ is to be called the _longitude-represented wavefunction_ for the state $|\psi\rangle$. It may be viewed as a function defined on a circle since $\phi \in (-\pi, \pi)$. An important remark is in order here. It follows from the definition (3.3.24) that

$$|\phi\rangle_s = e^{-i(\phi+\pi)\hat{S}_z}|-\pi\rangle_s. \tag{3.3.31}$$

Hence,[62]

$$|\pi\rangle_s = e^{-2\pi i\hat{S}_z}|-\pi\rangle_s = e^{-2S\pi i}|-\pi\rangle_s, \tag{3.3.32}$$

$$\psi_s(\pi) = e^{2S\pi i}\psi_s(-\pi). \tag{3.3.33}$$

---

[62] To be rigorous, the π on the left-hand side is $\pi - 0$, and the $-\pi$ on the right-hand side is $-\pi + 0$.

For brevity, this condition is to be called the <u>Möbius boundary condition</u>;[63] ψ_s is a continuous function on the circle if S is an integer, but, like the Möbius strip, it changes sign on completion of a round-the-circle trip if S is a half-integer.[64] The subscript s in $|\phi\rangle_s$ and ψ_s is meant to be a reminder of this condition.[65]

Incidentally, the longitude basis constitutes an orthogonal set in the limit of $S \uparrow \infty$ in the following formal sense:

$$\lim_{S\uparrow\infty} {}_s\langle\phi'|\phi\rangle_s = 2\pi\delta(\phi' - \phi). \tag{3.3.34}$$

In order to understand the nature of the longitude representation, let us consider the longitude-represented wavefunction for the spin coherent state $|\mathbf{n}\rangle$. Suppose that $\mathbf{n}$ lies in the xy-plane:

$$\mathbf{n} = \mathbf{e}_x \cos\phi_\mathbf{n} + \mathbf{e}_y \sin\phi_\mathbf{n}. \tag{3.3.35}$$

Putting $\theta_\mathbf{n} = \pi/2$ in Eq. (D.4.9) in Appendix D, one finds that[66]

$$|\mathbf{n}\rangle = 2^{-S} \exp(-i\phi_\mathbf{n}\hat{S}_3) \exp(\hat{S}_-)|S\rangle, \tag{3.3.36}$$

$${}_s\langle\phi|\mathbf{n}\rangle = \sum_m b_{Sm}^{1/2} \, e^{im\varphi}, \quad \varphi \equiv \phi - \phi_\mathbf{n}, \tag{3.3.37}$$

where

$$b_{Sm} := |\langle m|\mathbf{n}\rangle|^2 = \frac{(2S)!}{(S-m)!(S+m)!}\left(\frac{1}{2}\right)^{2S} \tag{3.3.38}$$

is the (normalized) binomial coefficient. In particular, for $S \gg 1$, the above expression reduces to

$$b_{Sm} \simeq (\pi S)^{-1/2} \exp(-S\mu^2), \tag{3.3.39}$$

where

$$\mu \equiv m/S. \tag{3.3.40}$$

In other words, when $|\mathbf{n}\rangle$ is expanded in terms of the standard basis (3.3.20), the dominant terms are those of $|m| \lesssim S^{1/2}$ (or $\mu \lesssim S^{-1/2}$) as seen from Fig. 3.14. Hence,

---

[63] It is the *periodic boundary condition* for an integral S and the *anti-periodic boundary condition* for a half-integral S.

[64] $S \in \mathbf{N}_{1/2}$, namely 1/2 times a positive odd integer; a more rigorous nomenclature would be "half positive-odd integer".

[65] If the phase factor $\exp(-im\phi)$ on the right-hand side of the definition (3.3.24) is replaced by $\exp(-i(m+1/2)\phi)$ in the case of a half-integral S, the ket $|\phi\rangle_s$ becomes periodic. This replacement is equivalent to the transformation (3.3.80) introduced below. The ket $|\phi\rangle_s$ modified in this way is sometimes called the "periodic coherent state". See, e.g. T. Kashiwa, Y. Ohnuki, and M. Suzuki, *Path Integral Methods*, Oxford (1997), Section 3.4.

[66] The overall phase factor $\exp(i\phi_\mathbf{n}S)$ on the right-hand side is omitted for brevity.

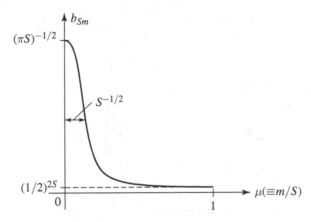

Fig. 3.14. Binomial coefficient for $S \gg 1$.

$$
\begin{aligned}
{}_s\langle \phi | \mathbf{n} \rangle &\simeq (\pi S)^{-1/4} S \sum_m \frac{1}{S} \exp\left\{ -\frac{S}{2}\left(\frac{m}{S}\right)^2 + iS\frac{m}{S}\varphi \right\} \\
&\simeq (\pi S)^{-1/4} S \int_{-1}^{1} d\mu \exp\left(-\frac{S}{2}\mu^2 + iS\mu\varphi \right) \\
&\simeq (\pi S)^{-1/4} S^{1/2} \int_{-\infty}^{\infty} dx \exp\left(-\frac{1}{2}x^2 + iS^{1/2}\varphi x \right) \\
&= (4\pi S)^{1/4} \exp\left\{ -\frac{S}{2}(\phi - \phi_\mathbf{n})^2 \right\}.
\end{aligned}
\tag{3.3.41}
$$

This wavefunction is sharply peaked at $\phi = \phi_\mathbf{n}$ with width $S^{-1/2}$, thereby quantitatively confirming the picture (3.3.22). To put it differently, this wavefunction shows that the picture is exact only in the limit $S \uparrow \infty$; for a generic S ($\gg 1$), quantum fluctuation of $\mathcal{O}(S^{-1/2})$ has to be taken into account (Fig. 3.15). In summary, the longitude representation is an intuitive representation that is in line with the classical picture for a spin. Of course, the above-mentioned quantum fluctuation is a familiar fact in the standard quantum mechanics of a spin. The magnitude of the fluctuation can be estimated without recourse to the longitude-represented wavefunction; noting the rotational symmetry around the $\mathbf{n}$-axis and the identity $\hat{\mathbf{S}}^2 = S(S+1)$, one can show for an arbitrary unit vector $\mathbf{l}$ perpendicular to $\mathbf{n}$ that

$$
\langle \mathbf{n} | (\mathbf{l} \cdot \hat{\mathbf{S}})^2 | \mathbf{n} \rangle = \frac{1}{2}\{S(S+1) - S^2\} = \frac{1}{2}S.
\tag{3.3.42}
$$

Stated in classical terms, the spin points to the $\mathbf{n}$-direction on average, but fluctuates within the cone of vertical angle of order $S^{-1/2}$ around $\mathbf{n}$.

Although $\psi_s(\phi)$ satisfies the normalization condition (3.3.30), a certain caution is needed to apply probability interpretation to it as is always the case with

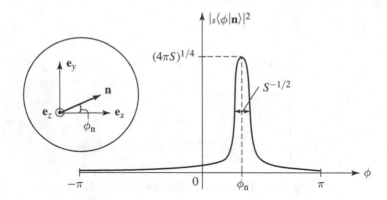

Fig. 3.15. Longitude-represented wavefunction for spin coherent state.

representations on an over-complete basis; even if $\phi' \neq \phi$, the event for the spin
to be found at ϕ' and the event for the spin to be found at ϕ are not mutually ex-
clusive, because $|\phi'\rangle_s$ is not orthogonal to $|\phi\rangle_s$. In this connection, readers should
recall the following: in the quantum mechanics of a particle, one concludes from
the normalization condition

$$\int d\mathbf{r}\,|\psi(\mathbf{r})|^2 = 1 \tag{3.3.43}$$

that $|\psi(\mathbf{r})|^2$ is the probability density for the particle to be found at $\mathbf{r}$, but this conclu-
sion is warranted only on the premise that the event for the particle to be found at $\mathbf{r}'$
and the event for the particle to be found at $\mathbf{r}$ are mutually exclusive (namely, $\mathbf{r}' \neq \mathbf{r}$
implies $\langle \mathbf{r}'|\mathbf{r}\rangle = 0$). However, the above-mentioned events are approximately ex-
clusive if $|\phi' - \phi| > S^{-1/2}$. Hence, in the case of $S \gg 1$, the usual probability
interpretation may be applied to $\psi_s(\phi)$ more or less safely.

3.3.3 Longitude representation of spin operator

The aim of the present section is to cast the Schrödinger equation into the
form of a differential equation with respect to ϕ. For this purpose we need the
longitude representation of the Hamiltonian. Consider first the operator $\hat{S}_z$. By use
of Eqs. (3.3.24) and (3.3.29), one finds for an arbitrary state $|\psi\rangle$ that

$$\begin{aligned}
{}_s\langle \phi|\hat{S}_z|\psi\rangle &= \sum_m e^{im\phi}\langle m|\hat{S}_z|\psi\rangle \\
&= \sum_m e^{im\phi} m\langle m|\psi\rangle = \frac{1}{i}\frac{\partial}{\partial \phi}\psi_s(\phi).
\end{aligned} \tag{3.3.44}$$

Hence, the differential operator $-i\partial/\partial\phi$ is to be called the longitude representation of $\hat{S}_z$. Next, by use of the standard formula

$$\hat{S}_+|m\rangle = S(m)|m+1\rangle, \tag{3.3.45}$$

$$S(m) := \{S(S+1) - m(m+1)\}^{1/2}, \tag{3.3.46}$$

one finds

$$\begin{aligned} {}_s\langle\phi|\hat{S}_-|\psi\rangle &= \sum_m e^{im\phi}\langle m|\hat{S}_-|\psi\rangle \\ &= \sum_m e^{im\phi}S(m)\langle m+1|\psi\rangle \\ &= e^{-i\phi}\sum_m e^{im\phi}S(m-1)\langle m|\psi\rangle. \end{aligned} \tag{3.3.47}$$

Although m is restricted to the integers between $-S$ and S, the above manipulation is justified by the following property:

$$S(S) = S(-S-1) = 0. \tag{3.3.48}$$

Similarly, one finds

$${}_s\langle\phi|\hat{S}_+|\psi\rangle = e^{i\phi}\sum_m e^{im\phi}S(m)\langle m|\psi\rangle. \tag{3.3.49}$$

Hereafter, we make the following assumption on $|\psi\rangle$:

Assumption $\mathcal{A}1$: $|\psi\rangle$ is a state in which the spin lies almost in the xy-plane, that is,

$$|\langle m|\psi\rangle|^2 \sim 0 \quad : |m|\gtrsim S^{1/2}. \tag{3.3.50}$$

A typical example of $|\psi\rangle$ is the spin coherent state $|\mathbf{n}\rangle$ with $\mathbf{n}$ lying in the xy-plane; see Eq. (3.3.39). The assumption $\mathcal{A}1$ is a formal statement for $\mathcal{A}0$ in Section 3.3.1. Under this assumption, only those m satisfying the inequality $|m| \lesssim S^{1/2}$ contribute to the sum on the right-hand side of Eq. (3.3.49). Hence, we may effectively write that

$$S(m) \simeq S + \frac{1}{2} - \frac{1}{2S}\left(m+\frac{1}{2}\right)^2, \tag{3.3.51}$$

where the approximate equality $\simeq$ stipulates that terms of $\mathcal{O}(S^{-1/2})$ have been retained whereas those of $\mathcal{O}(S^{-1})$ have been discarded. It follows that

$$\begin{aligned} {}_s\langle\phi|\hat{S}_+|\psi\rangle &\simeq e^{i\phi}\left\{ S + \frac{1}{2} - \frac{1}{2S}\left(\frac{1}{i}\frac{\partial}{\partial\phi}+\frac{1}{2}\right)^2 \right\}\psi_s(\phi) \\ &= \left\{\left(S+\frac{1}{2}\right)e^{i\phi} - \frac{1}{2S}\left(\frac{1}{i}\frac{\partial}{\partial\phi}-\frac{1}{2}\right)e^{i\phi}\left(\frac{1}{i}\frac{\partial}{\partial\phi}+\frac{1}{2}\right)\right\}\psi_s(\phi). \end{aligned} \tag{3.3.52}$$

The differential operator on the right-hand side is to be called the longitude representation of $\hat{S}_+$. Similarly, it follows from Eq. (3.3.47) that

$$_s\langle\phi|\hat{S}_-|\psi\rangle \simeq e^{-i\phi}\left\{S + \frac{1}{2} - \frac{1}{2S}\left(\frac{1}{i}\frac{\partial}{\partial\phi} - \frac{1}{2}\right)^2\right\}\psi_s(\phi)$$

$$= \left\{\left(S + \frac{1}{2}\right)e^{-i\phi} - \frac{1}{2S}\left(\frac{1}{i}\frac{\partial}{\partial\phi} + \frac{1}{2}\right)e^{-i\phi}\left(\frac{1}{i}\frac{\partial}{\partial\phi} - \frac{1}{2}\right)\right\}\psi_s(\phi). \quad (3.3.53)$$

The differential operator on the right-hand side is Hermitian conjugate to the one in Eq. (3.3.52), as it should be. As is clear from the above argument, we may make a formal statement that

$$\frac{1}{iS}\frac{\partial}{\partial\phi} = \mathcal{O}\left(S^{-1/2}\right). \quad (3.3.54)$$

Of course, this estimate is only valid with the proviso that the operator on the left-hand side acts on a wavefunction representing a state that satisfies the condition (3.3.50).

Let us now be content with the lowest-order result neglecting terms of $\mathcal{O}(S^{-1/2})$ as well, and stipulate this with the approximate equality $\approx$, then

$$_s\langle\phi|\frac{1}{S}\hat{S}_\pm|\psi\rangle \approx \left\{\left(1 + \frac{1}{2S}\right)e^{\pm i\phi} + \frac{1}{2}\frac{1}{S}\frac{\partial}{\partial\phi}e^{\pm i\phi}\frac{1}{S}\frac{\partial}{\partial\phi}\right\}\psi_s(\phi), \quad (3.3.55)$$

$$_s\langle\phi|\frac{1}{S}\hat{S}_y|\psi\rangle = \frac{1}{2iS}\{_s\langle\phi|\hat{S}_+|\psi\rangle - {}_s\langle\phi|\hat{S}_-|\psi\rangle\}$$

$$\approx \left\{\left(1 + \frac{1}{2S}\right)\sin\phi + \frac{1}{2}\frac{1}{S}\frac{\partial}{\partial\phi}\sin\phi\frac{1}{S}\frac{\partial}{\partial\phi}\right\}\psi_s(\phi). \quad (3.3.56)$$

Thus, we have found the longitude representation of $\hat{S}_y/S$ up to $\mathcal{O}(S^{-1})$. Accordingly, it follows with the same degree of accuracy that

$$_s\langle\phi|\frac{1}{S^2}\hat{S}_y^2|\psi\rangle \approx \left\{\left(1 + \frac{1}{S}\right)(\sin\phi)^2 + \frac{1}{S}\frac{\partial}{\partial\phi}(\sin\phi)^2\frac{1}{S}\frac{\partial}{\partial\phi}\right\}\psi_s(\phi). \quad (3.3.57)$$

Collecting the results obtained so far, we find the longitude representation of $H(\hat{\mathbf{S}})$, to be denoted by H_S:

$$_s\langle\phi|H(\hat{\mathbf{S}})|\psi\rangle = H_S\,\psi_s(\phi),$$

$$H_S \approx -\frac{1}{2}\frac{1}{S}\frac{\partial}{\partial\phi}\frac{1}{M(\phi)}\frac{1}{S}\frac{\partial}{\partial\phi} - \frac{\mathcal{B}_\parallel}{iS}\frac{\partial}{\partial\phi} + \tilde{U}(\phi), \quad (3.3.58)$$

$$M(\phi) := \frac{1}{2K_\parallel}\{1 + a(\sin\phi)^2 + ab'\cos\phi + ab\sin\phi\}^{-1}, \quad (3.3.59)$$

$$\tilde{U}(\phi) := K_\perp\left\{\left(1 + \frac{1}{S}\right)(\cos\phi)^2 - \left(1 + \frac{1}{2S}\right)2b'\cos\phi - \left(1 + \frac{1}{2S}\right)2b\sin\phi\right\},$$

$$(3.3.60)$$

where a, b' and b are the positive constants defined by Eq. (3.3.13). It is noteworthy that the quantity $M(\phi)$, which may be interpreted as the mass,[67] depends on ϕ. This circumstance is analogous to that for the nucleation theory developed in the preceding section.

3.3.4 Longitude-represented Schrödinger equation and its consequences

The time evolution of the spin governed by the Hamiltonian (3.3.1) is thus seen to be described approximately by the following Schrödinger equation:

$$i\hbar \frac{\partial}{\partial t} \psi_s(\phi, t) \approx H_S \, \psi_s(\phi, t), \qquad (3.3.61)$$

where ψ_s is to obey the boundary condition (3.3.33),

Recall that we adopted q, instead of R/R_0, as our working variable in dealing with nucleation theory. In the same spirit, we dispose of the ϕ-dependence of $M(\phi)$ by use of the following transformation of ϕ to φ:

$$\varphi = -\pi + \gamma \int_{-\pi}^{\phi} d\phi' \{M(\phi')\}^{1/2} =: F^{-1}(\phi), \qquad (3.3.62)$$

$$\gamma^{-1} \equiv \frac{1}{2\pi} \int_{-\pi}^{\pi} d\phi \{M(\phi)\}^{1/2}. \qquad (3.3.63)$$

The new variable φ, which is a monotonically increasing function of ϕ, is restricted to the range $(-\pi, \pi)$ as ϕ is. The function specifying the one-to-one correspondence between ϕ and φ is denoted by F. With the aid of this function, we define

$$\mu(\varphi) := \{M(F(\varphi))\}^{1/4}, \qquad (3.3.64)$$

$$\Psi_s(\varphi, t) := \{\mu(\varphi)\}^{-1} \gamma^{-1/2} \, \psi_s(F(\varphi), t), \qquad (3.3.65)$$

$$U(\varphi) := \tilde{U}(F(\varphi)) - \frac{1}{2}\mathcal{B}_{\parallel}^2 \{\mu(\varphi)\}^4. \qquad (3.3.66)$$

Then, Eq. (3.3.61) can be rewritten as

$$
\begin{aligned}
i\hbar \frac{\partial}{\partial t} \Psi_s(\varphi, t) &\approx \left\{ -\frac{\gamma^2}{2S^2}\mu(\varphi)\frac{\partial}{\partial\varphi}\frac{1}{\{\mu(\varphi)\}^2}\frac{\partial}{\partial\varphi}\mu(\varphi) - \mathcal{B}_{\parallel}\frac{\gamma}{iS}\mu(\varphi)\frac{\partial}{\partial\varphi}\mu(\varphi) \right. \\
&\qquad \left. + \frac{1}{2}\mathcal{B}_{\parallel}^2\{\mu(\varphi)\}^4 + U(\varphi) \right\} \Psi_s(\varphi, t) \\
&\approx \left[\frac{\gamma^2}{2S^2}\left\{ \frac{\partial}{i\partial\varphi} - 2\pi A_{\parallel}(\varphi) \right\}^2 + U(\varphi) \right] \Psi_s(\varphi, t), \qquad (3.3.67)
\end{aligned}
$$

$$A_{\parallel}(\varphi) := \frac{S\mathcal{B}_{\parallel}}{2\pi\gamma}\{\mu(\varphi)\}^2, \qquad (3.3.68)$$

---

[67] The actual dimension of $M(\phi)$ is [energy]$^{-1}$, of course.

where the difference between the first and the second lines of Eq. (3.3.67), being of $\mathcal{O}(S^{-2})$, has been neglected since we are working with the accuracy of $\mathcal{O}(S^{-1})$ only. We also have

$$\int_{-\pi}^{\pi} \frac{d\varphi}{2\pi} |\Psi_s(\varphi, t)|^2 = \int_{-\pi}^{\pi} \frac{d\phi}{2\pi} |\psi_s(\phi, t)|^2 = 1, \qquad (3.3.69)$$

$$\Psi_s(\pi, t) = e^{2S\pi i} \Psi_s(-\pi, t). \qquad (3.3.70)$$

In the situation of strong longitudinal anisotropy and weak magnetic field where

$$a \ll 1, \quad ab' = \mathcal{B}'/2K_\parallel \ll 1, \quad ab = \mathcal{B}/2K_\parallel \ll 1, \quad \mathcal{B}_\parallel/\sqrt{K_\parallel K_\perp} \ll 1, \qquad (3.3.71)$$

we have

$$\gamma^2 \simeq 2K_\parallel, \quad F(\varphi) \sim \varphi, \quad U(\varphi) \simeq \tilde{U}(\varphi). \qquad (3.3.72)$$

In this case, therefore, the potential in Eq. (3.3.67) is the same as the naive spin potential apart from corrections of $\mathcal{O}(1/S)$ to the coefficients $K_\perp$, $\mathcal{B}'$ and $\mathcal{B}$. For an actual single-domain ferromagnet with $S \sim 10^4$, say, these corrections are entirely negligible. Thus, the result of the naive quantum theory has been justified apart from the effect of $A_\parallel(\varphi)$ to be discussed shortly. It should be emphasized, however, that there do not exist a "Hermitian <u>longitude operator</u> $\hat{\phi}$" and the associated eigenket $|\phi\rangle$ such that

$$\hat{\phi}|\phi\rangle = \phi|\phi\rangle, \quad \langle\phi'|\phi\rangle = 2\pi\delta(\phi' - \phi). \qquad (3.3.73)$$

Equation (3.3.61) is to be understood as the Schrödinger equation in the longitude representation, an important consequence of which is the Möbius boundary condition (3.3.70) at the <u>dateline</u> $\varphi = \pm\pi$.

Let us apply Eq. (3.3.67) to the MQC-situation without the assumption of strong longitudinal anisotropy. Suppose for the sake of simplicity that $\mathcal{B}_\parallel = b = b' = 0$. One may follow the standardization procedure of Section 2.1 to find

$$h = \left\{ \frac{2}{S(S+1)} \right\}^{1/2} \left\{ \int_0^{\pi/2} \frac{d\phi}{\sqrt{a^{-1} + (\sin\phi)^2}} \right\}^{-1}. \qquad (3.3.74)$$

In particular, consider the case of $a \gg 1$. The most dominant contribution to the above integral comes from the region $\phi \sim 0$. Hence, h should be proportional to $(\log a)^{-1}$. A little more involved calculation gives

$$h \simeq \left\{ \frac{8}{S(S+1)} \right\}^{1/2} \frac{1}{\log 16a} \quad : a \gg 1. \qquad (3.3.75)$$

It can also be shown that $I \simeq \sqrt{8} \log 4a / \log 16a$. The ground-state energy splitting is thus found as

$$\Delta \simeq \{1 + (-1)^{2S}\} \tilde{A} \left(\frac{1}{4a}\right)^{\sqrt{S(S+1)}}. \tag{3.3.76}$$

At first sight, it might look strange that the right-hand side is not manifestly proportional to an exponential function to the base e. No mystery, however, is involved here; the above expression merely reflects the fact that h is proportional to $(\log a)^{-1}$.

Incidentally, in a situation where quantum tunneling may occur, the kinetic energy is of the same order as the zero-point energy associated with the well:

$$h^2 \frac{\partial^2}{\partial \varphi^2} \sim h\omega_0. \tag{3.3.77}$$

It follows that

$$\frac{\partial}{\partial \varphi} \sim h^{-1/2} \sim S^{1/2}, \tag{3.3.78}$$

which justifies Eq. (3.3.54) or equivalently the assumption $\mathcal{A}1$.

The foremost factor $\{1 + (-1)^{2S}\}$ on the right-hand side of Eq. (3.3.76), which appears regardless of the value of a, embodies the interference effect originating from the Möbius boundary condition. This effect may also be understood in analogy with the Aharonov–Bohm effect as follows. Let $A_s(\varphi)$ be an arbitrary periodic function such that

$$\int_{-\pi}^{\pi} d\varphi \, A_s(\varphi) = S, \quad A_s(\varphi) \in \mathbf{R}. \tag{3.3.79}$$

With the aid of this function, one may effect the transformation

$$\Psi(\varphi, t) := \Psi_s(\varphi, t) \exp\left\{2\pi i \int_{-\pi}^{\varphi} d\varphi' A_s(\varphi')\right\} \tag{3.3.80}$$

to find

$$i\hbar \frac{\partial}{\partial t} \Psi(\varphi, t) \approx \left[\frac{\gamma^2}{2S^2} \left\{\frac{1}{i} \frac{\partial}{\partial \varphi} - 2\pi A(\varphi)\right\}^2 + U(\varphi)\right] \Psi(\varphi, t), \tag{3.3.81}$$

$$A(\varphi) := A_s(\varphi) + A_{\parallel}(\varphi), \tag{3.3.82}$$

where the new wavefunction satisfies the periodic boundary condition:

$$\Psi(\pi, t) = \Psi(-\pi, t). \tag{3.3.83}$$

Equation (3.3.81) is formally equivalent to the Schrödinger equation of a charged particle moving along a ring threaded through by a <u>magnetic flux</u>[68] Φ, where

$$
\Phi := \int_{-\pi}^{\pi} d\varphi \, A(\varphi) = S + \frac{SB_{\|}}{2\pi\gamma} \int_{-\pi}^{\pi} d\varphi \, \{\mu(\varphi)\}^2
$$

$$
= S + \frac{SB_{\|}}{2\pi} \int_{-\pi}^{\pi} d\phi \, M(\phi). \tag{3.3.84}
$$

The function $A_s(\varphi)$, which is arbitrary so long as the condition (3.3.79) is fulfilled, turns out to be that part of the vector potential which corresponds to the <u>magnetic flux</u> S. As is well-known in the context of the Aharonov–Bohm effect, physical phenomena governed by the above Schrödinger equation are periodic in Φ with period 1. For example, in the presence of $B_{\|}$, the foremost factor $\{1 + (-1)^{2S}\}$ on the right-hand side of Eq. (3.3.76) is replaced[69] by $|2\cos(\pi\Phi)|$. With $b = b' = 0$, the integral in Eq. (3.3.84) may be evaluated as

$$
\frac{1}{2\pi} \int_{-\pi}^{\pi} d\phi \, M(\phi) = \frac{1}{2K_{\|}} \int_{-\pi}^{\pi} \frac{d\phi}{2\pi} \frac{1}{1 + a(\sin\phi)^2} = \frac{1}{2K_{\|}\sqrt{1+a}}. \tag{3.3.85}
$$

Hence, the factor in question is expressed explicitly as[70]

$$
\left| 2\cos\left\{ \pi \left(1 + \frac{B_{\|}}{2\sqrt{K_{\|}(K_{\|} + K_{\perp})}} \right) S \right\} \right|. \tag{3.3.86}
$$

3.3.5 Symmetry and interference effect

This subsection summarizes those rigorous results which are valid independently of approximations introduced above (and, hence, regardless of the value of S); they follow directly from the symmetry of the Hamiltonian.

(i) *Time-reversal symmetry*

For a half-integral S in the absence of a magnetic field, the ground-state energy splitting vanishes; $\Delta = 0$ due to the foremost factor in Eq. (3.3.76). This property, which implies that the ground state of $H(\hat{\mathbf{S}})$ is degenerate, is an example of the *Kramers degeneracy*. In other words, it is a result of the time-reversal symmetry

$$
H(-\hat{\mathbf{S}}) = H(\hat{\mathbf{S}}), \tag{3.3.87}
$$

which holds for the Hamiltonian (3.3.1) in the absence of a magnetic field.

[68] Recall that the mechanical momentum of a charged particle with a charge q is represented by $-i\hbar\nabla - q\mathbf{A}_{\text{conv}}$, which is proportional to $-i\nabla - 2\pi\mathbf{A}$ if the vector potential is measured in units of the flux quantum; $\mathbf{A} := (q/2\pi\hbar)\mathbf{A}_{\text{conv}}$.

[69] When it comes to actually solving this equation, it is convenient to eliminate $A(\varphi)$ by a (singular) gauge transformation of the form of Eq. (3.3.80) with $\Psi_s(\varphi, t)$ and $A_s(\varphi')$ replaced by $\tilde{\Psi}(\varphi, t)$ and $A(\varphi')$, respectively, and solve the resulting equation for $\tilde{\Psi}(\varphi, t)$ under the boundary condition $\tilde{\Psi}(\pi, t) = \exp(-2\pi\Phi i)\,\tilde{\Psi}(-\pi, t)$.

[70] For the original derivation of this "topological factor", see D. Loss, D. P. DiVincenzo and G. Grinstein, *Phys. Rev. Lett.* **69** (1992), 3232; J. von Delft and C. L. Henley, *Phys. Rev. Lett.* **69** (1992), 3236; A. Garg, *Europhys. Lett.* **22** (1993), 205.

(ii) π-*rotation symmetry*

Let $|m; \mathbf{e}_y\rangle$ be the eigenstate of $\hat{S}_y$ with eigenvalue m:

$$\hat{S}_y |m; \mathbf{e}_y\rangle = m |m; \mathbf{e}_y\rangle. \tag{3.3.88}$$

Comparison of this equation with the definition (3.3.21) of the spin coherent state gives

$$|\pm S; \mathbf{e}_y\rangle = |\pm \mathbf{e}_y\rangle. \tag{3.3.89}$$

Let $\hat{R}$ ($\equiv \exp(-i\pi \hat{S}_y)$) be the operator of rotation by π around the y-axis, then

$$\hat{R} |m; \mathbf{e}_y\rangle = e^{-im\pi} |m; \mathbf{e}_y\rangle. \tag{3.3.90}$$

Since $\hat{R}$ is a unitary operator, the following identity holds for an arbitrary polynomial $f(H)$:

$$f(\hat{H}) = \hat{R}^\dagger f(\hat{R}\hat{H}\hat{R}^\dagger)\hat{R}. \tag{3.3.91}$$

It follows that

$$\mathcal{A}_{m'm} \equiv \langle m'; \mathbf{e}_y| f(\hat{H})|m; \mathbf{e}_y\rangle \tag{3.3.92}$$

$$= \langle m'; \mathbf{e}_y| \hat{R}^\dagger f(\hat{R}\hat{H}\hat{R}^\dagger)\hat{R}|m; \mathbf{e}_y\rangle$$

$$= e^{i(m'-m)\pi} \langle m'; \mathbf{e}_y| f(\hat{R}\hat{H}\hat{R}^\dagger)|m; \mathbf{e}_y\rangle. \tag{3.3.93}$$

Now that $\mathcal{A}_{m'm}$ has been expressed in two forms, one may average them to find

$$\mathcal{A}_{m'm} = \frac{1}{2}\big\{\langle m'; \mathbf{e}_y| f(\hat{H})|m; \mathbf{e}_y\rangle + e^{i(m'-m)\pi} \langle m'; \mathbf{e}_y| f(\hat{R}\hat{H}\hat{R}^\dagger)|m; \mathbf{e}_y\rangle\big\}. \tag{3.3.94}$$

It is for convenience of the ensuing discussion (i.e. comparison with the path-integral method) that the arithmetic mean is adopted for the average. If one makes the choice

$$f(H) = \exp(-iHt/\hbar), \tag{3.3.95}$$

then $\mathcal{A}_{m'm}$ expresses the transition amplitude from the state $|m; \mathbf{e}_y\rangle$ to $|m'; \mathbf{e}_y\rangle$. The first and the second terms on the right-hand side of Eq. (3.3.94) express the time evolution governed by $\hat{H}$ and $\hat{R}\hat{H}\hat{R}^\dagger$ (i.e. <u>reflection</u> of $\hat{H}$ with respect to the $\mathbf{e}_y$-axis), respectively. Hence, one may view the transition expressed by $\mathcal{A}_{m'm}$ as being made up of the two kinds of time evolution. The two kinds are on equal footing with each other. It is, however, to be noted that the second term carries the phase factor $e^{i(m'-m)\pi}$. The results so far are valid for any $\hat{H}$. Now, consider the case of $\mathcal{B}_\parallel = \mathcal{B}' = 0$. Regardless of the value of $\mathcal{B}$, the Hamiltonian $\hat{H} \equiv H(\hat{\mathbf{S}})$ given by Eq. (3.3.1) is invariant under π-rotation around the $\mathbf{e}_y$-axis:

$$\hat{R}\hat{H}\hat{R}^\dagger = \hat{H}. \tag{3.3.96}$$

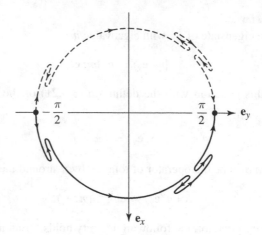

Fig. 3.16. A typical pair of paths appearing in the path integral for a spin.

Hence,

$$\mathcal{A}_{m'm} \propto 1 + e^{i(m'-m)\pi}. \tag{3.3.97}$$

Thus, one arrives at the following theorem:

$$\text{If } m' - m \text{ is odd, then } \mathcal{A}_{m'm} = 0. \tag{3.3.98}$$

As a result of this theorem, any transition from $|-S; \mathbf{e}_y\rangle$ to $|S; \mathbf{e}_y\rangle$ is forbidden for a half-integral S:

$$\langle \mathbf{e}_y | \exp(-i\hat{H}t/\hbar)| - \mathbf{e}_y \rangle = 0 \; : S \in \mathbf{N}_{1/2}, \; \forall t. \tag{3.3.99}$$

Quantum tunneling of a spin is often described with the terminology of a path integral. The above result would run as follows:

The transition amplitude from $|-\mathbf{e}_y\rangle$ to $|\mathbf{e}_y\rangle$ can be viewed as the sum of amplitudes[71] each of which is associated with one of infinitely many paths[72] connecting $\mathbf{S} = -S\mathbf{e}_y$ and $\mathbf{S} = S\mathbf{e}_y$. Decompose the first term on the right-hand side of Eq. (3.3.94) into contributions from infinitely many paths, and arbitrarily choose one of them. Suppose that the path chosen is the one depicted by the solid curve in Fig. 3.16. To this path, there corresponds a unique path obtained by reflecting the former with respect to the $\mathbf{e}_y$-axis (the broken curve in the figure). The latter path contributes to the second term of Eq. (3.3.94). Because of the symmetry of (3.3.96), the amount of this contribution is the same as that of the

[71] Amplitudes here refer to those in the context of path integrals. For a path integral for a spin, see e.g. Kashiwa, Ohnuki and Suzuki : ibid.

[72] In the spin-coherent-state path integral, for example, paths are those in the phase space S^2 for a spin. In a path integral in general, infinitely many paths are treated on equal footing (_Feynman democracy_); see e.g. R. P. Feynman and A. R. Hibbs, _Quantum Mechanics and Path Integrals_, McGraw-Hill (1965). It was to facilitate the comparison with this democracy that we adopted the arithmetic mean in Eq. (3.3.94).

former path to the first term. Between them, however, there is a phase difference of $2S\pi$. For a half-integral S, therefore, contributions from infinitely many paths cancel themselves out pair-wise.

Incidentally, Eq. (3.3.99) remains valid if $| \pm \mathbf{e}_y \rangle$ is replaced by $| \pm \pi/2 \rangle_s$:

$$_s\langle \pi/2 \,|\, f(\hat{H}) \,|\, -\pi/2 \rangle_s \propto 1 + e^{2S\pi i}. \tag{3.3.100}$$

It is too rash, however, to conclude on the basis of Eqs. (3.3.99) or (3.3.100) that the spin cannot undergo the tunneling in question if S is a half integer and $\mathcal{B}_{\parallel} = \mathcal{B}' = 0$; even though the transition (3.3.99) is forbidden, there is no reason for the transition amplitude to the neighboring state

$$\langle S - 1; \mathbf{e}_y | \exp(-i\hat{H}t/\hbar) \,|\, -S; \mathbf{e}_y \rangle \tag{3.3.101}$$

to vanish. For $S \gg 1$, it would be difficult to experimentally distinguish the state $|S; \mathbf{e}_y\rangle$ from $|S - 1; \mathbf{e}_y\rangle$. In an experiment, transition from $| - S; \mathbf{e}_y\rangle$ to either $|S; \mathbf{e}_y\rangle$ or $|S - 1; \mathbf{e}_y\rangle$ will be recorded as the change of the spin direction from $-\mathbf{e}_y$ to $\mathbf{e}_y$. If S is an integer, the transition amplitude in Eq. (3.3.99) does not vanish but (3.3.101) does. Thus, MQT is not drastically affected by the parity of $2S$.

Note added in the English edition: Two years after the publication of the Japanese edition, the interference effect represented by the factor (3.3.86) has been detected with a magnetic molecule Fe_8 for which $S = 10$ (see Ref. [18] in the bibliography).

Exercises

Exercise 3.1. For a SQUID of the size indicated in Fig. 3.1, typical values of constants are (cf. Ref.[11] in the bibliography):

$$C = 80\,\text{fF}, \quad \mathcal{R} = 5\,\text{k}\Omega, \quad \mathcal{L} = 210\,\text{pH}, \quad I_c = 0.78\,\mu\text{A}. \tag{3.3.102}$$

Estimate the values of I_{Φ_0} and γ.

Exercise 3.2. Cast the flux potential into the standard form $V(q)$ and evaluate the frequency ω_0 of the infinitesimal oscillation around its minimum.

Exercise 3.3. Use the constants quoted in Exercise 3. 1 above to estimate the values of U_0, τ_0, ω_0, h_0, h and η in the case of Fig. 3.3. Do the same with Fig. 3.4.

Exercise 3.4. The differential operator on the right-hand side of Eq. (3.2.35) might not look Hermitian. Show that it is in fact Hermitian with respect to the integration measure $\{M(R)\}^{1/2}dR \,(\propto dq)$. (Cf. The discussion following Eq. (3.3.64).)

Exercise 3.5.

(1) Adopt ϕ_0, λ_0, $\lambda_0/2c_0$ and ρ_0 as units of field, length, time and mass density, respectively, to render $I[\psi]$ given by Eq. (3.2.58) dimensionless.

(2) Let g be the Planck constant in these units, then

$$g = 4\hbar m_A c_0^2 / A\rho_0 \lambda_0^d. \tag{3.3.103}$$

(3) Consider the dimensionless field potential $\mathcal{U}(\phi) := \mathcal{U}_{\text{conv}}(\phi)/\rho_0 c_0^2$. Its simplest analytical form corresponding to Fig. 3.8 is the following:

$$\mathcal{U}(\phi) = \mathcal{U}^{\text{STDW}}(\varphi) := \mathcal{U}^{\text{SSDW}}(\varphi) - 4\varepsilon\varphi^3 + 3\varepsilon\varphi^4, \tag{3.3.104}$$

$$\mathcal{U}^{\text{SSDW}}(\varphi) := \frac{1}{2}\varphi^2(1-\varphi)^2, \quad \varphi \equiv (\phi_> - \phi)/\phi_0. \tag{3.3.105}$$

Use this form to determine the kink $b(x)$ and the constant $\bar{\sigma}$ in Eq. (3.2.79).

Exercise 3.6. It has been argued that the spin direction in a single-domain ferromagnet can change via the tunnel effect. Is this argument consistent with the law of conservation of angular momentum? Answer the corresponding question for a SQUID.

Exercise 3.7. Display graphically how the spin potential $U_S(\phi)/K_\perp$ depends on b and b'.

Exercise 3.8. By use of the naive WKB method, derive Eq. (3.3.16) assuming that the wavefunction obeys the periodic boundary condition. Show that the foremost factor 2 is replaced by 0 if the anti-periodic boundary condition is imposed (case of half-integral S). Also, evaluate the coefficient $\tilde{A}$.

Exercise 3.9. Consider the double bumpy slope depicted in Fig. 3.12. By use of the naive WKB method, derive the decay-rate formula, which would correspond to Eq. (2.4.5) in the single bump case, for the metastable state ($\phi \sim -\pi/2$).

Exercise 3.10. Show that the set $\{|\mathbf{n}\rangle \mid \mathbf{n} \in S^1\}$ does not constitute a basis.

Exercise 3.11. A wavefunction, being a complex number, may be represented by a point on the complex plane. Utilize this fact to display visually the Möbius boundary condition.

Exercise 3.12. Display b_{Sm} graphically for $S = 1/2, 1, 3/2, 2, \ldots$, and compare the result with Eq. (3.3.39).

Exercise 3.13. Display the wavefunction (3.3.37) graphically for $S = 1/2, 1, 3/2, 2, \ldots$. In which region of ϕ is the approximate form (3.3.41) valid?

Exercise 3.14. Pay due attention to the boundary condition (3.3.33) to prove that the differential operator on the right-hand side of Eq. (3.3.44) is Hermitian and that those in Eqs.(3.3.52) and (3.3.53) are Hermitian conjugate to each other.

Exercise 3.15. Show that the difference between the first and the second lines of Eq. (3.3.67) is given by

$$-\frac{\gamma^2}{2S^2}\{(\partial^2 \log \mu) - (\partial \log \mu)^2\} \quad : \partial \equiv \frac{\partial}{\partial\varphi}, \quad \mu \equiv \mu(\varphi). \tag{3.3.106}$$

Exercise 3.16. Compare $U(\varphi)$ with $\tilde{U}(\varphi)$ for some typical values of a, b and b'.

Exercise 3.17. Show that there do not exist $\hat{\phi}$ and $|\phi\rangle$ satisfying Eq. (3.3.73).

Exercise 3.18. Derive Eqs.(3.3.74) and (3.3.75). Readers may recognize that the integral on the right-hand side of Eq. (3.3.74) can be expressed as $\sqrt{\frac{a}{1+a}} K\left(\sqrt{\frac{a}{1+a}}\right)$, where $K(k)$ is the complete elliptic integral of the first kind of modulus k. The knowledge of the elliptic integral, however, is not necessary to derive Eq. (3.3.75).

Exercise 3.19. Under the π-rotation $\hat{R}$ around the y-axis, the states $|\pm \pi/2\rangle_s$ remain invariant apart from changes in phase:

$$\hat{R}|\pm \pi/2\rangle_s = e^{\mp iS\pi}|\pm \pi/2\rangle_s. \tag{3.3.107}$$

Derive this result and prove Eq. (3.3.100).

4

Environmental problems

Quantum-mechanical interference effects may be washed out by the influence of the environment. This can happen regardless of the system in question being macroscopic or not. According to the traditional view, as introduced in Chapter 1, this should happen inevitably for a macrosystem. This chapter surveys the mechanism of decoherence in order to familiarize readers with its essence.

4.1 Coherence and decoherence

The word *coherence* is used very frequently, and often in an ambiguous manner. For example, one sometimes encounters such an expression as "a coherent linear combination". Here, the adjective *coherent* is used rather emotionally for the purpose of emphasis, perhaps; it is logically redundant since there is no such concept as "an incoherent linear combination". Any quantum-mechanical state can be viewed as a linear combination of one sort or other, and can exhibit interference effects if a stage is set up appropriately. In this sense, any quantum-mechanical state is coherent.[1] Usually, one is interested in a specific interference effect; a situation is said to be coherent if the interference effect in question is visible and incoherent if not. In other words, it is meaningless just to ask whether a situation is coherent or not. One has to specify both the situation and which interference effect one is interested in. Analogous remarks apply to the word *mixture*. The adjective "incoherent" in the often-encountered phrase "incoherent mixture" is logically redundant since there is no such concept as "a coherent mixture".

Now, the topic of this chapter is *decoherence*[2] (i.e. the disappearance or the loss of coherence), that is, how interference effects may be affected by external influences.

---

[1] The so-called *coherent state* is a proper noun designating a particular state.
[2] This word is used also in the theory of consistent (or decohering or non-interfering) histories. This theory, which has been developed since the 1980s, concerns the fundamental framework of quantum mechanics. The concept of "decoherence" in this theory, however, is quite different from that in this book.

The mechanism of coherence may be classified into *dephasing* and *dissipation*:

$$\text{DDD}: \quad \text{Decoherence} = \text{Dephasing} \vee \text{Dissipation}. \qquad (4.1.1)$$

In general, dissipation means an irreversible transfer of a quasi-conserved quantity from a macrosystem in question to its environment. Here, a physical quantity associated with the macrosystem is said to be quasi-conserved if it is rigorously conserved when the macrosystem is isolated and if it changes slowly when the macrosystem interacts with the environment. For example, the momentum and the angular momentum of a body floating in a near vacuum are quasi-conserved. In practice, since environments are not so ideal as to allow the momentum or the angular momentum of a macrosystem to be quasi-conserved, the word dissipation is used synonymously with the irreversible transfer of energy from a macrosystem in question to its environment. Dephasing, on the other hand, means that the environment disturbs the phase of the quantum state of the macrosystem. This disturbance, to be described more precisely below, is conventionally regarded as "uncontrollable". It is noteworthy that dissipation is an important but only a partial mechanism of decoherence, for which dephasing is at least equally important. This has long since been recognized in theories of magnetic relaxation, for instance. The situation is the same with macroscopic quantum phenomena.

4.2 General discussion of dephasing in MQC

Consider the MQC-situation depicted in Fig. 2.1(a). Label the eigenstates associated with the macrosystem Hamiltonian $\hat{H}_S$ in ascending order with respect to the energy eigenvalue:

$$\hat{H}_S |n\rangle = E_n |n\rangle, \quad E_0 < E_1 < E_2 < \cdots. \qquad (4.2.1)$$

Following Section 2.2, the ground-state energy splitting (i.e. the energy difference between the ground state and the first excited state) is to be denoted by $h\Delta$:

$$\Delta := (E_1 - E_0)/h. \qquad (4.2.2)$$

The linear combinations

$$|\pm\rangle := \frac{1}{\sqrt{2}} (|0\rangle \pm |1\rangle) \qquad (4.2.3)$$

of the ground and the first excited states represent situations in which the <u>particle</u> is located in the right or the left well, respectively.

Let the initial state of the macrosystem be $|-\rangle$ (see Eq. (2.2.7)). Try to guess how an environment affects the proper time evolution (2.2.10) of the macrosystem. Perhaps, immediately conceivable effects are the following:

(1) The energy levels of the macrosystem will be shifted due to perturbations exerted by the environment. Let δE_n be the shift suffered by E_n.

(2) The first-excited-state component in the initial state will decay to the ground state, thereby increasing the ground-state component. Let T_1 be the lifetime of the first excited state.[3]

(3) Dephasing will occur. Let $\exp\{-i\vartheta_n(t)\}$ be the "uncontrollable phase factor" to be multiplied to the nth eigenstate $|n\rangle$ of $\hat{H}_S$.

Taking account of these three effects, one could guess that the time evolution (2.2.10) would be replaced by the following:[4]

$$|\psi(t)\rangle = \sqrt{P_0(t)}e^{-i(E_0+\delta E_0)t/h-i\vartheta_0(t)}|0\rangle - \sqrt{P_1(t)}e^{-i(E_1+\delta E_1)t/h-i\vartheta_1(t)}|1\rangle$$

$$\propto \sqrt{P_0(t)}e^{it\tilde{\Delta}/2+i\vartheta(t)/2}|0\rangle - \sqrt{P_1(t)}e^{-it\tilde{\Delta}/2-i\vartheta(t)/2}|1\rangle, \qquad (4.2.4)$$

where

$$P_1(t) \simeq \frac{1}{2}e^{-t/T_1}, \quad P_0(t) \simeq \frac{1}{2}+\left\{\frac{1}{2}-P_1(t)\right\} = 1-\frac{1}{2}e^{-t/T_1}, \quad (4.2.5)$$

$$\tilde{\Delta} \equiv \Delta + (\delta E_1 - \delta E_0)/h, \qquad (4.2.6)$$

$$\vartheta(t) \equiv \vartheta_1(t) - \vartheta_0(t). \qquad (4.2.7)$$

According to this guess, the probability $\mathcal{P}_\pm(t)$ for the macrosystem to be found in $|\pm\rangle$ at time t is given as

$$\mathcal{P}_\pm(t) = |\langle\pm|\psi(t)\rangle|^2$$

$$= \left|\frac{1}{\sqrt{2}}\left\{\sqrt{P_0(t)}e^{it\tilde{\Delta}/2+i\vartheta(t)/2} \mp \sqrt{P_1(t)}e^{-it\tilde{\Delta}/2-i\vartheta(t)/2}\right\}\right|^2$$

$$= \frac{1}{2}\left\{P_0(t) + P_1(t) \mp 2\sqrt{P_0(t)P_1(t)}\,\Re\left(e^{-i\vartheta(t)}e^{-it\tilde{\Delta}}\right)\right\}. \qquad (4.2.8)$$

This result implies that the phase of the quantum resonant oscillation gets shifted by a t-dependent amount $\vartheta(t)$. The behavior of the amplitude $\sqrt{P_0(t)P_1(t)}$ of the oscillation varies depending on temporal domain. In the short-time domain[5] ($t_c \ll t \ll T_1$) one may neglect quantities of $\mathcal{O}((t/T_1)^2)$ to find

$$\sqrt{P_0(t)} \simeq \frac{1}{\sqrt{2}}\left(1+\frac{t}{2T_1}\right) \simeq \frac{1}{\sqrt{2}}e^{t/2T_1}, \quad \sqrt{P_1(t)} \simeq \frac{1}{\sqrt{2}}e^{-t/2T_1}, \quad (4.2.9)$$

$$2\sqrt{P_0(t)P_1(t)} \simeq 1. \qquad (4.2.10)$$

---

[3] The subscript 1 in T_1 does not signify the first excited state; see the end of this section.

[4] The proportionality coefficient present in the last line of Eq. (4.2.4) is an overall phase factor, which can be dropped without any physical consequence.

[5] The constant t_c is a microscopic characteristic time (e.g. ω_c^{-1} in the next section) to characterize the interaction between the macrosystem and the environment. At any rate, as with the decay considered in Section 2.5, $P_1(t)$ behaves exponentially only after the quantum Zeno temporal domain.

Hence, the amplitude is essentially independent of time. In the <u>long-time domain</u> $(t \gtrsim T_1)$, on the other hand, one may neglect quantities of $\mathcal{O}(e^{-3t/2T_1})$ to find

$$2\sqrt{P_0(t)P_1(t)} \simeq \sqrt{2}e^{-t/2T_1}. \tag{4.2.11}$$

Accordingly, the expectation value $E_S(t)$ for the energy of the macrosystem is found as

$$E_S(t) = E_0 P_0(t) + E_1 P_1(t) = E_0 + \frac{1}{2}(E_1 - E_0)e^{-t/T_1}. \tag{4.2.12}$$

It follows that the characteristic time for the decay of the quantum resonant oscillation is twice the characteristic time T_1 for the energy dissipation.

The origin of the "uncontrollable phase" $\vartheta(t)$ might be understood as follows. Suppose for concreteness that the environment exerts a random <u>magnetic field</u> $\mathcal{B}(t)$ on the macrosystem. The macrosystem Hamiltonian is then augmented by[6]

$$\hat{H}'_S := h\mathcal{B}(t)\hat{S}_3 = \frac{1}{2}h\mathcal{B}(t)(|1\rangle\langle 1| - |0\rangle\langle 0|). \tag{4.2.13}$$

This additional term has the effect of increasing the instantaneous ground-state energy splitting by the amount of $h\mathcal{B}(t)$ at each t. Hence,

$$\vartheta(t) = \int_0^t dt'\mathcal{B}(t'). \tag{4.2.14}$$

Experimenters, not being in possession of means to measure the t-dependence of $\mathcal{B}(t)$, would hide their ignorance by regarding $\{\mathcal{B}(t)\}$ as a *stochastic process* and taking an appropriate stochastic average. Suppose for simplicity that the stochastic process in question is a *Gaussian process* and denote the averaging procedure with a bar, then

$$\overline{\mathcal{B}(t)} = 0, \quad \overline{\mathcal{B}(t_3)\mathcal{B}(t_2)\mathcal{B}(t_1)} = 0,$$
$$\overline{\mathcal{B}(t_4)\mathcal{B}(t_3)\mathcal{B}(t_2)\mathcal{B}(t_1)} = \overline{\mathcal{B}(t_4)\mathcal{B}(t_3)}\,\overline{\mathcal{B}(t_2)\mathcal{B}(t_1)} + \overline{\mathcal{B}(t_4)\mathcal{B}(t_2)}\,\overline{\mathcal{B}(t_3)\mathcal{B}(t_1)}$$
$$+ \overline{\mathcal{B}(t_4)\mathcal{B}(t_1)}\,\overline{\mathcal{B}(t_3)\mathcal{B}(t_2)}, \tag{4.2.15}$$
$$\cdots\cdots\cdots .$$

That is, a product of an odd number of $\mathcal{B}(t)$'s is averaged to 0, and a product of an even number of $\mathcal{B}(t)$'s is decomposed into the sum of products of averages of all possible pairs. It follows that

$$\overline{e^{-i\vartheta(t)}} = e^{-\frac{1}{2}\Theta(t)}, \tag{4.2.16}$$

$$\Theta(t) \equiv \int_0^t dt_2 \int_0^t dt_1 \overline{\mathcal{B}(t_2)\mathcal{B}(t_1)}. \tag{4.2.17}$$

---

[6] The operator $\hat{S}_3$ is defined by Eq. (2.2.4).

Accordingly,

$$\overline{\mathcal{P}_{\pm}(t)} = \frac{1}{2}\left\{ P_0(t) + P_1(t) \mp 2\sqrt{P_0(t)P_1(t)}\,e^{-\frac{1}{2}\Theta(t)}\cos(t\tilde{\Delta}) \right\}. \quad (4.2.18)$$

Furthermore, if $\{\mathcal{B}(t)\}$ is a *stationary process*, that is, if $\overline{\mathcal{B}(t_2)\mathcal{B}(t_1)}$ depends on the time difference $t_2 - t_1$ alone, then

$$\Theta(t) \simeq \mathcal{B}_0^2 \tau_{\mathrm{c}} t \quad : t \gg \tau_{\mathrm{c}}, \quad (4.2.19)$$

where[7]

$$\mathcal{B}_0^2 \equiv \overline{\mathcal{B}(t)\mathcal{B}(t)}, \quad (4.2.20)$$

$$\tau_{\mathrm{c}} \equiv \mathcal{B}_0^{-2} \int_{-\infty}^{\infty} \mathrm{d}t_2 \, \overline{\mathcal{B}(t_2)\mathcal{B}(t_1)}. \quad (4.2.21)$$

Thus, one may conclude that dephasing tends to damp the quantum resonant oscillation.

Plausible as it may seem, the above argument is not satisfactory at all; it pretends to deal with the effects of an environment, but the dynamical degrees of freedom of the environment do not make their appearance. The environment should not merely be referred to implicitly, but should be treated quantum-mechanically. Let $\hat{H}_{\mathcal{E}}$ be the Hamiltonian for the environment,[8] $\hat{H}_{S\mathcal{E}}$ be the Hamiltonian describing the interaction between the macrosystem and the environment, and $\hat{H}$ be the total Hamiltonian for the entire system composed of the macrosystem and the environment:

$$\hat{H} := \hat{H}_S + \hat{H}_{\mathcal{E}} + \hat{H}_{S\mathcal{E}}. \quad (4.2.22)$$

We denote with $|\Psi(t)\rangle\rangle$ the state[9] of the entire system at time t. Also, we denote with $\mathcal{H}_S$ and $\mathcal{H}_{\mathcal{E}}$ the Hilbert spaces to describe the macrosystem and the environment, respectively. Of course, the former is the space spanned by the basis

$$\{|n\rangle \mid n = 0, 1, 2, \ldots\}. \quad (4.2.23)$$

The state $|\Psi(t)\rangle\rangle$ belongs to the direct-product Hilbert space $\mathcal{H}_S \otimes \mathcal{H}_{\mathcal{E}}$.

Suppose that the initial state of the macrosystem is $|-\rangle$ and that of the environment is $|\chi\rangle$. The time evolution of the entire system is dictated by

$$i\hbar\frac{\mathrm{d}}{\mathrm{d}t}|\Psi(t)\rangle\rangle = \hat{H}|\Psi(t)\rangle\rangle, \quad (4.2.24)$$

$$|\Psi(0)\rangle\rangle = |-, \chi\rangle\rangle \equiv |-\rangle|\chi\rangle. \quad (4.2.25)$$

---

[7] The constant τ_{c} is called the correlation time of the noise $\overline{\mathcal{B}(t_2)\mathcal{B}(t_1)}$.
[8] The subscript $\mathcal{E}$ stands for the environment.
[9] The double-ket symbol $|\quad\rangle\rangle$ is to be used to emphasize that the state in question refers to the entire system composed of the macrosystem and the environment.

This equation may be solved formally as

$$|\Psi(t)\rangle\rangle = \hat{U}(t)|\Psi(0)\rangle\rangle, \quad \hat{U}(t) := \exp(-i\hat{H}t/h). \tag{4.2.26}$$

This state can be expanded in terms of the basis (4.2.23) of $\mathcal{H}_S$ with expansion coefficients being t-dependent states belonging to $\mathcal{H}_{\mathcal{E}}$:

$$|\Psi(t)\rangle\rangle = \sum_{n=0}^{\infty} e^{-iE_n t/h}|n\rangle|\widetilde{\chi_n(t)}\rangle \quad : |\widetilde{\chi_n(t)}\rangle \in \mathcal{H}_{\mathcal{E}}. \tag{4.2.27}$$

Thus, in general, once interaction takes place between the macrosystem and the environment, the state $|\Psi(t)\rangle\rangle$ is no longer a direct product, such as the one in Eq. (4.2.25), but an entangled state. Let $P_n(t)$ be the probability for the macrosystem to be found in the state $|n\rangle$ at time t, then

$$P_n(t) = \||\widetilde{\chi_n(t)}\rangle\|^2. \tag{4.2.28}$$

By use of these probabilities, the expectation value $E_S(t)$ for the energy of the macrosystem is expressed as

$$E_S(t) = \sum_{n=0}^{\infty} E_n P_n(t). \tag{4.2.29}$$

This quantity is to be called the _macrosystem energy_ for brevity.

Focus attention on the terms of $n = 0$ and $n = 1$ in Eq. (4.2.27), which may be rewritten as

$$|\Psi(t)\rangle\rangle = e^{-i(E_0+E_1)t/2h}\{|+\rangle|\widetilde{\chi_+(t)}\rangle + |-\rangle|\widetilde{\chi_-(t)}\rangle\}$$

$$+ \sum_{n=2}^{\infty} e^{-iE_n t/h}|n\rangle|\widetilde{\chi_n(t)}\rangle, \tag{4.2.30}$$

$$|\widetilde{\chi_\pm(t)}\rangle := \frac{1}{\sqrt{2}}\left\{e^{it\Delta/2}|\widetilde{\chi_0(t)}\rangle \pm e^{-it\Delta/2}|\widetilde{\chi_1(t)}\rangle\right\}. \tag{4.2.31}$$

It is clear that

$$|\widetilde{\chi_-(0)}\rangle = |\chi\rangle, \quad |\widetilde{\chi_+(0)}\rangle = 0, \tag{4.2.32}$$

$$|\widetilde{\chi_n(t)}\rangle = 0 \quad : n \geq 2. \tag{4.2.33}$$

It is convenient for the ensuing discussion to normalize $|\widetilde{\chi_n(t)}\rangle$:

$$|\chi_n(t)\rangle := \{P_n(t)\}^{-1/2}|\widetilde{\chi_n(t)}\rangle. \tag{4.2.34}$$

The probability $\mathcal{P}_\pm(t)$ for the macrosystem to be found in the state $|\pm\rangle$ at time t may be calculated as

$$\mathcal{P}_\pm(t) = \||\widetilde{\chi_\pm(t)}\rangle\|^2 = \frac{1}{2}\{P_0(t) + P_1(t) \mp \mathcal{I}(t)\}, \tag{4.2.35}$$

where

$$\mathcal{I}(t) := \Re\{e^{-it\Delta}\mathcal{C}(t)\}, \tag{4.2.36}$$

$$\mathcal{C}(t) := -2\langle\widetilde{\chi_0(t)}|\widetilde{\chi_1(t)}\rangle = 2\sqrt{P_0(t)P_1(t)}\{-\langle\chi_0(t)|\chi_1(t)\rangle\}. \tag{4.2.37}$$

Note the interference term $\mathcal{I}$, which does not appear in the macrosystem energy (4.2.29). The factor $\mathcal{C}(t)$ is a measure of the extent to which coherence is retained: coherence in this case refers to the states $|0\rangle$ and $|1\rangle$ of the macrosystem. $\mathcal{C}(t)$ consists of two factors. The first factor $2\{P_0(t)P_1(t)\}^{1/2}$ reflects the time-dependence of the probability for the macrosystem to be found in $|0\rangle$ or $|1\rangle$. The absolute value of the second factor $-\langle\chi_0(t)|\chi_1(t)\rangle$, which is unity at $t = 0$ since $|\chi_0(0)\rangle = -|\chi_1(0)\rangle$, decreases with time in general; the direction[10] of $|\chi_0(t)\rangle$ tends to deviate from that of $|\chi_1(t)\rangle$ as time goes on. Comparing Eq. (4.2.35) with Eq. (4.2.8), one may conclude that both the effect (4.2.16) and the energy-level shift must be hidden in this second factor. We somewhat abuse language to call $-\langle\chi_0(t)|\chi_1(t)\rangle$ the *dephasing factor*. The above consideration shows the following:

For a quantum-mechanical description of dephasing effects, it is necessary to treat the environment quantum-mechanically as well.

The dephasing factor is expected to behave typically as

$$-\langle\chi_0(t)|\chi_1(t)\rangle \sim \exp\left\{-\left(\frac{1}{T_2'} + i\Delta'\right)t + i\theta_0\right\} \quad : t \gg t_c, \tag{4.2.38}$$

where T_2' is the characteristic time of the dephasing, and Δ' and θ_0 express the phase shift.[11] Comparison of the above expression with Eqs. (4.2.18) and (4.2.19) reveals that T_2' and Δ' correspond to $2/\mathcal{B}_0^2\tau_c$ and the energy shift $(\delta E_1 - \delta E_0)/h$, respectively. If $P_n(t)$ is assumed to behave as in Eq. (4.2.5), then

$$E_S(t) \simeq E_0 + \frac{h\Delta}{2}e^{-t/T_1} \quad : t \gg t_c, \tag{4.2.39}$$

$$\mathcal{I}(t) \simeq \begin{cases} e^{-t/T_2'}\cos(t\tilde{\Delta} - \theta_0) & : t_c \ll t \ll T_1, \\ \sqrt{2}\,e^{-t/T_2}\cos(t\tilde{\Delta} - \theta_0) & : t \gtrsim T_1, \end{cases} \tag{4.2.40}$$

where

$$\frac{1}{T_2} \equiv \frac{1}{2T_1} + \frac{1}{T_2'}, \quad \tilde{\Delta} \equiv \Delta + \Delta'. \tag{4.2.41}$$

The characteristic time T_2 characterizing the decrease of $\mathcal{I}(t)$ in the long-time domain consists of two terms; the term $1/2T_1$ originating from the lifetime of the

---

[10] Direction in the Hilbert space.
[11] The former is proportional to t and the latter is independent of t.

state $|1\rangle$ and the term $1/T_2'$ originating from the dephasing. The frequency Δ of the quantum resonant oscillation suffers a shift by the amount Δ'. In the language of the fictitious spin introduced in Section 2.2, the energy dissipation corresponds to the decrease of $\langle \hat{S}_3 \rangle(t)$ and the damping of the interference term corresponds to the decrease of $|\langle \hat{S}_- \rangle(t)|$ (i.e. the damping of the precession of spin). For this reason, T_1 and T_2 are called the *longitudinal relaxation time* and the *transverse relaxation time*, respectively. The shift Δ' corresponds to the *frequency shift* in magnetic resonance. These terminologies have been used traditionally in the theory of magnetic relaxation (or magnetic resonance).[12]

4.3 Dissipative environment

Let us introduce models for the environment with simple properties[13] to facilitate understanding of the essential features of decoherence. This section considers a model which focuses on dissipation.

Suppose that the macrosystem Hamiltonian is specified by Eq. (4.2.1) with the origin of energy now chosen at E_0, and that the environment consists of bosons, namely, harmonic oscillators:

$$\hat{H}_{\mathcal{E}} := \sum_{\alpha} \hbar \omega_\alpha \hat{b}_\alpha^\dagger \hat{b}_\alpha, \tag{4.3.1}$$

$$[\hat{b}_\alpha, \hat{b}_\beta^\dagger] = \delta_{\alpha\beta}, \quad [\hat{b}_\alpha, \hat{b}_\beta] = 0. \tag{4.3.2}$$

Let $|\text{vac}\rangle$ be the ground state[14] of $\hat{H}_{\mathcal{E}}$, and $|\alpha\rangle$ be the state with a single boson α present:

$$\hat{b}_\alpha |\text{vac}\rangle = 0, \quad |\alpha\rangle := \hat{b}_\alpha^\dagger |\text{vac}\rangle. \tag{4.3.3}$$

We envisage the interaction between the macrosystem and the environment as follows: a boson is created when the macrosystem makes a transition from the first excited state to the ground state. Since the interaction Hamiltonian $\hat{H}_{S\mathcal{E}}$ must be Hermitian, a boson should be annihilated when the macrosystem makes a transition from the ground state to the first excited state:[15]

$$\hat{H}_{S\mathcal{E}} := i\hbar \sum_{\alpha} \kappa_\alpha (\hat{S}_- \hat{b}_\alpha^\dagger - \hat{S}_+ \hat{b}_\alpha), \tag{4.3.4}$$

$$\hat{S}_- := \hat{S}_1 - i\hat{S}_2 = -i|0\rangle\langle 1|, \quad \hat{S}_+ := \hat{S}_-^\dagger, \tag{4.3.5}$$

---

[12] See, e.g. C. P. Slichter, *Principles of Magnetic Resonance*, Harper and Row (1963).

[13] The nature of the interaction of the environment with the macrosystem is to be included in the property of environment; an isolated environment by itself is of no interest.

[14] The state in which bosons are absent, namely, the vacuum state.

[15] The operator $\hat{S}$ is defined in Section 2.2. In the literature of magnetic relaxation or quantum optics, the interaction of the form of Eq. (4.3.4) is often called the *rotating-wave approximant* to the more general interaction.

where κ_α are constants representing the strength of the interaction.[16] The time evolution of the entire system is governed by the Schrödinger equation (4.2.24). Obviously, the interaction (4.3.4) conserves the <u>total particle number</u> $\hat{N}_{tot}$:

$$[\hat{H}, \hat{N}_{tot}] = 0, \tag{4.3.6}$$

where

$$\hat{N}_{tot} := \left(\frac{1}{2} + \hat{S}_3\right) + \sum_\alpha \hat{b}_\alpha^\dagger \hat{b}_\alpha. \tag{4.3.7}$$

Therefore, one can decompose the entire Hilbert space into subspaces each of which corresponds to an eigenvalue of $\hat{N}_{tot}$, and consider the time evolution in each of the subspaces separately.

The state $|0, \text{vac}\rangle\rangle$ ($\equiv |0\rangle|\text{vac}\rangle$) by itself spans the subspace of $\hat{N}_{tot} = 0$. Hence, this state remains unchanged as time goes on. The subspace of $\hat{N}_{tot} = 1$ is spanned by the states $|1, \text{vac}\rangle\rangle$ ($\equiv |1\rangle|\text{vac}\rangle$) and $\{|0, \alpha\rangle\rangle$ ($\equiv |0\rangle|\alpha\rangle)\}$. Hence, the time evolution of the state $|1, \text{vac}\rangle\rangle$ should be of the following form:

$$c(t)|1, \text{vac}\rangle\rangle + \sum_\alpha c_\alpha(t)|0, \alpha\rangle\rangle. \tag{4.3.8}$$

Putting this into the Schrödinger equation, one finds

$$i\dot{c}(t)|1, \text{vac}\rangle\rangle + i\sum_\alpha \dot{c}_\alpha(t)|0, \alpha\rangle\rangle = c(t)\Delta|1, \text{vac}\rangle\rangle + \sum_\alpha \{\omega_\alpha c_\alpha(t)|0, \alpha\rangle\rangle$$
$$+ c(t)\kappa_\alpha|0, \alpha\rangle\rangle + c_\alpha(t)\kappa_\alpha|1, \text{vac}\rangle\rangle\}. \tag{4.3.9}$$

Hence

$$\dot{c}(t) = -i\left(c(t)\Delta + \sum_\alpha c_\alpha(t)\kappa_\alpha\right), \tag{4.3.10}$$

$$\dot{c}_\alpha(t) = -i\left(c_\alpha(t)\omega_\alpha + c(t)\kappa_\alpha\right). \tag{4.3.11}$$

Taking account of the initial condition

$$c(0) = 1, \quad c_\alpha(0) = 0 \tag{4.3.12}$$

by use of the Laplace transformation[17]

$$\tilde{f}(s) := \int_0^\infty dt\, e^{-st} f(t) \quad : \Re s > 0, \tag{4.3.13}$$

[16] We assume without loss of generality that $\kappa_\alpha > 0$; this is ensured by an appropriate adjustment of the phase of $\hat{b}_\alpha$.

[17] In general, a linear differential equation may be solved by use of either the Fourier transformation or the Laplace transformation. The latter incorporates the initial values automatically, and hence is suitable for an initial-value problem.

one obtains

$$s\tilde{c}(s) - 1 = -i\left(\tilde{c}(s)\Delta + \sum_\alpha \tilde{c}_\alpha(s)\kappa_\alpha\right), \tag{4.3.14}$$

$$s\tilde{c}_\alpha(s) = -i(\tilde{c}_\alpha(s)\omega_\alpha + \tilde{c}(s)\kappa_\alpha). \tag{4.3.15}$$

Elimination of $\tilde{c}_\alpha(s)$ leads to

$$\tilde{c}(s) = \left(s + i\Delta + \sum_\alpha \frac{\kappa_\alpha^2}{s + i\omega_\alpha}\right)^{-1}. \tag{4.3.16}$$

With the aid of the single-particle density of states weighted with the interaction strength[18]

$$J_B(\omega) := \sum_\alpha \kappa_\alpha^2 \delta(\omega - \omega_\alpha), \tag{4.3.17}$$

the above expression may be rewritten as

$$\tilde{c}(s) = \left(s + i\Delta + \int_0^\infty d\omega \frac{J_B(\omega)}{s + i\omega}\right)^{-1}. \tag{4.3.18}$$

These results remain unchanged if the bosons are replaced by fermions,[19] or if $\hat{b}_\alpha$ is replaced by the spin-1/2 lowering operator $\hat{s}_-^{(\alpha)}$; the statistics of particles do not matter in the subspace of $\hat{N}_{tot} = 1$.

Given a closed form of $J_B(\omega)$, the ω-integration in the above equation may be performed and the resulting $\tilde{c}(s)$ may be converted back to $c(t)$ via the inverse Laplace transformation. If the interaction is weak ($J_B(\Delta) \ll \Delta$), the following approximate expression is valid in the temporal domain[20] $t \gg \omega_c^{-1}$:

$$c(t) \propto \exp(-it\tilde{\Delta} - \pi J_B(\tilde{\Delta})t), \tag{4.3.19}$$

$$\tilde{\Delta} = \Delta + \mathcal{P}\int_0^\infty d\omega \frac{J_B(\omega)}{\tilde{\Delta} - \omega} \quad (\mathcal{P} \text{ denotes the principal value}), \tag{4.3.20}$$

and accordingly

$$T_1 \simeq 1/2\pi J_B(\tilde{\Delta}). \tag{4.3.21}$$

Putting the details of these calculations aside, consider the following initial condition:

$$|\Psi(0)\rangle\rangle = |-, \text{vac}\rangle\rangle \equiv |-\rangle|\text{vac}\rangle. \tag{4.3.22}$$

---

[18] This density of states is essentially the same as the environmental frequency distribution $J(\omega)$ in the next chapter.

[19] To ensure that the model continues to be a physical one, the operator $\hat{S}$ should also be replaced by another fermion operator.

[20] The constant ω_c is the characteristic frequency such that "$J_B(\omega) \simeq 0 : \omega > \omega_c$" (see the next chapter).

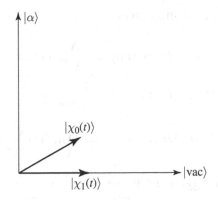

Fig. 4.1. Change of environmental states.

This corresponds to choosing $|\text{vac}\rangle$ as the environmental initial state $|\chi\rangle$ in Eq. (4.2.25). The state evolving from (4.3.22) may be cast into the form of (4.2.27) with

$$|\widetilde{\chi_0(t)}\rangle = \frac{1}{\sqrt{2}}\left\{|\text{vac}\rangle - \sum_\alpha c_\alpha(t)|\alpha\rangle\right\}, \tag{4.3.23}$$

$$|\widetilde{\chi_1(t)}\rangle = -\frac{1}{\sqrt{2}}\widetilde{c(t)}|\text{vac}\rangle, \quad \widetilde{c(t)} \equiv e^{it\Delta}c(t), \tag{4.3.24}$$

$$P_1(t) = \frac{1}{2}|c(t)|^2, \tag{4.3.25}$$

$$P_0(t) = \frac{1}{2}\left(1 + \sum_\alpha |c_\alpha(t)|^2\right) = 1 - P_1(t), \tag{4.3.26}$$

$$-\langle\chi_0(t)|\chi_1(t)\rangle = \frac{\widetilde{c(t)}/2}{\sqrt{P_0(t)P_1(t)}} = \frac{1}{\sqrt{2-|c(t)|^2}}\frac{\widetilde{c(t)}}{|c(t)|}. \tag{4.3.27}$$

Dissipation necessarily changes the state of the environment, too. In the present model, $|\chi_1(t)\rangle$ remains to be $|\text{vac}\rangle$, but $|\chi_0(t)\rangle$ deviates from the latter (Fig. 4.1). In the long-time domain ($t \gtrsim T_1$), in particular, the absolute value of the dephasing factor is approximately equal to $1/\sqrt{2}$. Hence, the amplitude of the quantum resonant oscillation is $1/\sqrt{2}$ times $2\sqrt{P_0(t)P_1(t)}$, with the latter being the amplitude expected from the naive consideration of dissipative effects in Section 4.2. Although readers might argue as to whether or not it is appropriate to regard this factor of $1/\sqrt{2}$ as representing a kind of dephasing effect, it should be clear that the factor is found only by treating the environment quantum-mechanically. Incidentally, the energy possessed initially by the macrosystem is transferred eventually to the environment. At intermediate times, however, a part of the energy is temporarily stored in the interaction; note that $\hat{H}_S + \hat{H}_\mathcal{E}$ does not commute with $\hat{H}$ and is not conserved. An explicit calculation gives

$$E_S(t) \equiv \langle\langle\Psi(t)|\hat{H}_S|\Psi(t)\rangle\rangle = \frac{h\Delta}{2}|c(t)|^2, \tag{4.3.28}$$

$$E_\mathcal{E}(t) \equiv \langle\langle\Psi(t)|\hat{H}_\mathcal{E}|\Psi(t)\rangle\rangle = \frac{1}{2}\sum_\alpha h\omega_\alpha|c_\alpha(t)|^2, \tag{4.3.29}$$

$$E_{S\mathcal{E}}(t) \equiv \langle\langle\Psi(t)|\hat{H}_{S\mathcal{E}}|\Psi(t)\rangle\rangle = \Re\left\{hc^*(t)\sum_\alpha \kappa_\alpha c_\alpha(t)\right\}. \tag{4.3.30}$$

It should be noted that $E_{S\mathcal{E}}(t)$ can be negative. The upper bound of its absolute value may be estimated by use of the Schwarz inequality as

$$|E_{S\mathcal{E}}(t)| \leq h|c(t)|\{1-|c(t)|^2\}^{1/2}\left\{\sum_\alpha \kappa_\alpha^2\right\}^{1/2} \leq \frac{h}{2}\left\{\sum_\alpha \kappa_\alpha^2\right\}^{1/2}. \tag{4.3.31}$$

Hence, if the interaction is weak, the sum $E_S(t) + E_\mathcal{E}(t)$ is almost independent of t in the timescale of the order of T_1 (the van Hove limit).

4.4 Non-dissipative environment

Let us now turn to the model to be called the *Coleman–Hepp environment*,[21] which elucidates the essence of dephasing most succinctly. We choose $(E_0 + E_1)/2$ as the origin of energy and rewrite the macrosystem Hamiltonian as

$$\hat{H}_S = \frac{1}{2}h\Delta(\hat{\Pi}_1 - \hat{\Pi}_0) + \sum_{n=2}^{\infty} E_n \hat{\Pi}_n, \tag{4.4.1}$$

where

$$\hat{\Pi}_n := |n\rangle\langle n| \tag{4.4.2}$$

is the projector onto the state $|n\rangle$. Suppose that the environment consists of N spins each of magnitude $1/2$:

$$\text{Environment} = \{\hat{s}^{(j)}|\ j = 1, \ldots, N\} \ : (\hat{s}^{(j)})^2 = \frac{1}{2}\left(\frac{1}{2}+1\right). \tag{4.4.3}$$

Furthermore, suppose for the sake of simplicity that the Hamiltonian of the environment by itself is 0 whereas the interaction between the macrosystem and the

---

[21] Originally proposed as a model for a detector in discussing quantum-mechanical measurement processes, and is known in this context as the *Coleman–Hepp model*.

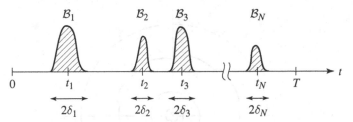

Fig. 4.2. Pulsed magnetic field: interaction between the macrosystem and the Coleman–Hepp environment.

environment is described by

$$\hat{H}_{S\mathcal{E}} := \hat{\Pi}_1 \sum_{j=1}^{N} h\mathcal{B}_j(t - t_j)\hat{s}_1^{(j)}, \tag{4.4.4}$$

where $\mathcal{B}_j(t)$ is a function of the shape of a pulse with width $2\delta_j$ (Fig. 4.2). In other words, the macrosystem exerts a pulsed magnetic field in the "x-direction" on each of the environmental spins successively if and only if the macrosystem is in the state $|1\rangle$.[22] For the sake of simplicity, it is assumed that the neighboring pulses do not overlap:

$$0 < t_1 < t_2 < \cdots < t_N = T, \qquad \delta_j + \delta_{j+1} < t_{j+1} - t_j, \tag{4.4.5}$$

where T is a given constant. For each j (i.e. for each of the environmental spins), we define the spin-up state $|\uparrow\rangle^{(j)}$, the spin-down state $|\downarrow\rangle^{(j)}$ and the spin-θ state $|\theta\rangle^{(j)}$:

$$\hat{s}_3^{(j)}|\uparrow\rangle^{(j)} = \frac{1}{2}|\uparrow\rangle^{(j)}, \qquad \hat{s}_3^{(j)}|\downarrow\rangle^{(j)} = -\frac{1}{2}|\downarrow\rangle^{(j)}, \tag{4.4.6}$$

$$|\theta\rangle^{(j)} := \cos(\theta/2)|\downarrow\rangle^{(j)} - i\sin(\theta/2)|\uparrow\rangle^{(j)}. \tag{4.4.7}$$

The state in which all the spins point downwards is to be denoted by $|\text{vac}\rangle$:

$$|\text{vac}\rangle \equiv |\downarrow, \downarrow, \ldots, \downarrow\rangle \equiv |\downarrow\rangle^{(1)}|\downarrow\rangle^{(2)} \cdots |\downarrow\rangle^{(N)}. \tag{4.4.8}$$

The following notation is also to be used:

$$|\theta_1, \theta_2, \ldots, \theta_N\rangle \equiv |\theta_1\rangle^{(1)}|\theta_2\rangle^{(2)} \cdots |\theta_N\rangle^{(N)}. \tag{4.4.9}$$

Now, let us study the time evolution starting from the initial state (4.3.22). In the first place, note that

$$[\hat{\Pi}_n, \hat{H}] = 0. \tag{4.4.10}$$

---

[22] The ensuing discussion remains essentially unchanged if $\hat{\Pi}_1$ is replaced by $\hat{\Pi}_0$.

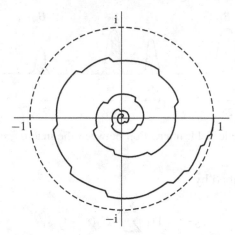

Fig. 4.3. The measure of coherence $\mathrm{e}^{-\mathrm{i}t\Delta}\mathcal{C}(t)$ in the case of Coleman–Hepp environment.

Hence, it suffices to consider the time evolution in each subspace $\mathcal{H}_n$ separately, where $\mathcal{H}_n$ is the subspace in which the macrosystem is in the state $|n\rangle$. Let $\hat{H}_n$ be the operator obtained by restricting $\hat{H}$ to $\mathcal{H}_n$, then

$$\hat{H}_0 = -\frac{1}{2}h\Delta, \tag{4.4.11}$$

$$\hat{H}_1 = \frac{1}{2}h\Delta + \sum_{j=1}^{N} h\mathcal{B}_j(t - t_j)\hat{s}_1^{(j)}. \tag{4.4.12}$$

These are operators acting on the environment alone. Let $\hat{U}_n(t)$ be the corresponding time-evolution operator, then[23]

$$\hat{U}_0(t) = \exp(\mathrm{i}t\Delta/2), \tag{4.4.13}$$

$$\hat{U}_1(t) = \exp\left\{-\mathrm{i}t\Delta/2 - \mathrm{i}\sum_{j=1}^{N}\theta_j(t)\hat{s}_1^{(j)}\right\}, \tag{4.4.14}$$

where

$$\theta_j(t) \equiv \int_0^t \mathrm{d}t'\mathcal{B}_j(t' - t_j). \tag{4.4.15}$$

Accordingly

$$\hat{U}_0(t)|\mathrm{vac}\rangle = \mathrm{e}^{\mathrm{i}t\Delta/2}|\mathrm{vac}\rangle, \tag{4.4.16}$$

$$\hat{U}_1(t)|\mathrm{vac}\rangle = \mathrm{e}^{-\mathrm{i}t\Delta/2}|\theta_1(t), \theta_2(1), \dots, \theta_N(t)\rangle. \tag{4.4.17}$$

---

[23] Since the Hamiltonian in this section depends on time, $\hat{U}(t)$ cannot be expressed in such a simple form as in Eq. (4.2.26). Nevertheless, it can be integrated in the following form because the operators $\{\hat{s}_1^{(j)}\}$ in $\hat{H}_1$ commute with each other.

Thus, if the macrosystem is in the state $|1\rangle$, the direction of the j-th spin at time t is given by $\theta_j(t)$; the spins which pointed downwards initially have been rotated by the effect of the pulsed magnetic field, with $\theta_j(t)$ defined by Eq. (4.4.15) being the rotation angle of the j-th spin. Collecting the above results, one finds

$$
\begin{aligned}
|\Psi(t)\rangle\rangle &= \hat{U}(t)|\Psi(0)\rangle\rangle \\
&= \frac{1}{\sqrt{2}}\{|0\rangle\hat{U}_0(t)|\text{vac}\rangle - |1\rangle\hat{U}_1(t)|\text{vac}\rangle\} \\
&= \frac{1}{\sqrt{2}}\{e^{it\Delta/2}|0,\text{vac}\rangle\rangle - e^{-it\Delta/2}|1, \theta_1(t), \theta_2(t), \ldots, \theta_N(t)\rangle\rangle\}.
\end{aligned}
$$

$$(4.4.18)$$

This implies that

$$
|\widetilde{\chi_0(t)}\rangle = \frac{1}{\sqrt{2}}|\text{vac}\rangle, \tag{4.4.19}
$$

$$
|\widetilde{\chi_1(t)}\rangle = -\frac{1}{\sqrt{2}}|\theta_1(t), \theta_2(t), \ldots, \theta_N(t)\rangle. \tag{4.4.20}
$$

Equations (4.2.35) and (4.2.37) then gives

$$
\mathcal{P}_\pm(t) = \frac{1}{2}[1 \mp \Re\{e^{-it\Delta}\mathcal{C}(t)\}], \tag{4.4.21}
$$

$$
\mathcal{C}(t) = \langle\text{vac}|\theta_1(t), \theta_2(t), \ldots, \theta_N(t)\rangle = \prod_{j=1}^{N}\cos\{\theta_j(t)/2\}. \tag{4.4.22}
$$

If the interaction between the macrosystem and the environment were absent, all $\theta_j(t)$ would vanish and $\mathcal{C}(t)$ would be unity at any t. Suppose that the interaction is such that the pulse strength

$$
\theta_j \equiv \int_{-\infty}^{\infty} dt\, \mathcal{B}_j(t) \tag{4.4.23}
$$

is not an integral multiple of 2π. As time goes on, each $\theta_j(t)$ attains the value θ_j, and $|\mathcal{C}(t)|$ decreases at least globally. In particular, if

$$
\mathcal{B}_j(t) > 0, \quad 0 < \theta_j < \pi \quad : \forall t, \; \forall j, \tag{4.4.24}
$$

then $|\mathcal{C}(t)|$ decreases monotonically. If $N \gg 1$ in addition, $|\mathcal{C}(t)|$ tends to 0 as t tends to T; the quantum resonant oscillation damps away, that is, the coherence is lost asymptotically (Fig. 4.3).

Note incidentally that the macrosystem energy is conserved in the present model:

$$
[\hat{H}_\mathcal{S}, \hat{H}] = 0. \tag{4.4.25}
$$

This property implies that energy dissipation cannot occur and hence that the decoherence described above is due purely to dephasing.

4.5 Quantum Zeno effect

The effect in question was proposed about two decades ago under the title "The Zeno's paradox in quantum theory",[24] where it was argued roughly as follows: "Prepare a system in a metastable state, which, if left to itself, will decay to a stable state. Continuously observe the system to check if it remains in the metastable state, then the system will not decay because the state of the system continuously collapses back to the metastable state due to the effect of the observation." Several years ago, even an experimental paper was published to claim[25] that the "collapse of wavepacket" has been demonstrated through the detection of quantum Zeno effect. After some controversy, it is now generally accepted that the result of the above experiment can be explained without resort to the "collapse of wavepacket". Indeed, phenomena analogous to the Zeno paradox that an arrow cannot move are predicted quite naturally within the framework of quantum mechanics; it is a direct consequence of the general structure of the Schrödinger equation that these phenomena arise if the system in question interacts <u>strongly</u> with its environment (inclusive of measurement apparatus). Accordingly, the phrase "quantum Zeno effect"[26] is more appropriate than "quantum Zeno paradox".

Quantum Zeno effect is sometimes invoked in discussing environmental effects on a macrosystem. However, an environment giving rise to quantum Zeno effect is rather special and rare in practice; it must interact <u>strongly</u> (in the sense to be explained below) with the macrosystem and is quite different from the more realistic harmonic environment introduced in the next chapter. All the same, we spare this section to explain about quantum Zeno effect, since it sheds light on an aspect of decoherence. We adopt a simple model,[27] where the environment is the Coleman–Hepp environment as in the preceding section. Let its initial state be $|\text{vac}\rangle$. The macrosystem, on the other hand, is arbitrary. Its initial state is to be denoted by $|\psi\rangle$. Accordingly

$$|\Psi(0)\rangle\rangle = |\psi, \text{vac}\rangle\rangle \equiv |\psi\rangle|\text{vac}\rangle. \tag{4.5.1}$$

Suppose that interaction between the macrosystem and the environment has the

---

[24] B. Misra and E. C. G. Sudarshan, *J. Math. Phys.* **18** (1977), 756.
[25] This claim was withdrawn soon after, though.
[26] Also called the "watched-pot effect".
[27] In what follows, we inherit the terminology of the preceding section to call the system in question the macrosystem. The ensuing argument, however, is valid regardless of the system being macroscopic or microscopic; quantum Zeno effect is quite general.

following special form:

$$\hat{H}_{S\mathcal{E}} := \hat{\Pi} \sum_{j=1}^{N} h\mathcal{B}_j(t - t_j)\hat{s}_1^{(j)}, \quad \hat{\Pi} \equiv 1 - |\psi\rangle\langle\psi|, \quad (4.5.2)$$

$$\int_{-\infty}^{\infty} dt\, \mathcal{B}_j(t) = \pi \quad : \forall j. \quad (4.5.3)$$

This interaction is similar to (4.4.4), but is different from the latter in that it is effective only if the macrosystem is in a state orthogonal to the initial state and that the pulse is a π-*pulse*. We suppose for simplicity that the pulses are separated by a constant interval τ and that the pulse width δ is sufficiently narrow:

$$t_j = j\tau, \quad \delta_j = \delta \ll \tau_S, \quad (4.5.4)$$

where the width δ is defined as in Fig. 4.2, and τ_S is the characteristic timescale associated with the intrinsic time evolution of the macrosystem in isolation from the environment; for example, if the macrosystem is described by the two states $\{|0\rangle, |1\rangle\}$ as in Section 4.2, then $\tau_S \sim \Delta^{-1}$.

Now, the intrinsic time evolution of the macrosystem can be expressed in the following form:

$$\exp(-it\hat{H}_S/h)|\psi\rangle = c(t)|\psi\rangle + |\phi(t)\rangle, \quad (4.5.5)$$

where $|\phi(t)\rangle$, which is not normalized in general, is orthogonal to the state $|\psi\rangle$:

$$\langle\psi|\phi(t)\rangle = 0. \quad (4.5.6)$$

Let us follow the time evolution of the entire system step by step.

(1) $0 < t < \tau - \delta$:
Since $\hat{H} = \hat{H}_S$ in this interval, it follows from Eq. (4.5.5) that

$$|\Psi(\tau - \delta)\rangle\rangle = c(\tau)|\psi, \text{vac}\rangle\rangle + |\phi_1, \text{vac}\rangle\rangle, \quad |\phi_1\rangle := |\phi(\tau)\rangle. \quad (4.5.7)$$

To be rigorous, the argument of c and ϕ on the right-hand side should be $\tau - \delta$, which has been approximated by τ in view of the condition (4.5.4).

(2) $\tau - \delta < t < \tau + \delta$:
In this interval, $\hat{H} \simeq \hat{H}_{S\mathcal{E}}$ since the condition (4.5.4) ensures that the effect of $\hat{H}_S$ is negligible. Hence, the time-evolution operator $\hat{U}(t, \tau - \delta)$ from time $\tau - \delta$ to time t may be expressed with the aid of the spin rotation angle $\theta_j(t)$ defined by Eq. (4.4.15) as

$$\hat{U}(t, \tau - \delta) = \exp\left\{-i\hat{\Pi}\theta_1(t)\hat{s}_1^{(1)}\right\} = (1 - \hat{\Pi}) + \hat{\Pi}\exp\left\{-i\theta_1(t)\hat{s}_1^{(1)}\right\}. \quad (4.5.8)$$

It follows that

$$\hat{U}(\tau + \delta, \tau - \delta) = (1 - \hat{\Pi}) + \hat{\Pi} \exp\left\{-i\pi \hat{s}_1^{(1)}\right\}, \tag{4.5.9}$$

$$|\Psi(\tau + \delta)\rangle\rangle = \hat{U}(\tau + \delta, \tau - \delta)|\Psi(\tau - \delta)\rangle\rangle$$

$$= c(\tau)|\psi, \text{vac}\rangle\rangle + |\phi_1, \uparrow\downarrow \cdots \downarrow\rangle\rangle. \tag{4.5.10}$$

(3) $\tau + \delta < t < 2\tau - \delta$:

Since $\hat{H} = \hat{H}_S$ again in this interval, it follows from Eq. (4.5.5) that

$$|\Psi(2\tau - \delta)\rangle\rangle = c(\tau)\{c(\tau)|\psi, \text{vac}\rangle\rangle + |\phi_1, \text{vac}\rangle\rangle\} + |\tilde{\phi}, \uparrow\downarrow \cdots \downarrow\rangle\rangle, \tag{4.5.11}$$

where

$$|\tilde{\phi}\rangle := \exp(-i\tau \hat{H}_S/h)|\phi_1\rangle =: c_1(\tau)|\psi\rangle + |\phi_2\rangle, \quad \langle\psi|\phi_2\rangle = 0. \tag{4.5.12}$$

The operator $\hat{U}(2\tau + \delta, 2\tau - \delta)$ should be expressible analogously to $\hat{U}(\tau + \delta, \tau - \delta)$, thus:

$$|\Psi(2\tau + \delta)\rangle\rangle = \hat{U}(2\tau + \delta, 2\tau - \delta)|\Psi(2\tau - \delta)\rangle\rangle$$

$$= \{c(\tau)\}^2|\psi, \text{vac}\rangle\rangle + c(\tau)|\phi_1, \downarrow\uparrow\downarrow \cdots \downarrow\rangle\rangle$$

$$+ c_1(\tau)|\psi, \uparrow\downarrow\downarrow \cdots \downarrow\rangle\rangle + |\phi_2, \uparrow\uparrow\downarrow \cdots \downarrow\rangle\rangle. \tag{4.5.13}$$

Continuing the above step-by-step consideration, one may conclude by induction that the state at time $T \ (\equiv N\tau + \delta)$ is of the following form:

$$|\Psi(T)\rangle\rangle = \{c(\tau)\}^N|\psi, \text{vac}\rangle\rangle + \sum_{n \geq 0} |\tilde{\psi}_n, \star_n\rangle\rangle, \tag{4.5.14}$$

where $|\tilde{\psi}_n\rangle$ is a state of the macrosystem, and $|\star_n\rangle$ are environmental states all orthogonal to $|\text{vac}\rangle$:

$$\langle\text{vac}|\star_n\rangle = 0 \quad : \forall n. \tag{4.5.15}$$

On the other hand, $|\tilde{\psi}_n\rangle$ is not orthogonal to $|\psi\rangle$ in general, as may readily be seen by examining the case of $N = 2$:

$$|\tilde{\psi}_0\rangle = c_1(\tau)|\psi\rangle, \quad |\tilde{\psi}_1\rangle = c(\tau)|\phi_1\rangle, \quad |\tilde{\psi}_2\rangle = |\phi_2\rangle,$$

$$|\tilde{\psi}_n\rangle = 0 \quad : n \geq 3,$$

$$|\star_0\rangle = |\uparrow\downarrow\downarrow \cdots \downarrow\rangle, \quad |\star_1\rangle = |\downarrow\uparrow\downarrow \cdots \downarrow\rangle, \quad |\star_2\rangle = |\uparrow\uparrow\downarrow \cdots \downarrow\rangle.$$

Equation (4.5.14) gives the probability $P_{\text{tot}}(t)$ for the entire system to be found in the initial state at time T as

$$P_{\text{tot}}(T) = |\langle\langle\Psi(0)|\Psi(T)\rangle\rangle|^2 = p^N, \quad p := |c(\tau)|^2. \tag{4.5.16}$$

On the other hand, it follows from Eq. (4.5.5) that

$$c(\tau) = \langle\psi| \exp(-i\tau \hat{H}_S/h)|\psi\rangle, \tag{4.5.17}$$

which implies

$$p = 1 - (\tau/\tau_Q)^2 + \mathcal{O}(\tau^4), \quad \tau_Q := h/\Delta E, \tag{4.5.18}$$

where ΔE is the indeterminacy in the energy of the macrosystem in the initial state, namely ,

$$(\Delta E)^2 := \langle \psi | (\hat{H}_S - \langle \psi | \hat{H}_S | \psi \rangle)^2 | \psi \rangle, \tag{4.5.19}$$

and τ_Q is the corresponding characteristic time to be called the *quantum indeterminacy time*.

Now, we let N tend to a large number while keeping T constant so that

$$\tau/\tau_Q \simeq T/N\tau_Q \ll 1. \tag{4.5.20}$$

It follows that

$$P_{\text{tot}}(T) > 1 - N^{-1}(T/\tau_Q)^2. \tag{4.5.21}$$

Usually, experimenters are not concerned with which state the environment is found in. Hence, the quantity of their interest is not $P_{\text{tot}}(T)$ but the probability $P_S(T)$ for the macrosystem to be found in its initial state:

$$P_S(T) = |\langle\langle \psi, \text{vac} | \Psi(T) \rangle\rangle|^2 + \sum_{\star} |\langle\langle \psi, \star | \Psi(T) \rangle\rangle|^2, \tag{4.5.22}$$

where we have chosen

$$\{|\text{vac}\rangle, |\star\rangle \ | \ \langle \star | \text{vac} \rangle = 0\}$$

as the basis of $\mathcal{H}_\mathcal{E}$. The first term on the right-hand side is nothing but $P_{\text{tot}}(T)$. Since a probability is necessarily less than or equal to unity, we find

$$1 \geq P_S(T) \geq P_{\text{tot}}(T). \tag{4.5.23}$$

Combining this inequality with (4.5.21), we finally arrive at the important inequality:

$$0 \leq 1 - P_S(T) \leq N^{-1}(T/\tau_Q)^2. \tag{4.5.24}$$

Hence, if we let $N \to \infty$ while keeping T constant, then $P_S(T) \to 1$, which implies that, at time T, the macrosystem is surely found in the initial state. In practice, the mathematical limit of $N \to \infty$ is not attainable, but it should be in principle possible to set up a situation such that the right-hand side of (4.5.24) is sufficiently small for a given T. Thus, we are led to the following conclusion:

If a macrosystem interacts with the environment both <u>frequently enough</u> (Eq. (4.5.20)) and <u>strongly enough</u> (Eq. (4.5.3)), then it can happen that the macrosystem remains virtually in the initial state (Eq. (4.5.24)).

Consider, for instance, a SQUID set up in a MQT-situation, and suppose that its initial state can be represented by a wavepacket localized in a metastable well. If the interaction between the SQUID and the environment were pulse-like and effective only outside the well, a quantum Zeno effect would arise; in such a special situation, decay of the metastable state (i.e. leakage of the macrosystem wavefunction out of the well) would be suppressed. However, an actual environment for a SQUID is unlikely to be of this sort.

Incidentally, one may interpret the macrosystem as a particle and each of the environmental spins as a detector with the spin-up and the spin-down states corresponding to the on- and off-state of the detector, respectively, where the detector is turned into the on-state if it has actually detected the particle and remains in the off-state otherwise. Then, the discussion in this section can be adapted to the problem of successive measurements repeated N times. The above result implies that the probability for the measurement apparatus (i.e. the set of all detectors) to have been turned on vanishes in the limit of $N \to \infty$. It is interesting to note that the apparatus has achieved the act of measurement although it has never been turned on.[28]

Exercises

Exercise 4.1. As an example of a stationary noise with correlation time τ_c, consider

$$\overline{\mathcal{B}(t_2)\mathcal{B}(t_1)} = \mathcal{B}_0^2 \exp(-2|t_2 - t_1|/\tau_c), \tag{4.5.25}$$

and derive Eq. (4.2.19). Incidentally, for the purpose of deriving this equation alone, it would suffice to consider the *white noise*

$$\overline{\mathcal{B}(t_2)\mathcal{B}(t_1)} = \mathcal{B}_0^2 \tau_c \delta(t_2 - t_1), \tag{4.5.26}$$

which, however, is a formal object. A measurable physical quantity can not rigorously be proportional to a delta function. Rather, a physical noise may effectively look like a white noise in the timescale longer than τ_c. Evaluate correction terms to Eq. (4.2.19) for the example (4.5.25).

Exercise 4.2. Consider the spin operator (2.2.14) for a two-state system, and evaluate its expectation value with respect to the state (4.2.30). Express the result in terms of $P_0(t)$, $P_1(t)$ and $\mathcal{C}(t)$.

Exercise 4.3. Supposing that the closed-form expression for $J_B(\omega)$ is given by a Lorentzian peaked at ω_c with width W

$$J^{\text{Lorentz}}(\omega) := \kappa^2 \frac{W/2\pi}{(\omega - \omega_c)^2 + (W/2)^2}, \tag{4.5.27}$$

---

[28] This kind of measurement process is sometimes referred to as a "negative-result measurement". It is, however, a subjective matter whether a measurement result is regarded as negative or positive.

or by (5.3.7) in the next chapter, compute $c(t)$ and $c_\alpha(t)$ in Eq. (4.3.8). Use the result to study the behavior of $E_S(t)$, $E_\mathcal{E}(t)$ and $E_{S\mathcal{E}}(t)$.

Exercise 4.4. How is the result of Section 4.3 modified if the environment is initially in thermal equilibrium? Study the same problem with the environmental bosons replaced by fermions or spins of magnitude $1/2$.

Exercise 4.5. Under what circumstances does the absolute value of $\mathcal{C}(t)$ given by Eq. (4.4.22) decrease exponentially in time?

Exercise 4.6. In the model considered in Sections 4.4–4.5, it is somewhat artificial that the interaction Hamiltonian (4.4.4) depends on t. Modify the model so as to make the Hamiltonian t-independent while preserving the essential features described in the text. (Hint: In the original Coleman–Hepp model,[29] the environmental spins are distributed on a line and the macrosystem moves along the line.)

[29] K. Hepp, *Helv. Phys. Acta* **45** (1972), 237.

5

Harmonic environment

A frequently employed model for an environment is a set of harmonic oscillators (to be called the environmental oscillators) such that the Hamiltonian describing its interaction with the macrosystem is linear with respect to the position variables of the oscillators. Many actual environments are expected to be represented fairly well by this model. This chapter aims to cultivate a physical picture for the harmonic environment by studying its properties in classical situations.

5.1 Linearly-coupled harmonic-environment model

Supposing that the macrosystem Hamiltonian $\hat{H}_S$ is given by Eq. (1.4.18), we assume that the Hamiltonian $\hat{H}$ for the entire system composed of the macrosystem and the environment is given as[1]

$$\hat{H} := \hat{H}_S + \sum_{\alpha} \left[\frac{1}{2m_\alpha} \hat{p}_\alpha^2 + \frac{1}{2} m_\alpha \omega_\alpha^2 \{\hat{x}_\alpha - f_\alpha(\hat{R})\}^2 \right] - \frac{1}{2} \sum_{\alpha} \hbar \omega_\alpha, \quad (5.1.1)$$

$$[\hat{R}, \hat{P}] = i\hbar, \quad [\hat{x}_\alpha, \hat{p}_\beta] = i\hbar \delta_{\alpha\beta}, \text{ with all the other commutators being 0.} \quad (5.1.2)$$

The system described by this Hamiltonian is to be called the *linearly-coupled harmonic-environment model*. The function $f_\alpha(R)$ is arbitrary for the moment. The last constant term, which merely displaces the origin of energy, has been added for later convenience. The above Hamiltonian may be rewritten as

$$\hat{H} = \hat{H}_S + \hat{H}_\varepsilon + \hat{H}_{S\varepsilon}, \quad (5.1.3)$$

$$\hat{H}_\varepsilon := \sum_{\alpha} \left(\frac{1}{2m_\alpha} \hat{p}_\alpha^2 + \frac{1}{2} m_\alpha \omega_\alpha^2 \hat{x}_\alpha^2 - \frac{1}{2} \hbar \omega_\alpha \right), \quad (5.1.4)$$

$$\hat{H}_{S\varepsilon} := -\sum_{\alpha} m_\alpha \omega_\alpha^2 f_\alpha(\hat{R}) \hat{x}_\alpha + \frac{1}{2} \sum_{\alpha} m_\alpha \omega_\alpha^2 \{f_\alpha(\hat{R})\}^2. \quad (5.1.5)$$

---

[1] In the summation, α runs from 1 to N with N being the total number of environmental oscillators.

117

It might seem more reasonable to incorporate the last term in the last equation into $\hat{H}_S$, since it depends on $\hat{R}$ alone but not on $\hat{x}_\alpha$. The reason for its inclusion in the interaction Hamiltonian $\hat{H}_{S\mathcal{E}}$ will become clear in Section 5.6.

Let us cast $\hat{H}_S$ into the standard form following the procedure of Section 2.1. The constants R_0, U_0 and τ_0 introduced there are to be adopted as units. Accordingly, x_α, p_α, ω_α, f_α and H in Eqs. (5.1.1)–(5.1.5) are temporarily to be designated by $x_{\alpha\text{conv}}$, $p_{\alpha\text{conv}}$, $\omega_{\alpha\text{conv}}$, $f_{\alpha\text{conv}}$ and H_{conv}, respectively, and the dimensionless quantities x_α, p_α, ω_α, f_α and H are to be defined anew:

$$x_\alpha := (m_\alpha/M)^{1/2} x_{\alpha\text{conv}}/R_0, \quad p_\alpha := (m_\alpha U_0)^{-1/2} p_{\alpha\text{conv}}, \tag{5.1.6}$$

$$\omega_\alpha := \omega_{\alpha\text{conv}} \tau_0, \quad f_\alpha(q) := \left(m_\alpha/U_0\tau_0^2\right)^{1/2} f_{\alpha\text{conv}}(R_0 q + R_\text{m}), \tag{5.1.7}$$

$$H := H_{\text{conv}}/U_0 =: H_S + H_\mathcal{E} + H_{S\mathcal{E}}. \tag{5.1.8}$$

In terms of these quantities, we have

$$\hat{H}_S = \frac{1}{2}\hat{p}^2 + V(\hat{q}), \tag{5.1.9}$$

$$\hat{H}_\mathcal{E} = \sum_\alpha \left(\frac{1}{2}\hat{p}_\alpha^2 + \frac{1}{2}\omega_\alpha^2 \hat{x}_\alpha^2 - \frac{1}{2}h\omega_\alpha\right), \tag{5.1.10}$$

$$\hat{H}_{S\mathcal{E}} = -\sum_\alpha \omega_\alpha^2 f_\alpha(\hat{q})\hat{x}_\alpha + \frac{1}{2}\sum_\alpha \omega_\alpha^2 \{f_\alpha(\hat{q})\}^2, \tag{5.1.11}$$

$$[\hat{q}, \hat{p}] = ih, \quad [\hat{x}_\alpha, \hat{p}_\beta] = ih\delta_{\alpha\beta}, \tag{5.1.12}$$

where h is of course the dimensionless Planck constant defined by Eq. (2.1.8).

The physical situation embodied by this model may be viewed in two ways. In the first view, the macrosystem exerts a force $\omega_\alpha^2 f_\alpha(q)$ on each environmental oscillator α. In other words, the origin of the environmental oscillator α with the spring constant ω_α^2 is displaced by $f_\alpha(q)$, as is seen manifestly from Eq. (5.1.1) or

$$H_\mathcal{E} + H_{S\mathcal{E}} = \sum_\alpha \left[\frac{1}{2}p_\alpha^2 + \frac{1}{2}\omega_\alpha^2 \{x_\alpha - f_\alpha(q)\}^2 - \frac{1}{2}h\omega_\alpha\right]. \tag{5.1.13}$$

If $f_\alpha(q)$ happens to be independent of α, the model is said to be separable (the separable linearly-coupled harmonic-environment model, or the *separable model* in short):[2]

$$\text{Separable model:} \quad f_\alpha(q) = \gamma_\alpha f(q), \quad f(0) = 0, \tag{5.1.14}$$

where $f(q)$ is an arbitrary function of q and γ_α is a positive constant.[3]

---

[2] In this case, the dependences of the interaction on α and q are separated.
[3] Without loss of generality, one may impose the condition $f(0) = 0$ as well as $\gamma_\alpha > 0$; the transformation $(\hat{p}_\alpha, \hat{x}_\alpha) \rightarrow (-\hat{p}_\alpha, -\hat{x}_\alpha)$, which can be effected without violation of Eq. (5.1.12), allows one to change the sign of γ_α at will.

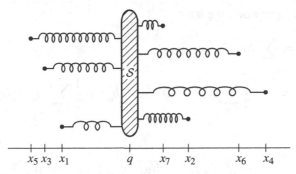

Fig. 5.1. A particle (the macrosystem) coupled to environmental oscillators via springs.

The second view is the one applicable if $f(q)$ in Eq. (5.1.14) is proportional to q. In this case, since the first term of $\hat{H}_{S\mathcal{E}}$ is linear with respect both to $\hat{x}_\alpha$ and to $\hat{q}$, the model is said to be bilinear (the bilinearly-coupled harmonic-environment model or the *bilinear model* in short):

$$\text{Bilinear model:} \quad f_\alpha(q) = \gamma_\alpha q. \tag{5.1.15}$$

The Hamiltonian may be rewritten as

$$H = \frac{1}{2}p^2 + \bar{V}(q) + \sum_\alpha \left\{ \frac{1}{2}p_\alpha^2 + \frac{1}{2}(1 - \gamma_\alpha)\omega_\alpha^2 x_\alpha^2 - \frac{1}{2}h\omega_\alpha \right\}$$

$$+ \frac{1}{2}\sum_\alpha \gamma_\alpha \omega_\alpha^2 (x_\alpha - q)^2, \tag{5.1.16}$$

$$\bar{V}(q) \equiv V(q) - \frac{1}{2}\left\{ \sum_\alpha \gamma_\alpha(1 - \gamma_\alpha)\omega_\alpha^2 \right\} q^2. \tag{5.1.17}$$

In other words, each environmental oscillator α with spring constant $(1 - \gamma_\alpha)\omega_\alpha^2$ is coupled to the particle q via the spring of spring constant $\gamma_\alpha \omega_\alpha^2$ (Fig. 5.1). In this view, the proper potential felt by the particle q is not $V(q)$ but $\bar{V}(q)$. It is, however, not possible in general to interpret q and $\{x_1, x_2, \ldots, x_N\}$ as the center-of-mass position and the set of relative positions, respectively, for a system of $(N + 1)$ particles $\{\tilde{x}_1, \tilde{x}_2, \ldots, \tilde{x}_{N+1}\}$ under an external potential V. If this interpretation is viable, V can be a function of q alone only in the case of $V(q) \propto q$ (e.g. the case of a uniform electric or gravitational field). As mentioned in Chapter 1, the situation envisaged in this book is such that q is a generic collective degree of freedom and $\{x_1, x_2, \ldots, x_N\}$ is the set of remaining degrees of freedom representing fluctuations around q.

It is convenient for the ensuing discussion to introduce the lowering and raising operators for the environmental oscillators:

$$\hat{b}_\alpha := \frac{1}{\sqrt{2}}\left\{ (\omega_\alpha/h)^{1/2}\hat{x}_\alpha + i(h\omega_\alpha)^{-1/2}\hat{p}_\alpha \right\}, \tag{5.1.18}$$

$$[\hat{b}_\alpha, \hat{b}_\beta^\dagger] = \delta_{\alpha\beta}, \quad [\hat{b}_\alpha, \hat{b}_\beta] = 0. \tag{5.1.19}$$

In terms of these operators, we have

$$\hat{H}_{\mathcal{E}} = \sum_\alpha h\omega_\alpha \hat{b}_\alpha^\dagger \hat{b}_\alpha, \tag{5.1.20}$$

$$\hat{H}_{S\mathcal{E}} = -\sqrt{\frac{h}{2}} \sum_\alpha \omega_\alpha^{3/2} f_\alpha(\hat{q})(\hat{b}_\alpha + \hat{b}_\alpha^\dagger) + \frac{1}{2} \sum_\alpha \omega_\alpha^2 \{f_\alpha(\hat{q})\}^2, \tag{5.1.21}$$

that is,

$$\hat{H}_{\mathcal{E}} + \hat{H}_{S\mathcal{E}} = \sum_\alpha h\omega_\alpha \left\{ \hat{b}_\alpha^\dagger - \sqrt{\frac{\omega_\alpha}{2h}} f_\alpha(\hat{q}) \right\} \left\{ \hat{b}_\alpha - \sqrt{\frac{\omega_\alpha}{2h}} f_\alpha(\hat{q}) \right\}. \tag{5.1.22}$$

5.2 Heisenberg equation of motion: part 1

Let us begin our study of the linearly-coupled harmonic-environment model by examining the Heisenberg equation of motion, which is suited to comparing the quantum-mechanical description with the classical one. Make an arbitrary choice of a fiducial time t_0 (to be called the initial time), and define the Heisenberg-picture operator $\hat{A}(t)$ corresponding to the Schrödinger-picture[4] operator $\hat{A}$ as

$$\hat{A}(t) := \exp\{\mathrm{i}(t - t_0)\hat{H}/h\}\hat{A}\exp\{-\mathrm{i}(t - t_0)\hat{H}/h\}. \tag{5.2.1}$$

Then, $\hat{q}(t)$ obeys the following Heisenberg equation of motion:

$$\frac{\mathrm{d}^2}{\mathrm{d}t^2}\hat{q}(t) = \frac{\mathrm{d}}{\mathrm{d}t}\hat{p}(t) = -\tilde{V}'(\hat{q}(t)) + \sum_\alpha \omega_\alpha^2[f_\alpha'(\hat{q}(t))\hat{x}_\alpha(t)]_+, \tag{5.2.2}$$

$$\tilde{V}(q) := V(q) + \frac{1}{2}\sum_\alpha \omega_\alpha^2\{f_\alpha(q)\}^2, \tag{5.2.3}$$

where the prime $'$ in $\tilde{V}'(q)$ and $f_\alpha'(q)$ denotes q-differentiation. The symbol

$$[\hat{A}\hat{B}]_+ := \frac{1}{2}(\hat{A}\hat{B} + \hat{B}^\dagger\hat{A}^\dagger) \tag{5.2.4}$$

is a device to express a product of two operators in a manifestly Hermitian form; it is not actually needed in Eq. (5.2.2) where $f_\alpha'(\hat{q}(t))$ and $\hat{x}_\alpha(t)$ referring to a common time t are commutative, but will be useful when a product of operators referring to different times is involved (e.g. in Eq. (5.2.12) below).

---

[4] The Schrödinger (or Heisenberg) picture is traditionally called the "Schrödinger (or Heisenberg) representation", which however is confusing with "representations of operators", for instance.

On the other hand, $\hat{b}(t)$ obeys the equation

$$\left(\frac{d}{dt} + i\omega_\alpha\right)\hat{b}_\alpha(t) = i\sqrt{\frac{\omega_\alpha^3}{2h}} f_\alpha(\hat{q}(t)), \tag{5.2.5}$$

which may be derived most conveniently by use of the expression (5.1.21). This equation is nothing but the quantum-mechanical version of the familiar equation of motion for a forced harmonic oscillator expressed in terms of $\hat{b}_\alpha(t)$ instead of $\hat{x}_\alpha(t)$. Although this equation may be solved easily with Lagrange's method of variation of constants, which is called the interaction-picture method in quantum mechanics, let us be somewhat pedantic and introduce the *retarded Green function* $\tilde{G}_\alpha$ associated with the differential operator on the left-hand side:

$$\tilde{G}_\alpha(t) := \theta(t)\exp(-i\omega_\alpha t), \tag{5.2.6}$$

$$\left(\frac{d}{dt} + i\omega_\alpha\right)\tilde{G}_\alpha(t - t') = \delta(t - t'), \tag{5.2.7}$$

where the step function $\theta(t - t')$ embodies the *retarded* nature. One may use this Green function to obtain the particular solution, which together with the initial condition

$$\hat{b}(t_0) = \hat{b} \tag{5.2.8}$$

that is required by the definition (5.2.1), gives the solution to Eq. (5.2.5) as

$$\hat{b}_\alpha(t) = e^{-i(t-t_0)\omega_\alpha}\hat{b}_\alpha + i\sqrt{\frac{\omega_\alpha^3}{2h}}\int_{t_0}^t dt'\, e^{-i(t-t')\omega_\alpha} f_\alpha(\hat{q}(t')). \tag{5.2.9}$$

Note that the first term (i.e. a solution to the homogeneous equation) on the right-hand side should be kept; otherwise, the above solution would not fulfill the commutation relation

$$[\hat{b}_\alpha(t), \hat{b}_\beta^\dagger(t)] = \delta_{\alpha\beta}, \quad [\hat{b}_\alpha(t), \hat{b}_\beta(t)] = 0, \tag{5.2.10}$$

which should hold at any time. It follows from Eq. (5.2.9) that

$$\begin{aligned}
\hat{x}_\alpha(t) &= \sqrt{\frac{h}{2\omega_\alpha}}\{\hat{b}_\alpha(t) + \hat{b}_\alpha^\dagger(t)\} \\
&= \sqrt{\frac{h}{2\omega_\alpha}}\{e^{-i(t-t_0)\omega_\alpha}\hat{b}_\alpha + \text{h.c.}\} \\
&\quad + \omega_\alpha\int_{t_0}^t dt'\,\sin((t - t')\omega_\alpha) f_\alpha(\hat{q}(t')). \tag{5.2.11}
\end{aligned}$$

By use of this result, one may eliminate $\hat{x}_\alpha(t)$ from Eq. (5.2.2) to obtain

$$\frac{d^2}{dt^2}\hat{q}(t) + \tilde{V}'(\hat{q}(t)) - \int_{t_0}^{t} dt' \sum_\alpha \omega_\alpha^3 [f_\alpha'(\hat{q}(t)) \sin((t-t')\omega_\alpha) f_\alpha(\hat{q}(t'))]_+$$

$$= \sqrt{\frac{h}{2}} \sum_\alpha \omega_\alpha^{3/2} [(e^{-i(t-t_0)\omega_\alpha} \hat{b}_\alpha + h.c.) f_\alpha'(\hat{q}(t))]_+, \qquad (5.2.12)$$

which is closed with respect to the variable $\hat{q}(t)$. It is the price paid for the elimination of the environmental degrees of freedom that this equation is "non-local in time"; the motion of $\hat{q}(t)$ is influenced by its own past history (the third term on the left-hand side). It is to be remembered that, although the environmental degrees of freedom have been eliminated, the motion of $\hat{q}(t)$ depends on the initial state of the environment through $\hat{b}_\alpha$ ($\equiv \hat{b}_\alpha(t_0)$) on the right-hand side.

What has been said so far might be too general to comprehend intuitively. Let us specialize to the separable model (5.1.14), then

$$\frac{d^2}{dt^2}\hat{q}(t) + \tilde{V}'(\hat{q}(t)) - \int_{t_0}^{t} dt' [f'(\hat{q}(t))\phi(t-t') f(\hat{q}(t'))]_+$$

$$= [\hat{\mathcal{R}}(t) f'(\hat{q}(t))]_+, \qquad (5.2.13)$$

$$\phi(t) := \sum_\alpha \bar{\gamma}_\alpha^2 \sin \omega_\alpha t, \qquad (5.2.14)$$

$$\hat{\mathcal{R}}(t) := \sqrt{\frac{h}{2}} \sum_\alpha \bar{\gamma}_\alpha (e^{-i(t-t_0)\omega_\alpha} \hat{b}_\alpha + h.c.), \qquad (5.2.15)$$

$$\bar{\gamma}_\alpha \equiv \gamma_\alpha \omega_\alpha^{3/2}. \qquad (5.2.16)$$

The physical meaning of each term in Eq. (5.2.13) should be clear. If the environmental oscillators were executing free oscillations independently of the macrosystem, the environment would exert on the macrosystem a force given by the right-hand side of Eq. (5.2.13), with the magnitude of the force being of $\mathcal{O}(h^{1/2})$ if the initial state of the environmental oscillators are limited to low-lying excited states. Actually, however, the macrosystem exerts a force on each of the environmental oscillators thereby displacing the position of the latter from that expected for a free oscillation. These displacements in turn change the force exerted by the environment on the macrosystem. This change of force is represented by the third term on the left-hand side of Eq. (5.2.13), and its magnitude is of $\mathcal{O}(h^0)$. Since the temporal duration of the order of ω_α^{-1} is needed for the environmental oscillator α to change its position appreciably in response to an external force, the environmental oscillator acted on by the macrosystem reacts back on the latter with a time delay of the same order (*retarded reaction*). The function $\phi(t-t')$ summarizes quantitatively the retarded reaction of all the environmental oscillators.

5.3 Environmental frequency distribution function

Let us label the frequencies $\{\omega_\alpha\}$ of the environmental oscillators in ascending order

$$0 < \omega_1 < \omega_2 < \cdots < \omega_N \equiv \omega_{max}, \tag{5.3.1}$$

and introduce a smooth function $\bar{\gamma}(\omega)$ such that

$$\bar{\gamma}(\omega_\alpha) = \bar{\gamma}_\alpha. \tag{5.3.2}$$

Keeping ω_{max} fixed, consider the situation of very large N. Then, the sum in Eq. (5.2.14) may be approximated by an integral, which is easier to handle than a sum. However, the summation symbol cannot simply be replaced by an integration symbol, since ω_α are not necessarily equally spaced. In order to make the transition from the sum to an integral transparent, it is convenient to define a function $J(\omega)$ as

$$J(\omega) := \frac{\pi}{2} \sum_\alpha \bar{\gamma}_\alpha^2 \delta(\omega - \omega_\alpha), \tag{5.3.3}$$

that is,

$$J(\omega) = \frac{\pi}{2}\{\bar{\gamma}(\omega)\}^2 D(\omega), \quad D(\omega) := \sum_\alpha \delta(\omega - \omega_\alpha). \tag{5.3.4}$$

The function $D(\omega)$ expresses the frequency distribution of the environmental oscillators, and $J(\omega)$ expresses the corresponding distribution weighted by the interaction strength $\{\bar{\gamma}(\omega)\}^2$. The function $J(\omega)$ is to be called simply the *environmental frequency distribution function*.[5]

Suppose that there exist characteristic frequencies ν_c and ω_c associated with the environment such that

$$\nu_c \ll \omega_0 \ll \omega_c, \tag{5.3.5}$$

where ω_0 is the characteristic frequency for the well of the potential $V(q)$ (Fig. 2.1). Suppose also that $J(\omega)$ behaves grossly as follows (Fig. 5.2):

$$J(\omega) \sim \begin{cases} 0 & : \omega \gtrsim \omega_c, \\ \eta\omega^s & : \nu_c \ll \omega \lesssim \omega_c, \\ \eta'(\omega - \nu_c)^{\sigma-1} & : \omega \gtrsim \nu_c, \\ 0 & : \omega < \nu_c, \end{cases} \tag{5.3.6}$$

where η and η' are positive constants which quantify the strength of the interaction between the environment and the macrosystem. The positive exponents s and σ vary depending on the environment. If the environment consists of phonons in a

---

[5] Also called "the spectral function" in specialized literature.

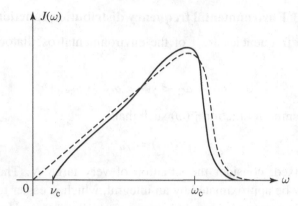

Fig. 5.2. Environmental frequency distribution function (the broken curve is that for the ohmic environment).

solid and if $\bar{\gamma}(\omega)$ happens to be a constant, then $\nu_c = 0$ and ω_c is of the order of the Debye frequency. If the environment consists of Bogoliubov quasiparticles in a superconductor, then $\sigma = 1/2$ and ν_c is of the order of Δ_s/h with Δ_s being the magnitude of the superconducting order-parameter field. And so on. In the case of $\nu_c = 0$, environments characterized by $J(\omega)$ of the above form are classified according to the value of s as follows:

$$0 < s < 1 : \text{ sub-ohmic environment,}$$
$$s = 1 : \quad \textit{ohmic environment,}$$
$$s > 1 : \quad \text{super-ohmic environment.}$$

For an explicit calculation, the following expression may be profitably employed:

$$J(\omega) = \eta\omega \left(\frac{\omega}{\omega_c}\right)^{s-1} \left(1 - \frac{\nu_c}{\omega}\right)^{\sigma-1} \exp(-\omega/\omega_c)\theta(\omega - \nu_c). \quad (5.3.7)$$

5.4 Retarded resistance function

For the ensuing discussion, it is also useful to define

$$\mathcal{T}(t) := \sum_\alpha \bar{\gamma}_\alpha^2 \omega_\alpha^{-1} \cos \omega_\alpha t. \quad (5.4.1)$$

This function, to be called the *retarded resistance function* (see the next section), is related to $\phi(t)$ as

$$\phi(t) = -\frac{d\mathcal{T}(t)}{dt}. \quad (5.4.2)$$

These functions may be expressed in terms of the environmental frequency distribution function (5.3.3) as

$$\phi(t) = \frac{2}{\pi} \int_0^\infty d\omega \, J(\omega) \sin \omega t, \tag{5.4.3}$$

$$T(t) = \frac{2}{\pi} \int_0^\infty \frac{d\omega}{\omega} J(\omega) \cos \omega t. \tag{5.4.4}$$

Let us survey properties of $T(t)$ under the condition (5.3.6). It is immediately obvious that

$$T(0) = \frac{2}{\pi} \int_0^\infty d\omega \frac{J(\omega)}{\omega} < \infty, \tag{5.4.5}$$

$$T(\infty) = 0, \tag{5.4.6}$$

$$\int_0^\infty dt \, T(t) = \left. \frac{J(\omega)}{\omega} \right|_{\omega \downarrow 0} = \begin{cases} \eta & : \quad s = 1, \quad v_c = 0, \\ 0 & : \quad s > 1 \text{ or } v_c > 0, \\ \infty & : \quad 0 < s < 1, \quad v_c = 0. \end{cases} \tag{5.4.7}$$

One may employ the model (5.3.7) to obtain an explicit expression for $T(t)$. In the case of $s = 1$ or $s = 2$, for example,

$$T(t) = \frac{2\eta\omega_c}{\pi} I_s(\omega_c t; v_c/\omega_c), \tag{5.4.8}$$

$$I_s(\tau; \epsilon) = \begin{cases} e^{-\epsilon} \left(\frac{1}{1+\tau^2} \cos \epsilon\tau - \frac{\tau}{1+\tau^2} \sin \epsilon\tau \right) & : s = 1, \\ e^{-\epsilon} \frac{(1-\tau^2) \cos \epsilon\tau - 2\tau \sin \epsilon\tau}{(1+\tau^2)^2} + \epsilon I_1(\tau; \epsilon) & : s = 2. \end{cases} \tag{5.4.9}$$

It has been assumed for simplicity that $\sigma = 1$. The corresponding expression for $\phi(t)$ follows readily from the relation (5.4.2). The result for $s = 1$ is shown in Fig. 5.3.

According to the definition (5.4.1), the retarded resistance function $T(t)$ is quasiperiodic; the situation is the same as the sum (2.5.3) in Section 2.5. Therefore, Eq. (5.4.6) is not literally correct. As N is increased with $\omega_{\max}$ fixed, however, the τ_P for $T(t)$ increases rapidly[6] except for the special case that all the ratios among ω_α are simple rational numbers. Hence, times t of our practical concern satisfies the condition

$$t \ll \tau_P. \tag{5.4.10}$$

The symbol ∞ in Eq. (5.4.6) stands for a time which is extremely long compared to ω_c^{-1} and yet satisfies the above condition. In fact, in the transition from the sum

---

[6] It easily outruns the present age of the universe.

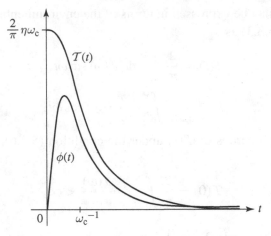

Fig. 5.3. The function $\phi(t)$ and the retarded resistance function $\mathcal{T}(t)$ (case of ohmic environment).

in Eq. (5.4.1) to the integral in Eq. (5.4.4), the following condition has been used:

$$\max_{\alpha}\{(\omega_{\alpha+1} - \omega_{\alpha})t\} \ll 1, \tag{5.4.11}$$

which is stronger than (5.4.10). Even if $\omega_{\alpha+1} - \omega_\alpha$ is infinitesimal, $|\sin(\omega_{\alpha+1}t) - \sin(\omega_\alpha t)|$ would not be infinitesimal if this condition were violated. Times t satisfying the inequality (5.4.11) are to be called the <u>ordinary times</u>. To sum up, $J(\omega)$ can be regarded as a continuous function of ω on the premise that only the <u>ordinary times</u> are considered. Furthermore, as far as the <u>ordinary times</u> are concerned, the effects of an environment are encapsulated in $J(\omega)$ regardless of the details[7] of its interaction with the macrosystem.

5.5 Heisenberg equation of motion: part 2

We now return to the main stream of this chapter. Equation (5.2.13) may be rewritten by use of the relation (5.4.2) as

$$\frac{\mathrm{d}^2}{\mathrm{d}t^2}\hat{q}(t) + \tilde{V}'(\hat{q}(t)) + \hat{\mathcal{M}}(t) = [\hat{\mathcal{R}}(t)f'(\hat{q}(t))]_+, \tag{5.5.1}$$

$$\hat{\mathcal{M}}(t) := \int_{t_0}^{t} \mathrm{d}t' \left[f'(\hat{q}(t))\frac{\mathrm{d}}{\mathrm{d}t}\mathcal{T}(t - t')f(\hat{q}(t')) \right]_+. \tag{5.5.2}$$

We work with $\mathcal{T}$ rather than ϕ, because the former allows the Markov approximation to be invoked as explained in the next section (see Fig. 5.3 to compare ϕ with $\mathcal{T}$).

---

[7] Namely, $\{(\omega_\alpha, \gamma_\alpha) \mid \alpha = 1, 2, \ldots, N\}$.

$\hat{\mathcal{M}}$ is the *memory term*, namely a term depending on the past history. It may be rearranged as

$$\hat{\mathcal{M}}(t) = -\int_{t_0}^{t} dt' \left[f'(\hat{q}(t)) \left(\frac{d}{dt'} T(t - t') \right) f(\hat{q}(t')) \right]_{+}$$

$$= \mathcal{V}'_{\text{ren}}(\hat{q}(t)) + [f'(\hat{q}(t))T(t - t_0)f(\hat{q}(t_0))]_{+}$$

$$+ \int_{t_0}^{t} dt' \left[f'(\hat{q}(t))T(t - t')\frac{df(\hat{q}(t'))}{dt'} \right]_{+}, \qquad (5.5.3)$$

$$\mathcal{V}_{\text{ren}}(q) \equiv -\frac{1}{2}T(0)\{f(q)\}^2. \qquad (5.5.4)$$

On the other hand,

$$\tilde{V}(q) - V(q) = \delta V(q) := \frac{1}{2}\sum_{\alpha} \omega_{\alpha}^2 \{f_{\alpha}(q)\}^2$$

$$= \frac{1}{2}\sum_{\alpha} \bar{\gamma}_{\alpha}^2 \omega_{\alpha}^{-1}\{f(q)\}^2 = \frac{1}{2}T(0)\{f(q)\}^2. \qquad (5.5.5)$$

It is seen that (5.5.4) is exactly cancelled out by (5.5.5) (this issue will be elaborated at the end of the next section). Hence,

$$\frac{d^2\hat{q}(t)}{dt^2} + V'(\hat{q}(t)) + \left[f'(\hat{q}(t)) \int_{t_0}^{t} dt' T(t - t')\frac{df(\hat{q}(t'))}{dt'} \right]_{+} = [f'(\hat{q}(t))\hat{\tilde{\mathcal{R}}}(t)]_{+}, \qquad (5.5.6)$$

$$\hat{\tilde{\mathcal{R}}}(t) := \hat{\mathcal{R}}(t) - T(t - t_0)f(\hat{q}) \qquad (5.5.7)$$

$$= \sqrt{\frac{\hbar}{2}} \sum_{\alpha} \bar{\gamma}_{\alpha} \{ e^{-i(t - t_0)\omega_{\alpha}} \hat{b}_{\alpha} + \text{h.c.} \}$$

$$- \sum_{\alpha} \frac{\bar{\gamma}_{\alpha}^2}{\omega_{\alpha}} \cos((t - t_0)\omega_{\alpha})f(\hat{q}) \qquad (5.5.8)$$

$$= \sqrt{\frac{\hbar}{2}} \sum_{\alpha} \bar{\gamma}_{\alpha} \left[e^{-i(t - t_0)\omega_{\alpha}} \left\{ \hat{b}_{\alpha} - \sqrt{\frac{\omega_{\alpha}}{2\hbar}} f_{\alpha}(\hat{q}) \right\} + \text{h.c.} \right]. \qquad (5.5.9)$$

Note that the combination of operators embraced by the curly bracket in the last line is the same as has been encountered in $\hat{H}_{\mathcal{E}} + \hat{H}_{S\mathcal{E}}$ of Eq. (5.1.22).

Hereafter, let us confine ourselves to the case of the bilinear model ($f(q) = q$). Then,

$$\frac{d^2\hat{q}(t)}{dt^2} + V'(\hat{q}(t)) + \int_{t_0}^{t} dt' T(t - t')\frac{d\hat{q}(t')}{dt'} = \hat{\tilde{\mathcal{R}}}(t). \qquad (5.5.10)$$

An equation of this form is called the *quantum Langevin equation*;[8] this reduces to a classical Langevin equation if $\hat{q}$ is replaced by a c-number variable. The third term on the left-hand side implies the existence of "frictional resistance retarded in time". It is because this retardation is quantitatively expressed by the function $\mathcal{T}(t)$ that we have borrowed the classical terminology to call the latter the retarded resistance function. The right-hand side, which depends only on initial values, may be interpreted as representing a fluctuating force (inclusive of a non-fluctuating systematic part).

Incidentally, the above integro-differential equation for $\hat{q}(t)$ can be transformed into a differential equation if desired. In the latter form, a retardation term disappears, but it does so only superficially; physical effects of retardation remain intact, of course. As a price for the elimination of the convolution term (i.e. the integration term), coefficients of the differential equation now depend explicitly on time, and the fluctuating force on the right-hand side is modified to a more complicated form. This transformation, which is cumbersome in general, is relatively straightforward in the special case of the macrosystem being a harmonic oscillator (i.e. $V(q) \propto q^2$).

5.6 Equation of motion in the classical regime

Let $|\Psi(0)\rangle\rangle$ be the initial state of the entire system, and denote an expectation value with respect to the state with the symbol $\langle\cdots\rangle$. By use of the notation

$$q(t) := \langle \hat{q}(t) \rangle \equiv \langle\langle \Psi(0) | \hat{q}(t) | \Psi(0) \rangle\rangle, \tag{5.6.1}$$

one may take the expectation value of both sides of equation of motion (5.5.10) to obtain

$$\ddot{q}(t) + \langle V'(\hat{q}(t)) \rangle + \tilde{\mathcal{M}}(t) = \langle \hat{\tilde{\mathcal{R}}}(t) \rangle, \tag{5.6.2}$$

$$\tilde{\mathcal{M}}(t) := \int_{t_0}^{t} dt'\, \mathcal{T}(t - t')\dot{q}(t') = \int_{-(t-t_0)}^{0} dt'\, \mathcal{T}(t')\dot{q}(t + t'). \tag{5.6.3}$$

Under the condition (5.3.5) that the timescale ω_0^{-1} characterizing the motion of the macrosystem falls between v_c^{-1} and ω_c^{-1}, the two factors in the integrand of $\tilde{\mathcal{M}}(t)$ are roughly of the form depicted in Fig. 5.4. Focusing attention on the temporal domain

$$\omega_c^{-1} \ll t - t_0 \ll v_c^{-1}, \tag{5.6.4}$$

---

[8] Sometimes, the phrase "Langevin equation" is reserved for "Langevin equation without retardation". In that case, the above equation is called a "generalized Langevin equation". See, e.g. R. Kubo, *Statistical Physics*, Vol. 2, Springer Verlag (1998).

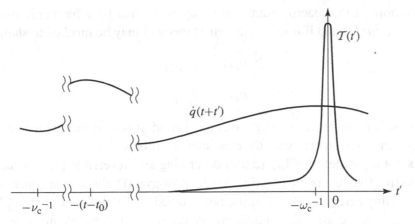

Fig. 5.4. Integrand of $\tilde{\mathcal{M}}(t)$.

one may evaluate $\tilde{\mathcal{M}}(t)$ as

$$\tilde{\mathcal{M}}(t) \simeq \dot{q}(t) \int_{-(t-t_0)}^{0} dt' \, T(t') \simeq \eta \dot{q}(t). \tag{5.6.5}$$

In short, the *Markov approximation* is valid. The last approximate equality follows from Eq. (5.4.7) in the case of the ohmic environment (i.e. $s = 1$). In the same temporal domain, the following approximation is also valid:

$$\hat{\tilde{\mathcal{R}}}(t) \simeq \hat{\mathcal{R}}(t). \tag{5.6.6}$$

If the state of the macrosystem is represented by a wavepacket localized within a region that is narrow compared to the characteristic scale of variation (i.e. q-dependence) of the potential, one has

$$\langle V'(\hat{q}(t)) \rangle \simeq V'(q(t)), \tag{5.6.7}$$

$$\ddot{q}(t) + \eta \dot{q}(t) + V'(q(t)) \simeq \langle \hat{\mathcal{R}}(t) \rangle. \tag{5.6.8}$$

The last equation reveals that η is the usual friction coefficient. Note that this equation is valid only if the environment is of the ohmic variety, which is the reason for the nomenclature "ohmic".

If the right-hand side happens to vanish, the macrosystem energy

$$E(t) := \frac{1}{2}\{\dot{q}(t)\}^2 + V(q(t)) \tag{5.6.9}$$

decreases in time in accordance with the equation

$$\dot{E}(t) = -\eta\{\dot{q}(t)\}^2. \tag{5.6.10}$$

Furthermore, if the macrosystem can be approximated by a harmonic oscillator $(V(q) \simeq \omega_0^2 q^2/2)$ and if $\eta \ll \omega_0$, the virial theorem may be invoked to show that

$$\frac{d}{dt}\overline{E(t)} \simeq -\eta\overline{E(t)}, \tag{5.6.11}$$

$$\overline{E(t)} \simeq \overline{E(0)}e^{-\eta t}, \tag{5.6.12}$$

where the bar denotes the average over one period. It is seen that the macrosystem energy damps on average with the characteristic time η^{-1}.

It is not a paradox that Eq. (5.6.8) describing an irreversible process has been derived from the reversible (i.e. time-reversal invariant) Schrödinger equation. The irreversibility here, namely the usual one typified by friction, does not imply that the entire system undergoes a unidirectional time evolution, but claims only that the macrosystem energy flows out unilaterally into the environment and that this energy can not return to the macrosystem within the ordinary time since the time of the order of τ_P is needed for it to do so. Equation (5.6.8) has been derived under the tacit assumption that $t > t_0$ for a given initial time t_0. This assumption is embodied in the retarded Green function. If one were to repeat the whole argument assuming that $t < t_0$, one would replace the *retarded solution* (5.2.9) with the corresponding *advanced solution*, which would eventually lead to Eq. (5.6.8) with η replaced by $-\eta$. In general, the time-reversal invariance of a fundamental equation is equivalent to the statement that the time-reversed version of a solution is another solution. The above-mentioned advanced solution is indeed the time-reversed version of the retarded solution. Thus, Eq. (5.6.8) does not contradict the time-reversal symmetry. What has been said in this paragraph applies quite generally to any theory whether quantum or classical.

In general, it is not an easy matter to determine the macrosystem potential $V(q)$ by theoretical means. In the search for macroscopic quantum tunneling, therefore, one will proceed as follows:

(1) Guess an approximate shape of $V(q)$ and express it with a function containing a few adjustable parameters.
(2) Prepare the macrosystem in question in the classical situation where Eq. (5.6.8) is expected to be applicable and observe its behavior. Compare the experimental result with the prediction of Eq. (5.6.8) to determine the value of η as well as the parameters introduced in the step (1) above.
(3) Use the obtained η and $V(q)$ to theoretically analyze quantum tunneling expected for the macrosystem.
(4) Prepare the macrosystem in a situation[9] where quantum tunneling is expected, and observe its behavior. Compare the experimental result with the theoretical prediction made

---

[9] That is, a quantum-mechanical situation (or, to be precise, the quasi-classical situation (2.1.11)).

in the step (3) above. If they agree with each other, it may be concluded that a macroscopic quantum tunneling has been detected.

As is clear from this procedure, it is $V(q)$ in Eq. (5.1.9) that can be determined experimentally. It is for this reason that the total Hamiltonian has been arranged in the form of Eqs. (5.1.8)–(5.1.11). The quantity (5.5.5) (i.e. the second term of Eq. (5.1.11)) is often called the counter term, and the cancellation of (5.5.4) by this term is called the potential renormalization. Such seemingly highbrow phrases as counter term or potential renormalization tend to make students worry whether or not they should include the counter term in the Hamiltonian. But, as should be clear by now, there is no need for such worry.

Incidentally, the second term on the left-hand side of Eq. (5.6.8) arises regardless of the value of ω_c provided that the condition (5.3.5) holds. In other words, contributions to the friction come from the low-frequency members ($\omega_\alpha \lesssim \omega_0$) of the environmental oscillators, while the high-frequency ones ($\omega_\alpha \gg \omega_0$) contribute to the "potential renormalization" only; indeed, under the condition (5.3.5), the dominant contribution to (5.5.4) comes from the high-frequency members.

Prepared in the classical situation, many macrosystems execute motions accompanied by usual friction, that is, they are likely to obey Eq. (5.6.8). Therefore, it is also likely to be a fairly general feature that actual environments are of the ohmic variety, although details may vary depending on the system in question.

Exercises

Exercise 5.1. Use Eq. (5.2.9) to confirm the commutation relation (5.2.10).

Exercise 5.2. Employ the expression (5.3.7) for $J(\omega)$ to study in detail the behavior of the retarded resistance function.

Exercise 5.3. Supposing that the macrosystem is in fact a harmonic oscillator, solve Eq. (5.6.10) and compare the result with Eq. (5.6.12).

Exercise 5.4. Derive the equation of motion obeyed by the fluctuation of $\hat{q}(t)$

$$\sigma(t) := \langle \{\hat{q}(t)\}^2 \rangle - \{\langle \hat{q}(t) \rangle\}^2, \tag{5.6.13}$$

where $\langle \cdots \rangle$ denotes an expectation value with respect to the initial state of the entire system as in Section 5.6.

Exercise 5.5. Study the properties of the *random force*

$$\delta\hat{\mathcal{R}}(t) := \hat{\mathcal{R}}(t) - \langle \hat{\mathcal{R}}(t) \rangle, \tag{5.6.14}$$

with the notation being the same as in Section 5.6, for typical initial states. In particular, consider the case

$$\langle \hat{b}_\alpha \rangle = 0, \quad \langle \hat{b}_\alpha^\dagger \hat{b}_\beta \rangle = n(\omega_\alpha)\delta_{\alpha\beta}, \quad \langle \hat{b}_\alpha \hat{b}_\beta \rangle = s(\omega_\alpha)\delta_{\alpha\beta}, \tag{5.6.15}$$

where $n(\omega)$ and $s(\omega)$ are given functions,

and show

$$\langle [\hat{\mathcal{R}}(t)\hat{\mathcal{R}}(t')]_+\rangle$$
$$= \frac{2h}{\pi} \int_0^\infty d\omega\, J(\omega) \left[\left\{ n(\omega) + \frac{1}{2} \right\} \cos((t-t')\omega) + \Re\{s(\omega)e^{-i(t+t'-2t_0)\omega}\} \right],$$

(5.6.16)

where the left-hand side is called the *symmetrized time-correlation function of random force* (to be abbreviated as the <u>noise</u>). This formula is valid if the initial states of the environmental oscillators are *squeezed states*. In this case, $s(\omega_\alpha)$ gives a measure of squeezing of the oscillator α and renders the noise non-stationary. However, the term containing $s(\omega)$ tends to 0 as $t_0 \downarrow -\infty$, that is, the noise is non-stationary only transiently, provided that

$$\int_0^\infty d\omega\, J(\omega)|s(\omega)| < \infty.$$

(5.6.17)

In the special case of

$$n(\omega) = \frac{1}{e^{h\omega/T_\star} - 1},$$

(5.6.18)

where $T_\star$ may be interpreted as the effective temperature of the environment (in the same spirit as one may infer the temperature of a star, for instance), one finds

$$\langle [\hat{\mathcal{R}}(t)\hat{\mathcal{R}}(t')]_+\rangle|_{t_0\downarrow-\infty} \simeq \frac{2}{\pi} \int_0^\infty d\omega \frac{J(\omega)}{\omega} \mathcal{E}(\omega; T_\star) \cos((t-t')\omega),$$

(5.6.19)

$$\mathcal{E}(\omega; T_\star) := \frac{h\omega}{e^{h\omega/T_\star} - 1} + \frac{1}{2}h\omega = \frac{h\omega}{2} \coth \frac{h\omega}{2T_\star}.$$

(5.6.20)

Study the $(t-t')$-dependence of this noise.

Discuss also the case where each of the environmental oscillators is initially in a coherent state.

6

Quantum resonant oscillation in the harmonic environment

In the last section of the preceding chapter, we considered a macrosystem prepared in the classical situation and studied the effect of the harmonic environment. The result, in a nutshell, is that a typical harmonic environment acts as a source of friction. Then, how about the situation where a macrosystem, if isolated from an environment, would behave quantum-mechanically? In this chapter we study the effect of the harmonic environment on a macrosystem prepared in the MQC-situation.

The main purpose of this book is a theoretical search for a possibility of macroscopic quantum phenomena in spite of the presence of an environment, that is, of overcoming the decoherence effects due to the environment. Therefore, we are interested in the case in which the interaction between the macrosystem and the environment is <u>weak</u> (to be stated quantitatively below).[1] Under such circumstances, it should be possible to treat the interaction by perturbation theory, which we now develop to elucidate the details of dephasing and dissipation effects.

6.1 Preliminary consideration on perturbation theory

Let us begin by recalling the standard elementary perturbation theory. Let $|\star\rangle$ and $\varepsilon_\star$ be a generic energy eigenstate and the associated energy eigenvalue, respectively, for the environment, with the ground state being denoted by $|\text{vac}\rangle$:

$$\hat{H}_\varepsilon \, |\star\rangle = \varepsilon_\star \, |\star\rangle, \qquad \hat{H}_\varepsilon |\text{vac}\rangle = 0. \tag{6.1.1}$$

Also, let $|\alpha\rangle$ be the state in which the environmental oscillator α is singly excited, and $|\alpha\beta\rangle$ be the one in which both α and β are singly excited:[2]

$$|\alpha\rangle := \hat{b}_\alpha^\dagger |\text{vac}\rangle, \qquad |\alpha\beta\rangle := \hat{b}_\alpha^\dagger \hat{b}_\beta^\dagger |\text{vac}\rangle. \tag{6.1.2}$$

---

[1] By contrast, if one were interested in the question why the majority of macrosystems seem to behave classically even though they should be governed by quantum mechanics, one would deal with the case where the interaction is so <u>strong</u> that even a faint trace of quantum-mechanical feature is washed out.

[2] If $\alpha = \beta$, $|\alpha\beta\rangle$ is the state with the oscillator being doubly excited and is not normalized to unity.

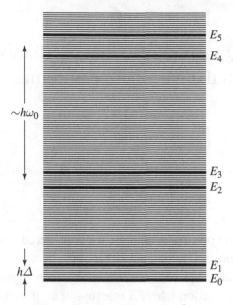

Fig. 6.1. Energy levels of $\hat{H}_0$.

As to the eigenstates and eigenvalues of the macrosystem Hamiltonian $\hat{H}_S$, we follow the notation introduced in Section 4.2. As in that section, a state for the entire system is to be written in such a form as

$$|n, \star\rangle\rangle \equiv |n\rangle|\star\rangle. \tag{6.1.3}$$

Accordingly, the state $|n, \mathrm{vac}\rangle\rangle$ is an eigenstate of

$$\hat{H}_0 := \hat{H}_S + \hat{H}_\mathcal{E}, \tag{6.1.4}$$

and the associated energy eigenvalue is E_n (cf. Fig. 6.1). Let δE_n be the shift suffered by this eigenvalue due to the perturbation $\hat{H}_{S\mathcal{E}}$, then the stationary perturbation theory (i.e. the perturbation theory applied to an energy eigenstate) gives

$$\delta E_n \simeq \delta E_n^{(1)} + {\sum_{m,\star}}' \frac{|\langle\langle m, \star|\hat{H}_{S\mathcal{E}}|n, \mathrm{vac}\rangle\rangle|^2}{E_n - (E_m + \varepsilon_\star)}, \tag{6.1.5}$$

where

$$\delta E_n^{(1)} := \langle\langle n, \mathrm{vac}|\hat{H}_{S\mathcal{E}}|n, \mathrm{vac}\rangle\rangle \tag{6.1.6}$$

is the contribution of the first-order perturbation. The approximate equality $\simeq$ stipulates that the perturbation has been treated up to the second order. The prime $'$ attached to the summation symbol indicates that the term of $(m, \star) = (n, \mathrm{vac})$ is to be omitted from the sum. The uncritical use of the stationary perturbation theory above is not quite warranted. In fact, the "state" with the energy shifted to $E_n + \delta E_n$

by perturbation is not stationary but decays with a finite "lifetime" Γ_n^{-1}. According to Fermi's golden rule, the latter is given as

$$\Gamma_n \simeq \frac{2\pi}{h} \sum_{m,\star}{}' |\langle\langle m, \star|\hat{H}_{S\mathcal{E}}|n, \text{vac}\rangle\rangle|^2 \delta(E_n - (E_m + \varepsilon_\star)). \qquad (6.1.7)$$

Combining Eqs. (6.1.5) and (6.1.7), one would obtain

$$\delta E_n - \delta E_n^{(1)} - \frac{1}{2}ih\Gamma_n \simeq \sum_{m,\star}{}' \frac{|\langle\langle m, \star|\hat{H}_{S\mathcal{E}}|n, \text{vac}\rangle\rangle|^2}{E_n - (E_m + \varepsilon_\star) + i0}, \qquad (6.1.8)$$

where $+0$ stands for an infinitesimal positive number. Now, let

$$\hat{H}' := \hat{H}_{S\mathcal{E}} - |n, \text{vac}\rangle\rangle \delta E_n^{(1)} \langle\langle n, \text{vac}|, \qquad (6.1.9)$$

then

$$\langle\langle n, \text{vac}|\hat{H}'|n, \text{vac}\rangle\rangle = 0. \qquad (6.1.10)$$

Hence, the above expression may be rewritten as

$$\delta E_n - \delta E_n^{(1)} - \frac{1}{2}ih\Gamma_n$$

$$\simeq \sum_{m,\star} \langle\langle n, \text{vac}|\hat{H}'|m, \star\rangle\rangle \frac{1}{E_n - (E_m + \varepsilon_\star) + i0} \langle\langle m, \star|\hat{H}'|n, \text{vac}\rangle\rangle \qquad (6.1.11)$$

$$= \langle\langle n, \text{vac}|\hat{H}' \frac{1}{E_n - \hat{H}_0 + i0} \hat{H}'|n, \text{vac}\rangle\rangle. \qquad (6.1.12)$$

The last expression, being concise, is suitable for some formal manipulation. For an explicit computation, however, one has to go back to Eq. (6.1.11).

The above argument, which is a more or less standard one encountered frequently, is somewhat dubious. For the delta function in Eq. (6.1.7) to make sense, it is necessary that the sum over $\star$ is able to be replaced by an integral. This in turn requires that the same is true for the sum in Eq. (6.1.5). As is well known, Eq. (6.1.8) is valid only if the integral to replace the sum in Eq. (6.1.5) is a principal-value integral. However, Eq. (6.1.5) does not provide any reason why it should be. In fact, a rational derivation of Eq. (6.1.12) is not possible with an argument combining the stationary perturbation theory and Fermi's golden rule, but is possible only if a perturbation theory is applied to the time evolution. We return to this issue in the next section. For the moment, we regard Eq. (6.1.12) as the definition of $\delta E_n - \delta E_n^{(1)} - \frac{1}{2}ih\Gamma_n$. Of course, this definition is valid only up to the second order in $\hat{H}_{S\mathcal{E}}$.

For brevity, we introduce the notation

$$\hat{f}_\alpha := \omega_\alpha^{3/2} f_\alpha(\hat{q}), \quad \Omega_{nm} := (E_n - E_m)/h. \qquad (6.1.13)$$

By use of $\hat{b}_\alpha|\text{vac}\rangle = 0$, we then find

$$\delta E_n^{(1)} = \frac{1}{2} \sum_\alpha \frac{1}{\omega_\alpha} \langle n|\hat{f}_\alpha^2|n\rangle. \tag{6.1.14}$$

Let us evaluate the right-hand side of Eq. (6.1.12) up to the second order in $\hat{f}_\alpha$. Since the second term of $\hat{H}'$, contributing to the third order at most, can be neglected, we may effectively put

$$\hat{H}'|n, \text{vac}\rangle\rangle \simeq -\sqrt{\frac{h}{2}} \sum_\alpha \hat{f}_\alpha |n, \alpha\rangle\rangle \tag{6.1.15}$$

in Eq. (6.1.12). Hence,

$$\begin{aligned}
\delta E_n - \delta E_n^{(1)} - \frac{1}{2}ih\Gamma_n &\simeq \frac{h}{2} \sum_{\alpha,\beta} \langle\langle n,\alpha|\hat{f}_\alpha \frac{1}{E_n - \hat{H}_0 + i0} \hat{f}_\beta|n,\beta\rangle\rangle \\
&= \frac{h}{2} \sum_{m,\alpha} \langle n|\hat{f}_\alpha|m\rangle \frac{1}{E_n - (E_m + h\omega_\alpha) + i0} \langle m|\hat{f}_\alpha|n\rangle \\
&= \frac{1}{2} \sum_{m,\alpha} |\langle m|\hat{f}_\alpha|n\rangle|^2 \left\{ \frac{1}{\Omega_{nm} - \omega_\alpha + i0} + \frac{1}{\omega_\alpha} \right\} \\
&\quad - \frac{1}{2} \sum_{m,\alpha} \frac{|\langle m|\hat{f}_\alpha|n\rangle|^2}{\omega_\alpha}.
\end{aligned} \tag{6.1.16}$$

The last line follows from the preceding one simply with ω_α^{-1} added and subtracted in each term of the summand. The last expression is convenient because dominant contributions from high-frequency members of the environmental oscillators cancel out[3] in the α-summation in the first term. The second term, on the other hand, reduces to the right-hand side of Eq. (6.1.14) if the m-summation is performed. It follows that

$$\delta E_n - \frac{1}{2}ih\Gamma_n \simeq \frac{1}{2} \sum_{m,\alpha} \frac{|\langle m|\hat{f}_\alpha|n\rangle|^2}{\omega_\alpha} \frac{\Omega_{nm}}{\Omega_{nm} - \omega_\alpha + i0}. \tag{6.1.17}$$

Hereafter, we confine ourselves to the separable model (5.1.14) and introduce another notation

$$f_{mn} := \langle m|f(\hat{q})|n\rangle. \tag{6.1.18}$$

Equation (6.1.17) may then be rewritten with the aid of the environmental frequency distribution function (5.3.3) as

$$\delta E_n - \frac{1}{2}ih\Gamma_n = \frac{1}{\pi} \sum_m |f_{mn}|^2 \Omega_{nm} \int_0^\infty \frac{d\omega}{\omega} \frac{J(\omega)}{\Omega_{nm} - \omega + i0}. \tag{6.1.19}$$

---

[3] For $\omega_\alpha \gg |\Omega_{nm}|$, the dominant part of the quantity embraced by the curly bracket is $-\omega_\alpha^{-1} + \omega_\alpha^{-1}$.

In other words,

$$\delta E_n = \frac{1}{\pi} \sum_m |f_{mn}|^2 \Omega_{mn} \mathcal{P} \int_0^\infty \frac{d\omega}{\omega} \frac{J(\omega)}{\omega + \Omega_{mn}}, \tag{6.1.20}$$

$$\Gamma_n = \frac{2}{h} \sum_m |f_{mn}|^2 \Omega_{nm} \int_0^\infty \frac{d\omega}{\omega} J(\omega) \delta(\Omega_{nm} - \omega)$$

$$= \frac{2}{h} \sum_m |f_{mn}|^2 J(\Omega_{nm}) \theta(\Omega_{nm}), \tag{6.1.21}$$

where the symbol $\mathcal{P}$ in Eq. (6.1.20) stipulates that the integral preceded by it is a principal-value integral, and $\theta(\Omega_{nm})$ in Eq. (6.1.21) is the step function, which indicates that the macrosystem initially in an excited state is allowed to make transition to the lower states only. In this sense, Γ_n reflects the dissipation effect. In particular,

$$\Gamma_0 = 0. \tag{6.1.22}$$

Now, suppose that δE_n and Γ_n have been evaluated. What can be concluded about MQC? To be specific, let us consider the problem posed in Section 4.2 assuming that the initial state of the entire system is given as

$$|\Psi(0)\rangle\rangle = |-, \text{vac}\rangle\rangle, \tag{6.1.23}$$

namely, that $|\chi\rangle$ in Eq. (4.2.25) is chosen to be $|\text{vac}\rangle$. What might be expected naively is the following:

$$|\Psi(t)\rangle\rangle = \{P_0(t)\}^{1/2}|0, \text{vac}\rangle\rangle \exp\{-i(E_0 + \delta E_0)t/h\}$$
$$- \{P_1(t)\}^{1/2}|1, \text{vac}\rangle\rangle \exp\{-i(E_1 + \delta E_1)t/h\}. \tag{6.1.24}$$

This is essentially the same as Eq. (4.2.4) with $\vartheta_n(t) = 0$. It should be clear that this expectation is inaccurate; in the presence of the interaction described by $\hat{H}_{S\mathcal{E}}$, the environmental oscillators can be excited, and as a consequence, environmental states other than $|\text{vac}\rangle$ should emerge as in Section 4.3. Thus, mere knowledge of δE_n and Γ_n is not sufficient to understand the influence of the environment on MQC. It is necessary to investigate the time evolution of the entire system in more detail. We do so in the next section, which treats the time evolution with perturbation theory.[4]

---

[4] This theory is conventionally called the "time-dependent perturbation theory", which is not quite an appropriate terminology.

6.2 Perturbation theory of time evolution

It is convenient to introduce the interaction-picture operators:

$$\hat{H}_{S\mathcal{E}}(t) := \mathrm{e}^{\mathrm{i}\hat{H}_0 t/h} \hat{H}_{S\mathcal{E}} \mathrm{e}^{-\mathrm{i}\hat{H}_0 t/h}, \quad \hat{q}(t) := \mathrm{e}^{\mathrm{i}\hat{H}_0 t/h} \hat{q} \mathrm{e}^{-\mathrm{i}\hat{H}_0 t/h}. \tag{6.2.1}$$

Consider the time-evolution operator

$$\hat{U}_I(t) := \exp(\mathrm{i}\hat{H}_0 t/h) \exp(-\mathrm{i}\hat{H}t/h), \tag{6.2.2}$$

and expand it up to the second order with respect to the interaction Hamiltonian to find

$$\hat{U}_I(t) \simeq 1 - \frac{\mathrm{i}}{h} \int_0^t \mathrm{d}t_1 \hat{H}_{S\mathcal{E}}(t_1) - \frac{1}{h^2} \int_0^t \mathrm{d}t_2 \int_0^{t_2} \mathrm{d}t_1 \hat{H}_{S\mathcal{E}}(t_2) \hat{H}_{S\mathcal{E}}(t_1). \tag{6.2.3}$$

As to the initial state of the entire system, we make a slightly more general assumption than (6.1.23):

$$|\Psi(0)\rangle\rangle = |\psi\rangle |\mathrm{vac}\rangle, \tag{6.2.4}$$

where $|\psi\rangle$ is an arbitrary state for the macrosystem. First, we compute

$$\hat{H}_{S\mathcal{E}}(t)|\mathrm{vac}\rangle = |\mathrm{vac}\rangle\langle\mathrm{vac}|\hat{H}_{S\mathcal{E}}(t)|\mathrm{vac}\rangle + \sum_\alpha |\alpha\rangle\langle\alpha|\hat{H}_{S\mathcal{E}}(t)|\mathrm{vac}\rangle \tag{6.2.5}$$

$$= \delta\hat{V}(t)|\mathrm{vac}\rangle - \sqrt{\frac{h}{2}} \sum_\alpha \hat{f}_\alpha(t)\mathrm{e}^{\mathrm{i}\omega_\alpha t} |\alpha\rangle, \tag{6.2.6}$$

where

$$\delta\hat{V}(t) := \delta V(\hat{q}(t)), \quad \hat{f}_\alpha(t) := \omega_\alpha^{3/2} f_\alpha(\hat{q}(t)), \tag{6.2.7}$$

with $\delta V(q)$ being defined by Eq. (5.5.5). Next, we compute

$$\hat{H}_{S\mathcal{E}}(t')\hat{H}_{S\mathcal{E}}(t)|\mathrm{vac}\rangle \simeq -\hat{H}_{S\mathcal{E}}(t')\sqrt{\frac{h}{2}} \sum_\alpha \hat{f}_\alpha(t)\mathrm{e}^{\mathrm{i}\omega_\alpha t} |\alpha\rangle$$

$$\simeq \frac{h}{2} \sum_\beta \hat{f}_\beta(t')(\hat{b}_\beta \mathrm{e}^{-\mathrm{i}\omega_\beta t'} + \mathrm{h.c.}) \sum_\alpha \hat{f}_\alpha(t)\mathrm{e}^{\mathrm{i}\omega_\alpha t} |\alpha\rangle$$

$$= \frac{h}{2} \left\{ \sum_\alpha \hat{f}_\alpha(t')\mathrm{e}^{-\mathrm{i}(t'-t)\omega_\alpha} \hat{f}_\alpha(t)|\mathrm{vac}\rangle \right.$$

$$\left. + \sum_{\alpha\beta} \hat{f}_\beta(t')\mathrm{e}^{\mathrm{i}\omega_\beta t' + \mathrm{i}\omega_\alpha t} \hat{f}_\alpha(t)|\alpha\beta\rangle \right\}. \tag{6.2.8}$$

These results may be summarized in the following form:

$$\hat{U}_I(t)|\mathrm{vac}\rangle \simeq \hat{U}_{\mathrm{vac}}(t)|\mathrm{vac}\rangle + \sum_\alpha \hat{U}_\alpha(t)|\alpha\rangle + \sum_{\alpha\beta} \hat{U}_{\alpha\beta}(t)|\alpha\beta\rangle, \tag{6.2.9}$$

where $\hat{\mathcal{U}}_{\text{vac}}(t)$, $\hat{\mathcal{U}}_\alpha(t)$ and $\hat{\mathcal{U}}_{\alpha\beta}(t)$ are operators acting on the macrosystem Hilbert space $\mathcal{H}_S$ such that

$$\hat{\mathcal{U}}_{\text{vac}}(t) := 1 - \frac{i}{h} \int_0^t dt_1 \delta \hat{V}(t_1)$$
$$- \frac{1}{2h} \sum_\alpha \int_0^t dt_2 \int_0^{t_2} dt_1 \hat{f}_\alpha(t_2) e^{-i(t_2-t_1)\omega_\alpha} \hat{f}_\alpha(t_1), \quad (6.2.10)$$

$$\hat{\mathcal{U}}_\alpha(t) := \frac{i}{\sqrt{2h}} \int_0^t dt_1 \hat{f}_\alpha(t_1) e^{i\omega_\alpha t_1}, \quad (6.2.11)$$

$$\hat{\mathcal{U}}_{\alpha\beta}(t) := -\frac{1}{2h} \int_0^t dt_2 \int_0^{t_2} dt_1 \hat{f}_\beta(t_2) e^{i\omega_\beta t_2 + i\omega_\alpha t_1} \hat{f}_\alpha(t_1). \quad (6.2.12)$$

Now, compare

$$\hat{U}(t)|\Psi(0)\rangle = \sum_n |n\rangle\langle n| \exp\{-i(\hat{H}_S + \hat{H}_\varepsilon)t/h\}\hat{U}_I(t)|\Psi(0)\rangle \quad (6.2.13)$$

with Eq. (4.2.27) to find

$$|\widetilde{\chi_n(t)}\rangle = \langle n| \exp(-i\hat{H}_\varepsilon t/h)\hat{U}_I(t)|\Psi(0)\rangle$$
$$= |\text{vac}\rangle\langle n|\hat{\mathcal{U}}_{\text{vac}}(t)|\psi\rangle + \sum_\alpha e^{-i\omega_\alpha t}|\alpha\rangle\langle n|\hat{\mathcal{U}}_\alpha(t)|\psi\rangle$$
$$+ \sum_{\alpha\beta} e^{-i(\omega_\alpha+\omega_\beta)t}|\alpha\beta\rangle\langle n|\hat{\mathcal{U}}_{\alpha\beta}(t)|\psi\rangle. \quad (6.2.14)$$

The problem has thus been reduced to the evaluation of matrix elements of $\hat{\mathcal{U}}_{\text{vac}}(t)$, $\hat{\mathcal{U}}_\alpha(t)$ and $\hat{\mathcal{U}}_{\alpha\beta}(t)$.

The diagonal element of $\hat{\mathcal{U}}_{\text{vac}}(t)$ is evaluated as

$$\langle n|\hat{\mathcal{U}}_{\text{vac}}(t)|n\rangle = 1 - \frac{i}{h}t\delta E_n^{(1)} + \frac{i}{2h} \sum_{m,\alpha} |\langle m|\hat{f}_\alpha|n\rangle|^2 \int_0^t dt' \frac{1 - e^{-i(\Omega_{mn}+\omega_\alpha)t'}}{\Omega_{mn} + \omega_\alpha}.$$
$$(6.2.15)$$

Hereafter, let us adopt the separable model again. Then, the third term on the right-hand side may be rewritten in terms of the environmental frequency distribution function as

$$\frac{i}{\pi h} \sum_m |f_{mn}|^2 \int_0^\infty d\omega J(\omega) \int_0^t dt' \frac{1 - e^{-i(\omega+\Omega_{mn})t'}}{\omega + \Omega_{mn}}. \quad (6.2.16)$$

By virtue of the numerator $1 - e^{-i(\omega+\Omega_{mn})t'}$, the integrand is regular at $\omega + \Omega_{mn} = 0$. Hence, the above integral over ω is well-defined even if Ω_{mn} is negative. Accordingly, the result of the integration remains unchanged if the integral is regarded as the principal-value integral, which allows one to perform the t'-integration term by term:

$$\langle n|\hat{\mathcal{U}}_{\text{vac}}(t)|n\rangle = 1 - \frac{i}{h}t\left\{\delta E_n^{(1)} - \frac{1}{\pi}\sum_m |f_{mn}|^2\,\mathcal{P}\int_0^\infty d\omega\,\frac{J(\omega)}{\omega + \Omega_{mn}}\right\}$$

$$- \frac{1}{\pi h}\sum_m |f_{mn}|^2\,\mathcal{P}\int_0^\infty d\omega\,J(\omega)\frac{1 - e^{-i(\omega+\Omega_{mn})t}}{(\omega + \Omega_{mn})^2}. \qquad (6.2.17)$$

The quantity embraced by the curly bracket in the second term on the right-hand side coincides with δE_n in Eq. (6.1.20). Hence, the following expression is valid up to the second order:

$$\langle n|\hat{\mathcal{U}}_{\text{vac}}(t)|n\rangle \simeq e^{-it\delta E_n/h}\left\{1 - \frac{1}{\pi h}\sum_m |f_{mn}|^2\,\mathcal{P}\int_0^\infty d\omega\,J(\omega)\frac{1 - e^{-i(\omega+\Omega_{mn})t}}{(\omega + \Omega_{mn})^2}\right\}$$

$$= e^{-it\delta E_n/h}\left\{1 - t\frac{1}{h}\sum_m |f_{mn}|^2\int_0^\infty d\omega\,J(\omega)\mathcal{D}_2(\omega + \Omega_{mn};t)\right.$$

$$\left. + \frac{i}{h}\sum_m |f_{mn}|^2\,\mathcal{P}\int_0^\infty d\omega\,J(\omega)\mathcal{D}_3(\omega + \Omega_{mn};t)\right\}, \qquad (6.2.18)$$

where $\mathcal{D}_2(\omega;t)$ and $\mathcal{D}_3(\omega;t)$ are defined as

$$\mathcal{D}_2(\omega;t) := \frac{1}{2\pi t}\left\{\frac{\sin(\omega t/2)}{\omega/2}\right\}^2, \qquad \mathcal{D}_3(\omega;t) := -\frac{1}{\pi}\frac{\sin\omega t}{\omega^2}. \qquad (6.2.19)$$

(These functions are related to the delta function; see below.) It is to be noted that Eq. (6.1.20), whose status was obscure in the framework of the stationary perturbation theory, has now emerged unambiguously.

Matrix elements of $\hat{\mathcal{U}}_\alpha(t)$ are evaluated as

$$\langle m|\hat{\mathcal{U}}_\alpha(t)|n\rangle = \frac{i}{\sqrt{2h}}\langle m|\hat{f}_\alpha|n\rangle\int_0^t dt'\,e^{i(\omega_\alpha+\Omega_{mn})t'}$$

$$= i\frac{2\pi}{\sqrt{2h}}\gamma_\alpha f_{mn}\mathcal{D}_1(\omega_\alpha + \Omega_{mn};t)e^{i(\omega_\alpha+\Omega_{mn})t/2}, \qquad (6.2.20)$$

$$\mathcal{D}_1(\omega;t) := \frac{1}{2\pi}\frac{\sin(\omega t/2)}{\omega/2}. \qquad (6.2.21)$$

Hereafter, we make a simplifying assumption that $f(q)$ is an odd function:

$$f(-q) = -f(q), \quad \delta V(-q) = \delta V(q). \qquad (6.2.22)$$

(The bilinear model (5.1.15) is an example which has this property.) Recall that, the macrosystem potential $V(q)$ in the MQC-situation being an even function, the macrosystem energy eigenstate $|n\rangle$ has the following parity:

$$\langle -q|n\rangle = (-)^n \langle q|n\rangle. \tag{6.2.23}$$

Therefore, one finds the selection rule

$$\langle m|\hat{f}(t)|n\rangle = 0 \quad : \ m - n \text{ is even,} \tag{6.2.24}$$
$$\langle m|\delta\hat{V}(t)|n\rangle = 0 \quad : \ m - n \text{ is odd.} \tag{6.2.25}$$

Accordingly, in particular,

$$\langle 0|\delta\hat{V}(t)|1\rangle = 0, \quad \langle 0|\hat{f}(t_2)\hat{f}(t_1)|1\rangle = 0. \tag{6.2.26}$$

Hence

$$\langle 0|\hat{\mathcal{U}}_{\text{vac}}(t)|1\rangle = 0. \tag{6.2.27}$$

Similarly, one finds

$$\langle 1|\hat{\mathcal{U}}_{\text{vac}}(t)|0\rangle = 0, \tag{6.2.28}$$
$$\langle 0|\hat{\mathcal{U}}_\alpha(t)|0\rangle = 0, \quad \langle 1|\hat{\mathcal{U}}_\alpha(t)|1\rangle = 0. \tag{6.2.29}$$

Now, recall that the frequency scale ω_c characterizing $J(\omega)$ is much higher than ω_0 (we assume that $\nu_c = 0$). So long as the temporal domain

$$t \gg \omega_0^{-1} \tag{6.2.30}$$

is concerned, therefore, the following approximation is valid in the integral involving $J(\omega)$ in Eq. (6.2.18):

$$\mathcal{D}_2(\omega;t) \sim \delta(\omega). \tag{6.2.31}$$

Here and in what follows, the symbol $\sim$ is to indicate an approximate equality valid in the temporal domain (6.2.30). Likewise, in the last integral in Eq. (6.2.18), the contribution of $\mathcal{D}_3(\omega;t)$ comes only from the region $\omega \sim 0$. These properties together with the spectral structure of $\hat{H}_S$ shown in Fig. 6.1[5] lead to

$$\langle 0|\hat{\mathcal{U}}_{\text{vac}}(t)|0\rangle \sim e^{-it\delta E_0/\hbar}\left\{1 + \frac{i}{\hbar}|f_{10}|^2 F_+(t)\right\}, \tag{6.2.32}$$

$$F_\pm(t) := \mathcal{P}\int_0^\infty d\omega\, J(\omega)\mathcal{D}_3(\omega \pm \Delta;t). \tag{6.2.33}$$

Note that terms involving $\mathcal{D}_2(\omega + \Omega_{m0};t)$ are all gone because of $\omega + \Omega_{m0} > 0$, and those involving $\mathcal{D}_3(\omega + \Omega_{m0};t)$ with $m \geq 3$ are also gone because of $\Omega_{m0} = \mathcal{O}(1) > 0$ (no terms with $m = 2$ are present to begin with). These terms

---

[5] The relevant feature in the present context is that $\{E_n \mid n \geq 2\}$ is separated from $\{E_0, E_1\}$ by the amount of $\hbar\omega_0$ at least.

contribute mainly to transient effects of characteristic time ω_0^{-1} (compare, however, Exercise 6. 7). Similarly, we find

$$\langle 1|\hat{\mathcal{U}}_{\text{vac}}|1\rangle \sim e^{-it\delta E_1/h}\left\{1 - \frac{1}{2}\Gamma_1 t + \frac{i}{h}|f_{01}|^2 F_-(t)\right\}, \tag{6.2.34}$$

where we have made use of the relation

$$\Gamma_1 = \frac{2}{h}|f_{01}|^2 J(\Delta) \sim \frac{2}{h}|f_{01}|^2 \int_0^\infty d\omega J(\omega)\mathcal{D}_2(\omega - \Delta;t) \tag{6.2.35}$$

which follows from Eq. (6.1.21).

Hereafter, let us have the ohmic environment in mind and envisage the following situation

$$\eta \ll h, \tag{6.2.36}$$

which is to be called the weak-interaction situation. Recall that the constant η is the friction coefficient in the classical regime. This fact, however, does not automatically imply that $\eta = \mathcal{O}(h^0) \gg h$; the constant η, which is a measure for the strength of the interaction between the macrosystem and the environment, is on equal footing with h, which is a measure for the quantum-mechanical nature of the macrosystem. The ratio between η and h can vary depending on the system. Indeed, the weak-interaction situation (6.2.36) can actually be set up (see, e.g. Eq. (3.1.49) or Chapter 9). In this situation, the inequality

$$\Gamma_1 = \frac{2\eta}{h}|f_{01}|^2\Delta \ll \Delta \ll \omega_0 \tag{6.2.37}$$

holds since $|f_{01}|$ is of $\mathcal{O}(1)$, and accordingly the following temporal domain exists:

$$\omega_0^{-1} \ll t \ll \Gamma_1^{-1}, \tag{6.2.38}$$

which is to be called the principal temporal domain.[6] Hereafter, let us focus our attention on this temporal domain, then,

$$\langle 0|\hat{\mathcal{U}}_{\text{vac}}(t)|0\rangle \sim \exp\left[-\frac{i}{h}\{t\delta E_0 - |f_{10}|^2 F_+(t)\}\right], \tag{6.2.39}$$

$$\langle 1|\hat{\mathcal{U}}_{\text{vac}}(t)|1\rangle \sim e^{-\Gamma_1 t/2}\exp\left[-\frac{i}{h}\{t\delta E_1 - |f_{01}|^2 F_-(t)\}\right], \tag{6.2.40}$$

where the estimate

$$|F_\pm(t)|/h = \mathcal{O}(\eta/h) \ll 1 \tag{6.2.41}$$

has been used (see Eq. (6.3.20) below).

---

[6] To be called so because it is the temporal domain of primary interest in this book. Compared to Γ_1^{-1}, it would be a short-time domain as described in Section 4. 2, while, compared to ω_0^{-1}, it would be a long-time domain.

6.3 Macroscopic quantum resonant oscillation

As in Section 4.2, let us choose the initial state $|\psi\rangle$ of the macrosystem as

$$|\psi\rangle = |-\rangle = \frac{1}{\sqrt{2}}(|0\rangle - |1\rangle), \qquad (6.3.1)$$

and compute $\mathcal{P}_{\pm}(t)$ in the principal temporal domain (6.2.38). The results of the preceding section may be summarized as

$$
\begin{aligned}
|\widetilde{\chi_0(t)}\rangle \sim{} & \frac{1}{\sqrt{2}}|\text{vac}\rangle \exp\left[-\frac{i}{\hbar}\{t\delta E_0 - |f_{10}|^2 F_+(t)\}\right] \\
& - i\frac{\pi}{\sqrt{\hbar}} f_{01} \sum_\alpha \bar{\gamma}_\alpha |\alpha\rangle e^{-i(\omega_\alpha + \Delta)t/2} \mathcal{D}_1(\omega_\alpha - \Delta; t) \\
& + \sum_{\alpha\beta} e^{-i(\omega_\alpha + \omega_\beta)t} |\alpha\beta\rangle\langle 0|\hat{\mathcal{U}}_{\alpha\beta}(t)|\psi\rangle,
\end{aligned} \qquad (6.3.2)
$$

$$
\begin{aligned}
|\widetilde{\chi_1(t)}\rangle \sim{} & -\frac{1}{\sqrt{2}}|\text{vac}\rangle e^{-\Gamma_1 t/2} \exp\left[-\frac{i}{\hbar}\{t\delta E_1 - |f_{01}|^2 F_-(t)\}\right] \\
& + i\frac{\pi}{\sqrt{\hbar}} f_{10} \sum_\alpha \bar{\gamma}_\alpha |\alpha\rangle e^{-i(\omega_\alpha - \Delta)t/2} \mathcal{D}_1(\omega_\alpha + \Delta; t) \\
& + \sum_{\alpha\beta} e^{-i(\omega_\alpha + \omega_\beta)t} |\alpha\beta\rangle\langle 1|\hat{\mathcal{U}}_{\alpha\beta}(t)|\psi\rangle.
\end{aligned} \qquad (6.3.3)
$$

Hence,

$$
\begin{aligned}
P_0(t) &= \|\,|\widetilde{\chi_0(t)}\rangle\,\|^2 \\
&\sim \frac{1}{2} + \frac{\pi^2}{\hbar}|f_{01}|^2 \sum_\alpha \bar{\gamma}_\alpha^2 \{\mathcal{D}_1(\omega_\alpha - \Delta; t)\}^2 \\
&= \frac{1}{2} + \frac{t}{\hbar}|f_{01}|^2 \int_0^\infty d\omega\, J(\omega)\mathcal{D}_2(\omega - \Delta; t), \qquad (6.3.4)
\end{aligned}
$$

where the contribution of the third term on the right-hand side of Eq. (6.3.2), being of $\mathcal{O}(\bar{\gamma}_\alpha^4)$ and hence beyond the second order, has been dropped in the present approximation. The first term in the last line of the above equation is the initial probability $P_0(0)\,(=1/2)$ for the macrosystem to be found in the ground state at $t = 0$, and the second term ($\propto |f_{01}|^2$) represents the effect of the first excited state making transition to the ground state due to the interaction with the environment. By virtue of (6.2.35), the above equation reduces to

$$P_0(t) \sim \frac{1}{2} + \frac{1}{2}\Gamma_1 t \sim 1 - \frac{1}{2}e^{-\Gamma_1 t}, \qquad (6.3.5)$$

where the last equality holds in the principal temporal domain under consideration. This result, having been derived by use of the perturbation theory, is valid only if $P_0(t)$ is not much different from $P_0(0)$. It is in the principal temporal domain that this condition is satisfied. Similarly,

$$P_1(t) = \| \, |\widetilde{\chi_1(t)}\rangle \|^2$$

$$\sim \frac{1}{2}e^{-\Gamma_1 t} + \frac{t}{\hbar}|f_{10}|^2 \int_0^\infty d\omega \, J(\omega)\mathcal{D}_2(\omega + \Delta; t), \qquad (6.3.6)$$

where the first and the second terms represent the effect of the first excited state making transition to the ground state, and that of the ground state making transition to the first excited state, respectively. The latter contribution is negligible since the "energy $H_S + H_\mathcal{E}$" cannot be conserved in this transition, that is, $\omega + \Delta$ cannot vanish in the above integrand. Hence,

$$P_1(t) \sim \frac{1}{2}e^{-\Gamma_1 t}. \qquad (6.3.7)$$

It has thus been shown that Eqs. (4.2.5), (4.2.10) and (4.2.12) with $T_1 = 1/\Gamma_1$ hold in the principal temporal domain up to the second order in perturbation. Accordingly,

$$P_0(t) + P_1(t) \sim 1, \quad P_0(t)P_1(t) \sim 1, \qquad (6.3.8)$$

$$C(t) \sim -\langle\chi_0(t)|\chi_1(t)\rangle. \qquad (6.3.9)$$

In other words, the effect of the environment on $\mathcal{P}_\pm(t)$ in the principal temporal domain manifests itself through the dephasing factor $C(t)$ rather than the dissipation effect (i.e. changes in $P_0(t)$ and $P_1(t)$). By use of the normalization factor (4.2.9) (see Eq. (4.2.34)), one finds

$$|\chi_0(t)\rangle \sim |\text{vac}\rangle e^{-\Gamma_1 t/2} \exp\left[-\frac{i}{\hbar}\{t\delta E_0 - |f_{10}|^2 F_+(t)\}\right]$$

$$- i\pi\sqrt{\frac{2}{\hbar}}f_{01}\sum_\alpha \bar{\gamma}_\alpha|\alpha\rangle e^{-i(\omega_\alpha+\Delta)t/2}\mathcal{D}_1(\omega_\alpha - \Delta; t) + \cdots, \qquad (6.3.10)$$

$$|\chi_1(t)\rangle \sim -|\text{vac}\rangle \exp\left[-\frac{i}{\hbar}\{t\delta E_1 - |f_{01}|^2 F_-(t)\}\right]$$

$$+ i\pi\sqrt{\frac{2}{\hbar}}f_{10}\sum_\alpha \bar{\gamma}_\alpha|\alpha\rangle e^{-i(\omega_\alpha-\Delta)t/2}\mathcal{D}_1(\omega_\alpha + \Delta; t) - \cdots, \qquad (6.3.11)$$

where $\cdots$ stands for terms involving $|\alpha\beta\rangle$. This result shows that the directions of $|\chi_0(t)\rangle$ and $|\chi_1(t)\rangle$, both of which are parallel to $|\text{vac}\rangle$ initially, deviate from each other as time goes on. This circumstance, which is analogous to that depicted in Fig. 4.1, is to be called the dephasing effect of the first kind due to the environment. This effect is succinctly revealed by the factor $e^{-\Gamma_1 t/2}$ accompanying $|\text{vac}\rangle$

in $|\chi_0(t)\rangle$; the $|vac\rangle$-component suffers a corresponding decrease, and is replaced by $|\alpha\rangle$-components in the second line. On the other hand, calculations leading to Eqs. (6.2.39) and (6.2.40) show that the process

$$\begin{cases} \text{macrosystem} & : \quad |0\rangle \to |1\rangle \to |0\rangle \\ \text{environment} & : \quad |vac\rangle \to |\alpha\rangle \to |vac\rangle \end{cases} \tag{6.3.12}$$

gives rise to the phase $|f_{10}|^2 F_+(t)/h$ in the coefficient of $|vac\rangle$ in $|\chi_0(t)\rangle$, and the process

$$\begin{cases} \text{macrosystem} & : \quad |1\rangle \to |0\rangle \\ \text{environment} & : \quad |vac\rangle \to |\alpha\rangle \end{cases} \tag{6.3.13}$$

gives rise to the phase $|f_{01}|^2 F_-(t)/h$ in the coefficient of $|vac\rangle$ in $|\chi_1(t)\rangle$. This is to be called the dephasing effect of the second kind.

The dephasing factor $\mathcal{C}(t)$ may be evaluated explicitly as

$$\mathcal{C}(t) = \mathcal{C}_{vac} + \mathcal{C}_{exc}(t), \tag{6.3.14}$$

where $\mathcal{C}_{vac}$ comes from the $|vac\rangle$-component which represents the environment remaining in the initial state, and $\mathcal{C}_{exc}$ comes from the excited-state components:

$$\mathcal{C}_{vac}(t) \sim e^{-\Gamma_1 t/2} \exp\left[-it\Delta' - \frac{i}{h}(|f_{10}|^2 F_+(t) - |f_{01}|^2 F_-(t)) \right], \tag{6.3.15}$$

$$\mathcal{C}_{exc}(t) \sim \frac{4\pi}{h}(f_{10})^2 e^{it\Delta} \int_0^\infty d\omega J(\omega) \mathcal{D}_1(\omega - \Delta; t) \mathcal{D}_1(\omega + \Delta; t), \tag{6.3.16}$$

$$\Delta' := (\delta E_1 - \delta E_0)/h. \tag{6.3.17}$$

Let us confine ourselves further to the temporal domain

$$\Delta^{-1} \ll t \ll \Gamma_1^{-1}, \tag{6.3.18}$$

which is a sub-domain of the principal temporal domain. In this sub-domain, we find

$$h(f_{10})^{-2} e^{-it\Delta} \mathcal{C}_{exc}(t) \sim \frac{J(\Delta)}{\Delta} \sin(t\Delta) - 2J'(0)\frac{1}{\pi(t\Delta)^2}, \tag{6.3.19}$$

$$F_+(t) \sim J'(0)\frac{\sin(t\Delta)}{\pi(t\Delta)^2}, \quad F_-(t) \sim -J'(\Delta) - J'(0)\frac{\sin(t\Delta)}{\pi(t\Delta)^2}, \tag{6.3.20}$$

where

$$J'(\omega) \equiv dJ(\omega)/d\omega. \tag{6.3.21}$$

Therefore, the interference term of $\mathcal{P}_{\pm}(t)$ is expressed in the form of the first line of Eq. (4.2.40):

$$\mathcal{P}_{\pm}(t) \sim \frac{1}{2}\{1 \mp e^{-\Gamma_1 t/2} \cos(t\tilde{\Delta} - \theta_0)\}, \qquad (6.3.22)$$

$$\tilde{\Delta} \equiv \Delta + \Delta', \quad \theta_0 \equiv |f_{01}|^2\{J(\Delta)/\Delta - J'(\Delta)\}/h. \qquad (6.3.23)$$

As mentioned above, $e^{-\Gamma_1 t/2}$ and θ_0 represent the dephasing effect of the first kind and the second kind, respectively.

We have thus arrived at the following conclusion: if the weak-interaction situation (6.2.36) is set up, the quantum resonant oscillation of frequency $\tilde{\Delta}$ with phase shifted by θ_0 occurs in the temporal domain (6.3.18), and the effect of energy dissipation (cf. (4.2.12)) on this oscillation is almost negligible.[7] This is a result that cannot be obtained from the mere knowledge of δE_n and Γ_n computed by use of the stationary perturbation theory.

Although the effect of the environment has been seen to appear in Δ', $F_{\pm}(t)$ and Γ_1 through the constant η, it is not warranted to assert that the environment influences MQC via friction. It is true that η plays the role of a friction coefficient in Eq. (5.6.8) in the classical situation, but, as mentioned above, the effect of the environment on $C(t)$ differs from that of dissipation of the macrosystem energy. Furthermore, the relaxation time for the macrosystem energy is Γ_1^{-1}, which is different from the corresponding relaxation time in the classical situation (see Eq. (5.6.12)). The constant η, being the measure of strength of the interaction between the macrosystem and the environment, plays various roles depending on the situation. It is because only a single type of interaction (of the form of $x_\alpha f(q)$) is involved in the model treated above that the environmental effect is encapsulated in the single constant η; this will not be the case if a more complicated model[8] is considered.

In dealing with ordinary macroscopic phenomena, the temporal domain of usual interest is the long-time domain $\eta t \gtrsim 1$ where the effect of friction becomes conspicuous in Eq. (5.6.8). It is then no longer possible to treat the interaction with a simple perturbation theory. A formal perturbation expansion may be used only if the summation is extended to infinite orders. Various highbrow techniques to do so have been devised. Of course, such a calculation necessarily involves an approximation, the validity of which is often hard to assess precisely. Thus, it is a considerably difficult matter to deal with a situation where quantum-mechanical features are almost lost and classical features are expected to take their place. By contrast, in this book we are mainly interested in the situation in which quantum-mechanical features are

---

[7] Of course, if the strong-interaction situation ($\eta \gg h$) were set up, Γ_1 would be much larger than Δ, and the quantum resonant oscillation would damp away in less than a period.

[8] For example, a model involving an interaction of the form of $p_\alpha p$ as well.

retained well, that is, in the <u>principal temporal domain</u> (6.2.38). It is quite fortunate that a simple perturbation theory is adequate to deal with this situation.[9]

6.4 Estimation of quantum-resonance frequency

The result of the preceding section shows that the environmental effects shift the frequency of the quantum resonant oscillation from Δ by the amount of Δ'. It is in principle possible to calculate the value of Δ' by the stationary perturbation theory of Section 6.1. But this is quite a cumbersome task in practice. For simplicity, let us replace the macrosystem with a two-state system $\{|0\rangle, |1\rangle\}$ and sketch the calculation of Δ' for the latter. In this case, Eq. (6.1.20) gives

$$\Delta' \simeq -\frac{1}{\pi h}|f_{01}|^2 \Delta \, \mathcal{P} \int_0^\infty \frac{d\omega}{\omega} J(\omega) \left(\frac{1}{\omega - \Delta} + \frac{1}{\omega + \Delta} \right). \qquad (6.4.1)$$

Assuming that the environment is of the ohmic variety, one may compute the above integral to find

$$\Delta' \simeq -\alpha \Delta \left\{ \log\left(\frac{p\omega_c}{\Delta}\right) + \mathcal{O}\left(\frac{\Delta}{\omega_c} \log \frac{\omega_c}{\Delta}\right) \right\}, \qquad (6.4.2)$$

$$\alpha := (2\eta/\pi h)|f_{01}|^2, \qquad (6.4.3)$$

where p is a positive number of $\mathcal{O}(1)$. In view of the fact that

$$\frac{\Delta}{\omega_c} = \mathcal{O}\left(\frac{\omega_0}{\omega_c} h^{-1/2} e^{-1/h}\right), \qquad (6.4.4)$$

which follows from Eq. (2.4.1), the second term on the right-hand side of Eq. (6.4.2) may be discarded. On the other hand, the order of magnitude of the first term is estimated as

$$\log(\omega_c/\Delta) = \mathcal{O}(h^{-1}). \qquad (6.4.5)$$

Hence, for the condition $|\Delta'/\Delta| \ll 1$ to be satisfied to ensure the validity of the perturbation theory, η is required to obey a condition much stronger than (6.2.36), namely the condition $\eta \ll h^2$. If this condition fails to hold (e.g. in the case of $h^2 < \eta \ll h$), higher order terms of the perturbation series have to be taken into account. One may treat them in the spirit of the renormalized perturbation theory[10]

---

[9] This might not be fortunate for those readers who are fond of transcendental techniques.

[10] The ordinary (i.e. Rayleigh–Schrödinger) perturbation theory calculates perturbative corrections in terms of unperturbed quantities (e.g. Δ in the present example). By contrast, the renormalized (i.e. Brillouin–Wigner) perturbation theory expresses perturbative corrections in terms of quantities including the corrections (e.g. $\tilde{\Delta}$ in the present example) as in the second term on the right-hand side of Eq. (6.4.6), which is nothing but the first term of Eq. (6.4.2) with Δ replaced by $\tilde{\Delta}$.

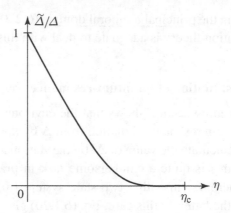

Fig. 6.2. η-dependence of quantum resonance frequency.

to argue that

$$\tilde{\Delta} \equiv \Delta + \Delta' \simeq \Delta - \alpha\tilde{\Delta}\log\left(\frac{p\omega_c}{\tilde{\Delta}}\right) + \cdots. \tag{6.4.6}$$

It then follows that

$$\Delta \simeq \left\{1 + \alpha\log\left(\frac{p\omega_c}{\tilde{\Delta}}\right) + \cdots\right\}\tilde{\Delta}$$

$$\simeq \tilde{\Delta}\exp\left\{\alpha\log\left(\frac{p\omega_c}{\tilde{\Delta}}\right)\right\} = \left(\frac{p\omega_c}{\tilde{\Delta}}\right)^{\alpha}\tilde{\Delta}. \tag{6.4.7}$$

This equation may be solved for $\tilde{\Delta}$ as

$$\tilde{\Delta} \simeq \left(\frac{\Delta}{p\omega_c}\right)^{\alpha/(1-\alpha)}\Delta. \tag{6.4.8}$$

If one believes this result, one arrives at the following conclusion:

There exists a critical value

$$\eta_c := \frac{\pi}{2|f_{01}|^2}h \tag{6.4.9}$$

for η such that $\tilde{\Delta}$ tends to 0 rapidly as η tends to η_c (Fig. 6.2).
This feature, which is characteristic to the ohmic environment, is brought about by the abundant presence of low-frequency oscillators in the environment. Although this is quite an interesting result,[11] we refrain from elaborating on the detailed derivation[12] of Eq. (6.4.8) since the situation of $\eta \sim \eta_c$, being the strong-interaction situation violating the condition (6.2.36), is outside the scope of this book.

---

[11] The critical value η_c is often regarded as marking the boundary between the quantum world and the classical world.

[12] The sketchy argument given in this section is so sloppy that it should not be taken literally. It is meant merely to convey a rough feel for the more careful argument. See, e.g. Refs. [21, 22]. Incidentally, the two-state approximation is too idealized to be applicable to actual systems where the states with $n \geq 2$ can play significant roles.

Exercises

Exercise 6.1. Compute f_{01} for the case of $f(q) = q$ and $V(q) = V^{\text{SSDW}}(q)$.

Exercise 6.2.

(1) One may sometimes encounter the following question:

> When does the environment influence the quantum resonant oscillation? Does the environment do so when the macrosystem is either in the left or the right well or when it is "in the process of tunneling"?

This sort of question is based on the classical picture of tunneling that a particle passes through a barrier. Hence, it is not quite clear how to formulate the question in quantum-mechanical language. A way to answer the question might be the following. Consider the case of

$$f(q) = q\,\theta(|q| - a) \quad : 0 < a \ll 1, \tag{6.4.10}$$

namely the case where interaction is absent in the region $|q| < a$. If the wavepacket representing the macrosystem is localized in the left (or right) well, $f(q)$ is effectively equal to q and Eq. (5.6.8) in Chapter 5 holds. Thus, the picture of the frictional motion in the classical regime is not affected by the constant a. Study how the quantum resonant oscillation depends on a.

(2) The following $f(q)$ also reduces effectively to q in the left (or right) well:

$$f(q) = \frac{q^2/a}{1 + |q|/a} \quad : 0 < a \ll 1. \tag{6.4.11}$$

This is an even function in contrast to the preceding case. Repeat the discussion of this chapter with this $f(q)$.

Exercise 6.3. How are the results of this chapter modified if the following initial state is chosen instead of (6.2.4)∧(6.3.1) ?

$$|\Psi(0)\rangle\rangle = |\Psi_-\rangle\rangle, \tag{6.4.12}$$

where

$$|\Psi_\pm\rangle\rangle \equiv \int dq \,\langle q|\pm\rangle \,|q\rangle|\{\xi(q)\}\rangle, \tag{6.4.13}$$

$$\{\xi(q)\} \equiv \{\xi_1(q),\, \xi_2(q),\, \ldots,\, \xi_N(q)\}, \quad \xi_\alpha(q) \equiv (\omega_\alpha/2h)^{1/2} f_\alpha(q), \tag{6.4.14}$$

with $|\{\xi\}\rangle$ denoting the coherent state for the environmental oscillators:

$$b_\alpha|\{\xi\}\rangle = \xi_\alpha|\{\xi\}\rangle. \tag{6.4.15}$$

Some insight into the state (6.4.13) may be gained by consideration of the state

$$|\Psi'_\pm\rangle\rangle = \int dq \,|q\rangle \,\langle q|\pm\rangle \,|\{\bar{\xi}\}\rangle = |\pm\rangle|\{\bar{\xi}\}\rangle, \tag{6.4.16}$$

which is the state (6.4.13) with $\xi(q)$ replaced by a q-independent constant $\bar{\xi}$: in the state (6.4.12), the macrosystem is localized in the left well, and the origin of the environmental oscillator α is shifted, depending on the position q of the macrosystem, by the amount of $f_\alpha(q)$ so as to minimize the potential for a given q (cf. Eq. (5.1.13)).

Exercise 6.4. Assume that the environment follows the macrosystem so faithfully that the state of the entire system is necessarily given by $|\Psi_+\rangle\rangle$ (or $|\Psi_-\rangle\rangle$) if the macrosystem is in the state $|+\rangle$ (or $|-\rangle$). (Such an environment is called the *adiabatic environment*.) Furthermore, approximate the Hilbert space for the entire system with the two-dimensional space spanned by $\{|\Psi_\pm\rangle\rangle\}$, then

$$\hat{H} \simeq \check{E}_+|\Psi_+\rangle\rangle\langle\langle\Psi_+| + \check{E}_-|\Psi_-\rangle\rangle\langle\langle\Psi_-| - \frac{1}{2}h\check{\Delta}\left(|\Psi_+\rangle\rangle\langle\langle\Psi_-| + |\Psi_-\rangle\rangle\langle\langle\Psi_+|\right). \quad (6.4.17)$$

Evaluate $\check{E}_\pm$ and $\check{\Delta}$. Note that

$$\frac{\check{\Delta}}{\Delta} \simeq \frac{|\langle\langle\Psi_-|\hat{H}|\Psi_+\rangle\rangle|}{|\langle-|\hat{H}_S|+\rangle|} < 1, \quad (6.4.18)$$

since the centers of wavepackets for the environmental oscillators are different between $|\Psi_+\rangle\rangle$ and $|\Psi_-\rangle\rangle$. The quantity on the left-hand side corresponds to the *Franck–Condon factor* in the formula for molecular spectral intensity.

Exercise 6.5. Suppose that $J(\omega)$ is given by Eq. (5.3.7) with $s = 1$. How is the result of Section 6.4 modified if $\nu_c \gg \Delta$? How about the case of the super-ohmic environment such that $s = 3$ and $\nu_c = 0$?

Exercise 6.6. Repeat the discussion of this chapter with a more general interaction (cf. p. 146, the second-to-last paragraph of Section 6.3).

Exercise 6.7. Since calculations leading from Eq. (6.2.31) to Eq. (6.3.23) are rather involved, we have discarded various terms, which do not essentially affect the final result, such as a correction term of $\mathcal{O}(1/t)$ to Eq. (6.2.31), similar correction terms to Eqs. (6.2.35), (6.3.4) and (6.3.6), and a term proportional to $\cos(t\Delta)$ in Eq. (6.3.19). Compute and recover these terms, and show that they modify Eq. (6.3.22) merely by the factor of $1 + \mathcal{O}(\eta/h)$ that multiplies each of its terms. Incidentally, to be precise, it is in the temporal domain (6.3.18) that those transient terms involving $\mathcal{D}_2(\omega + \Delta; t)$ are negligible.

7

Quantum decay in the harmonic environment

As in the preceding chapter, we consider the system described by the Hamiltonian (5.1.8)–(5.1.11). The situation envisaged in this chapter is that of MQT. Accordingly, the macrosystem potential $V(q)$ is supposed to be the bumpy slope depicted in Fig. 2.1(b).

7.1 Conjectures on decay rate

We have seen in Chapter 2 that the decay rate of the metastable state is given by the formula (2.4.5) if the macrosystem is isolated (i.e. not interacting with the environment). In that formula, the crucial factor is the exponent B/h, where B is independent of h. The coefficient A, on the other hand, is a gentle function of h. Although it is not obvious how the above formula is modified by the interaction of the macrosystem with the environment, let us make the following conjecture:

Conjecture 0: Express the decay rate Γ of the metastable state in the form

$$\Gamma = A \exp(-B/h), \tag{7.1.1}$$

then A is a gentle function of h and B is independent of h. Furthermore, if the effect of the environment can be encapsulated in a constant η, both A and B are functions of η.

It may be guessed that both A and B are <u>natural generalizations</u> of the corresponding quantities for the case of the single degree of freedom consisting of the macrosystem alone. However, the expression (2.4.7) for B involves the q-integration, which is specific to a system of a single degree of freedom. One is likely to be at a loss how to adapt it for the system of many degrees of freedom comprising both the macrosystem and the environment. Fortunately, the constant B can be expressed in the form of Eq. (2.6.11) as well and viewed as the bounce action. The latter

151

concept, being definable regardless of the number of degrees of freedom involved, may be utilized for the desired generalization. Thus, we are led to make an additional conjecture:

Conjecture 1 : The constant B is equal to the bounce action.

A considerable amount of preparation is needed to verify that these conjectures are correct (cf. Ref. [24]). Relegating this task to another book (to be published, perhaps, in future), we devote this chapter to study the consequence of these conjectures. In order to state Conjecture 1 precisely, it is necessary to begin by introducing the Euclidean action functional and the bounce defined as its non-trivial stationary point for the system of many degrees of freedom.

7.2 Euclidean action functional and bounce

In Chapter 2, the Euclidean action functional $S[q]$ corresponding to the Hamiltonian (2.1.6) was defined by Eq. (2.6.12). Similarly, the Euclidean action functional $S[q, x]$ corresponding to the Hamiltonian (5.1.8) is defined as

$$S[q, x] := S[q] + S_1[q, x], \tag{7.2.1}$$

$$S_1[q, x] := \int_{-\infty}^{\infty} d\tau \frac{1}{2} \sum_{\alpha} \{ (\dot{x}_\alpha(\tau))^2 + \omega_\alpha^2 (x_\alpha(\tau) - f_\alpha(q(\tau)))^2 \}, \tag{7.2.2}$$

with $S[q]$ being the one defined by Eq. (2.6.12). In the above, the position variables of the environmental oscillators are denoted collectively by x:

$$x := \{ x_\alpha \mid \alpha = 1, 2, \ldots, N \}. \tag{7.2.3}$$

The right-hand side of Eq.(7.2.1) may be interpreted as the action integral (with respect to the Euclidean time) for a fictitious particle, whose position variable is (q, x), moving in the $(N + 1)$-dimensional space under the potential

$$-V(q) - \frac{1}{2} \sum_{\alpha} \omega_\alpha^2 (x_\alpha - f_\alpha(q))^2 . \tag{7.2.4}$$

The domain $\mathcal{F}_\infty$ of definition of the above functional is to be taken as

$$\mathcal{F}_\infty := \{ (q, x) \mid q(\pm\infty) = 0, \ x(\pm\infty) = 0 \}. \tag{7.2.5}$$

This is the many-degrees-of-freedom counterpart of the function space (2.6.13) for the case of a single degree of freedom.[1]

---

[1] No confusion should arise if the two spaces are denoted by one and the same symbol.

Let the non-trivial stationary point of the functional $S[q, x]$ be denoted by (Q, χ) and be called the *bounce*:

$$\frac{\delta S[q, x]}{\delta q(\tau)}\bigg|_{(q,x)=(Q,\chi)} = 0, \qquad (7.2.6)$$

$$\frac{\delta S[q, x]}{\delta x_\alpha(\tau)}\bigg|_{(q,x)=(Q,\chi)} = 0, \qquad (7.2.7)$$

$$Q(0) > 0, \quad \dot{Q}(0) = 0. \qquad (7.2.8)$$

The last condition $\dot{Q}(0) = 0$, which is not needed logically, has been imposed so as to define the bounce uniquely; without this condition, the bounce is not unique because of the temporal translation invariance of the action functional as in the case of the system of a single degree of freedom. In accordance with the notation (7.2.3), the symbol χ stands for a set of N quantities, of course:

$$\chi := \{\chi_\alpha \mid \alpha = 1, 2, \ldots, N\}. \qquad (7.2.9)$$

By use of the quantities defined above, Conjecture 1 is stated formally as

$$B = S[Q, \chi]. \qquad (7.2.10)$$

7.2.1 Effective Euclidean action functional and a stationary point

The formal similarity of Eq. (7.2.10) to Eq. (2.6.11) for the system of a single degree of freedom may be made clear as follows. Given a function q belonging to $\mathcal{F}_\infty$, define $x^{[q]}$ with the following equation:

$$\frac{\delta S[q, x]}{\delta x_\alpha(\tau)}\bigg|_{x=x^{[q]}} = 0. \qquad (7.2.11)$$

Of course, $x^{[q]}$ is a functional of q. Furthermore, define

$$S_{\text{eff}}[q] := S[q, x^{[q]}], \qquad (7.2.12)$$

which, being a functional of q alone, is called the *effective Euclidean action functional*. In terms of this functional, the conjecture (7.2.10) is re-expressed as

$$B = S_{\text{eff}}[Q]. \qquad (7.2.13)$$

The Q in this expression is the same as the one in Eq. (7.2.10). It may be viewed as being defined by the following equation, too:

$$\frac{\delta S_{\text{eff}}[q]}{\delta q(\tau)}\bigg|_{q=Q} = 0, \quad Q(0) > 0, \quad \dot{Q}(0) = 0. \qquad (7.2.14)$$

In other words, Q is the non-trivial stationary point of $S_{\text{eff}}[q]$.

Equations (7.2.13) and (7.2.14) may be proven as follows. First, comparison of Eq. (7.2.7) with Eq. (7.2.11) shows

$$\chi = x^{[Q]}. \tag{7.2.15}$$

Next, differentiation of the effective Euclidean action functional gives (see Appendix C)

$$\frac{\delta S_{\text{eff}}[q]}{\delta q(\tau)} = \frac{\delta S[q, x]}{\delta q(\tau)}\bigg|_{x=x^{[q]}} + \int_{-\infty}^{\infty} d\tau' \sum_{\alpha} \frac{\delta S[q, x]}{\delta x_{\alpha}(\tau')}\bigg|_{x=x^{[q]}} \frac{\delta x_{\alpha}^{[q]}(\tau')}{\delta q(\tau)}, \tag{7.2.16}$$

which combined with Eq. (7.2.15) leads to

$$\frac{\delta S_{\text{eff}}[q]}{\delta q(\tau)}\bigg|_{q=Q} = \frac{\delta S[q, x]}{\delta q(\tau)}\bigg|_{(q,x)=(Q,\chi)}$$

$$+ \int_{-\infty}^{\infty} d\tau' \sum_{\alpha} \frac{\delta S[q, x]}{\delta x_{\alpha}(\tau')}\bigg|_{(q,x)=(Q,\chi)} \frac{\delta x_{\alpha}^{[q]}(\tau')}{\delta q(\tau)}\bigg|_{q=Q}. \tag{7.2.17}$$

Together with Eqs. (7.2.6) and (7.2.7), the above equation gives Eq. (7.2.14). To sum up, Q is determined as the non-trivial stationary point of $S_{\text{eff}}[q]$, and this Q in turn determines χ through Eq. (7.2.15). This two-step procedure determines the stationary point (Q, χ) of $S[q, x]$.

It is to be emphasized that the above analogy with the case of the system of a single degree of freedom is purely formal; the content is quite different. In particular, the effective Euclidean action functional, which is a result of partial elimination of degrees of freedom, is non-local in time (see the next subsection for details).

7.2.2 Effective Euclidean action functional in the harmonic-environment model

Let us return to the harmonic-environment model, and begin by determining $x^{[q]}$ on the basis of the general consideration in the preceding subsection. For the harmonic-environment model, Eq. (7.2.11) reduces to

$$-\ddot{x}_{\alpha}(\tau) + \omega_{\alpha}^2 x_{\alpha}(\tau) = 2\omega_{\alpha} \tilde{f}_{\alpha}(\tau), \tag{7.2.18}$$

$$\tilde{f}_{\alpha}(\tau) \equiv \frac{1}{2}\omega_{\alpha} f_{\alpha}(q(\tau)), \tag{7.2.19}$$

where $x_{\alpha}^{[q]}$ has been abbreviated as x_{α} for brevity. Thus, the problem to be solved is the following:

Given a function $\tilde{f}_\alpha(\tau)$ which tends to 0 as $\tau \to \pm\infty$, find the solution $x_\alpha(\tau)$ to Eq.(7.2.18) under the boundary condition

$$\lim_{\tau \to \pm\infty} x_\alpha(\tau) = 0. \tag{7.2.20}$$

The desired solution is found as

$$x_\alpha(\tau) = \int_{-\infty}^{\infty} d\tau' e^{-\omega_\alpha|\tau-\tau'|} \tilde{f}_\alpha(\tau'), \tag{7.2.21}$$

which can be rearranged in the following form as well:

$$x_\alpha(\tau) = \int_{0}^{\infty} d\tau' e^{-\omega_\alpha \tau'} \{\tilde{f}_\alpha(\tau - \tau') + \tilde{f}_\alpha(\tau + \tau')\}. \tag{7.2.22}$$

Substitution of the solution (7.2.21) for $x_\alpha(\tau)$ in Eq. (7.2.1) gives the effective Euclidean action functional. Let the result be expressed in the form

$$S_{\text{eff}}[q] = S[q] + S_{1\text{eff}}[q]. \tag{7.2.23}$$

Note that $S_{1\text{eff}}[q]$ is positive. Integrating once by parts, one finds

$$S_{1\text{eff}}[q] = \frac{1}{2} \sum_{\alpha} \int_{-\infty}^{\infty} d\tau \{-x_\alpha(\tau)\ddot{x}_\alpha(\tau) + \omega_\alpha^2 (x_\alpha(\tau) - f_\alpha(q(\tau)))^2\}, \tag{7.2.24}$$

where $x_\alpha(\tau)$ is given by Eq. (7.2.21), of course. The above expression may be simplified by use of Eq. (7.2.18) as

$$S_{1\text{eff}}[q] = \sum_{\alpha} \sigma_\alpha[q], \tag{7.2.25}$$

$$\sigma_\alpha[q] := \frac{1}{2}\omega_\alpha^2 \int_{-\infty}^{\infty} d\tau f_\alpha(q(\tau))\{f_\alpha(q(\tau)) - x_\alpha(\tau)\}. \tag{7.2.26}$$

This may be rewritten further by use of Eq.(7.2.21) as

$$\sigma_\alpha[q] = \sigma_{\alpha0}[q] + \sigma_{\alpha1}[q],$$

$$\sigma_{\alpha0}[q] := 2 \int_{-\infty}^{\infty} d\tau \{\tilde{f}_\alpha(\tau)\}^2,$$

$$\sigma_{\alpha1}[q] := -\omega_\alpha \int_{-\infty}^{\infty} d\tau \int_{-\infty}^{\infty} d\tau' \tilde{f}_\alpha(\tau) e^{-\omega_\alpha|\tau-\tau'|} \tilde{f}_\alpha(\tau'). \tag{7.2.27}$$

Now one may profitably invoke the <u>Caldeira–Leggett trick</u>; by use of the identities

$$fg = \frac{1}{2}(f^2 + g^2) - \frac{1}{2}(f - g)^2, \tag{7.2.28}$$

$$\omega_\alpha \int_{-\infty}^{\infty} d\tau \int_{-\infty}^{\infty} d\tau' \{\tilde{f}_\alpha(\tau)\}^2 e^{-\omega_\alpha|\tau-\tau'|} = \int_{-\infty}^{\infty} d\tau \{\tilde{f}_\alpha(\tau)\}^2 \int_{-\infty}^{\infty} dt\, e^{-|t|} = \sigma_{\alpha0}[q],$$

$$\tag{7.2.29}$$

one finds

$$\sigma_\alpha[q] = \frac{\omega_\alpha}{2} \int_{-\infty}^{\infty} d\tau \int_{-\infty}^{\infty} d\tau' e^{-\omega_\alpha|\tau-\tau'|} \{\tilde{f}_\alpha(\tau) - \tilde{f}_\alpha(\tau')\}^2$$

$$= \left(\frac{\omega_\alpha}{2}\right)^3 \int_{-\infty}^{\infty} d\tau \int_{-\infty}^{\infty} d\tau' e^{-\omega_\alpha|\tau-\tau'|} \{f_\alpha(q(\tau)) - f_\alpha(q(\tau'))\}^2. \quad (7.2.30)$$

The results (7.2.23), (7.2.25) and (7.2.30) give an explicit expression for the effective Euclidean action functional. It is to be noted that the right-hand side of the above equation is a double integral with respect to τ. In general, an action functional expressed by a multiple integral with respect to time is said to be *non-local in time*, whereas the usual action functional given by the Lagrangian integrated once over time is said to be local in time. The temporal non-locality is one of the important characteristics of the effective action functional.

Hereafter, we confine attention to the separable model (5.1.14) as in Chapter 5. Introducing a positive constant η, which is arbitrary for the moment and is to be specified shortly, we write

$$\mathcal{J}(\tau) := \frac{1}{2\eta} \sum_\alpha \gamma_\alpha^2 \omega_\alpha^3 e^{-\omega_\alpha|\tau|}. \quad (7.2.31)$$

In terms of this function, the second term of the effective action is expressed as

$$S_{1\text{eff}}[q] = \frac{\eta}{4} \int_{-\infty}^{\infty} d\tau \int_{-\infty}^{\infty} d\tau' \mathcal{J}(\tau - \tau')\{f(q(\tau)) - f(q(\tau'))\}^2. \quad (7.2.32)$$

In this form, $S_{1\text{eff}}[q]$ is called the *Caldeira–Leggett effective action*. Properties of the integral kernel $\mathcal{J}(\tau)$ are as follows. In terms of the environmental frequency distribution function $J(\omega)$ defined by Eq. (5.3.3), the kernel is expressed as

$$\mathcal{J}(\tau) = \frac{1}{\pi\eta} \int_0^{\infty} d\omega J(\omega) e^{-\omega|\tau|}. \quad (7.2.33)$$

Accordingly,

$$\int_{-\infty}^{\infty} d\tau \mathcal{J}(\tau) = \frac{2}{\pi\eta} \int_0^{\infty} \frac{d\omega}{\omega} J(\omega) = \frac{1}{\eta} \mathcal{T}(0) < \infty, \quad (7.2.34)$$

where $\mathcal{T}(t)$ is the retarded resistance function defined by Eq. (5.4.1). For the ohmic environment in particular, the expression (5.3.7) with $s = 1$ and $\nu_c = 0$ may be used for $J(\omega)$ with the result

$$\mathcal{J}^{\text{ohmic}}(\tau) = \frac{1}{\pi(|\tau| + \omega_c^{-1})^2}, \quad (7.2.35)$$

where the arbitrary constant η introduced in Eq. (7.2.31) has now been identified with the η in $J(\omega)$. It is characteristic of the ohmic environment that $\mathcal{J}(\tau)$ has a long-time tail proportional to τ^{-2}, which implies that the temporal non-locality is quite pronounced.

7.3 Bounce and decay rate in a bilinear model

The content of the preceding section is valid regardless of the choice of the macrosystem potential V. Let us consider the case that V is the bumpy slope. We also confine attention to the bilinear model (5.1.15). Then, Eq. (7.2.14) to determine the bounce may be written explicitly as

$$\ddot{Q}(\tau) = V'(Q(\tau)) + \eta \int_{-\infty}^{\infty} d\tau' J(\tau - \tau')\{Q(\tau) - Q(\tau')\}, \qquad (7.3.1)$$

$$Q(0) > 0, \quad \dot{Q}(0) = 0. \qquad (7.3.2)$$

Since J is a function of the time difference $\tau - \tau'$ alone, Eq. (7.3.1) is time-translation invariant as expected.

It is possible to solve this equation numerically for an arbitrary η. As mentioned in the preceding chapter, however, the main interest of this book is in the case of small η. Accordingly, we assume that the coupling constants $\{\gamma_\alpha\}$ for the interaction between the environment and the macrosystem are small, that is,

$$\eta \ll 1, \quad \tilde{\gamma}_\alpha := \gamma_\alpha/\eta^{1/2} = \mathcal{O}(1) \equiv \mathcal{O}(\eta^0), \qquad (7.3.3)$$

$$J(\tau) = \mathcal{O}(1). \qquad (7.3.4)$$

Under this assumption, Q may be obtained perturbatively. We expand it in powers of η as

$$Q = Q^{(0)} + \eta Q^{(1)} + \eta^2 Q^{(2)} + \cdots, \qquad (7.3.5)$$

and substitute this for Q in Eq. (7.3.1). Comparing coefficients of each power of η, we find up to the second order that

$$\ddot{Q}^{(0)}(\tau) = V'(Q^{(0)}(\tau)),$$

$$\mathcal{D}_* Q^{(1)}(\tau) = -\int_{-\infty}^{\infty} d\tau' J(\tau - \tau')\{Q^{(0)}(\tau) - Q^{(0)}(\tau')\}, \qquad (7.3.6)$$

$$\mathcal{D}_* Q^{(2)}(\tau) = -\frac{1}{2} V'''(Q^{(0)}(\tau))\{Q^{(1)}(\tau)\}^2$$
$$\qquad - \int_{-\infty}^{\infty} d\tau' J(\tau - \tau')\{Q^{(1)}(\tau) - Q^{(1)}(\tau')\},$$

where

$$\mathcal{D}_* := -\frac{d^2}{d\tau^2} + v_*(\tau), \quad v_*(\tau) := V''(Q^{(0)}(\tau)). \qquad (7.3.7)$$

Clearly, $\mathcal{Q}^{(0)}$ is nothing but the bounce in the system of a single degree of freedom introduced in Chapter 1. Let $\mathcal{G}_*(\tau, \tau')$ be the Green function associated with the differential operator $\mathcal{D}_*$ such that it obeys the Dirichlet and the Neumann boundary conditions at $\tau = -\infty$ and $\tau = 0$, respectively:

$$\mathcal{D}_*\mathcal{G}_*(\tau, \tau') = \delta(\tau - \tau'), \tag{7.3.8}$$

$$\mathcal{G}_*(\tau, \tau') = \mathcal{G}_*(\tau', \tau), \tag{7.3.9}$$

$$\mathcal{G}_*(-\infty, \tau') = \dot{\mathcal{G}}_*(0, \tau') = 0 \quad : \forall \tau' \in (-\infty, 0). \tag{7.3.10}$$

In terms of this Green function, Eq. (7.3.6) may be solved as

$$\mathcal{Q}^{(1)}(\tau) = -\int_{-\infty}^{0} d\tau' \int_{-\infty}^{\infty} d\tau'' \mathcal{G}_*(\tau, \tau') \mathcal{J}(\tau' - \tau'')\{\mathcal{Q}^{(0)}(\tau') - \mathcal{Q}^{(0)}(\tau'')\} \quad : \tau < 0. \tag{7.3.11}$$

Since the right-hand side of Eq. (7.3.6) is even in τ, the above result can be extended to an even function. Thus, we have obtained the first-order correction to the bounce. The second- and the higher-order corrections may be found similarly.

Now that the bounce has been seen to be obtainable with perturbation theory, the bounce action may be expanded in powers of η as well:

$$B = B^{(0)} + \eta B^{(1)} + \eta^2 B^{(2)} + \cdots. \tag{7.3.12}$$

The coefficients should be extracted from

$$
\begin{aligned}
B &\simeq S_{\text{eff}}\big[\mathcal{Q}^{(0)} + \eta\mathcal{Q}^{(1)} + \eta^2\mathcal{Q}^{(2)}\big] \\
&\simeq S\big[\mathcal{Q}^{(0)} + \eta\mathcal{Q}^{(1)} + \eta^2\mathcal{Q}^{(2)}\big] + S_{\text{1eff}}\big[\mathcal{Q}^{(0)} + \eta\mathcal{Q}^{(1)}\big],
\end{aligned} \tag{7.3.13}
$$

where the symbol $\simeq$ denotes approximate equality valid up to the second order. The first term can be expanded as

$$
\begin{aligned}
S\big[\mathcal{Q}^{(0)} &+ \eta\mathcal{Q}^{(1)} + \eta^2\mathcal{Q}^{(2)}\big] \\
&\simeq S[\mathcal{Q}^{(0)}] + \frac{\eta^2}{2}\int_{-\infty}^{\infty} d\tau \int_{-\infty}^{\infty} d\tau' \left.\frac{\delta^2 S[q]}{\delta q(\tau)\delta q(\tau')}\right|_{q=\mathcal{Q}^{(0)}} \mathcal{Q}^{(1)}(\tau)\mathcal{Q}^{(1)}(\tau') \\
&= S[\mathcal{Q}^{(0)}] + \frac{\eta^2}{2}\int_{-\infty}^{\infty} d\tau\, \mathcal{Q}^{(1)}(\tau)\mathcal{D}_*\mathcal{Q}^{(1)}(\tau),
\end{aligned} \tag{7.3.14}
$$

where Eqs. (2.6.14) and (C.0.17) have been used. Likewise, the second term can be expanded as

$$
\begin{aligned}
S_{\text{1eff}}\big[\mathcal{Q}^{(0)} &+ \eta\mathcal{Q}^{(1)}\big] \\
&\simeq S_{\text{1eff}}[\mathcal{Q}^{(0)}] + \frac{\eta^2}{2}\int_{-\infty}^{\infty} d\tau \int_{-\infty}^{\infty} d\tau'\{\mathcal{Q}^{(1)}(\tau) - \mathcal{Q}^{(1)}(\tau')\}\mathcal{J}(\tau - \tau')
\end{aligned}
$$

$$\times \left\{ \mathcal{Q}^{(0)}(\tau) - \mathcal{Q}^{(0)}(\tau') \right\}$$

$$= S_{1\mathrm{eff}}[\mathcal{Q}^{(0)}] - \eta^2 \int_{-\infty}^{\infty} \mathrm{d}\tau\, \mathcal{Q}^{(1)}(\tau) \mathcal{D}_* \mathcal{Q}^{(1)}(\tau), \qquad (7.3.15)$$

where Eq.(7.3.6) has been used. Hence,

$$B^{(0)} = S[\mathcal{Q}^{(0)}] \qquad (7.3.16)$$

is the same bounce action as for the system of a single degree of freedom, and

$$B^{(1)} = \frac{1}{4} \int_{-\infty}^{\infty} \mathrm{d}\tau \int_{-\infty}^{\infty} \mathrm{d}\tau'\, \mathcal{J}(\tau - \tau') \left\{ \mathcal{Q}^{(0)}(\tau) - \mathcal{Q}^{(0)}(\tau') \right\}^2 \qquad (7.3.17)$$

is a positive number of $\mathcal{O}(1)$. Finally,

$$B^{(2)} = -\frac{1}{2} \int_{-\infty}^{\infty} \mathrm{d}\tau\, \mathcal{Q}^{(1)}(\tau) \mathcal{D}_* \mathcal{Q}^{(1)}(\tau). \qquad (7.3.18)$$

Thus, the most important effect of the influence of the environment on the decay rate Γ of the metastable state of the macrosystem in the bilinearly-coupled harmonic environment is summarized succinctly by the *Caldeira–Leggett formula*:

$$\Gamma \propto \Gamma^{(0)} \exp\left(-B^{(1)} \eta / h\right), \qquad (7.3.19)$$

where $\Gamma^{(0)}$ is the decay rate (2.4.5) relevant to the macrosystem in isolation. In other words, the decay is suppressed by the interaction $\hat{H}_{S\mathcal{E}}$ (Eq. (5.1.11)). Note that one should be careful to say "the interaction $\hat{H}_{S\mathcal{E}}$" rather than "the interaction $-\sum_\alpha \omega_\alpha^2 f_\alpha(\hat{q}) \hat{x}_\alpha$"; these two expressions need to be distinguished as mentioned at the end of Section 5.6. An additional remark is in order: Although the effect of the environment is essentially encapsulated in the constant η, it may not quite be appropriate to call the effect "a friction effect" or "a dissipation effect" (cf. the remark at p. 146, the second-to-last paragraph of Section 6.3).

7.4 Remark on decay rate at finite temperatures

In order to determine the precise shape of the solid curve (or the dot-dashed one) depicted in Fig. 2.7 of Section 2.3, it is necessary to have a theory of decay at finite temperatures. Unfortunately, however, a well-founded theory is yet to be constructed. The following formula is often quoted for "the decay rate Γ^β at an arbitrary temperature β^{-1}":

$$\Gamma^\beta \sim \omega_0 \exp\{-S_\beta[\mathcal{Q}^\beta, \chi^\beta]/h\}, \qquad (7.4.1)$$

where $S_\beta[q, x]$ is the *Euclidean action functional at temperature* β^{-1}, and $(\mathcal{Q}^\beta, \chi^\beta)$ is its non-trivial stationary point.[2] Here, $S_\beta[q, x]$ is given by the same expression as $S[q, x]$ with the range of τ-integration restricted to the interval $(-\beta h/2, \beta h/2)$. It is a functional defined on the following space $\mathcal{F}_\beta$ of periodic functions:

$$\mathcal{F}_\beta := \{(q, x) \mid q(-\beta h/2) = q(\beta h/2), \; x(-\beta h/2) = x(\beta h/2)\}. \quad (7.4.2)$$

The purported derivation of the formula (7.4.1), however, suffers from various obscurities and difficulties, which would constitute a topic of another book (to be published, given the opportunity).

Exercises

Exercise 7.1. Consider the potential (7.2.4), which is defined in the $(N+1)$-dimensional configuration space, with $V = V^{SBS}$. Draw the contour diagram for its inverted version (i.e. the potential (7.2.4) multiplied by the overall factor -1), and infer how the metastable state would decay.

Exercise 7.2. How is Eq. (7.2.35) modified if $\nu_c \neq 0$ in the environmental frequency distribution function (5.3.7)?

Exercise 7.3. Show that $\mathcal{Q}(\tau)$ is an even function and accordingly so is $x_\alpha(\tau)$ by virtue of Eq. (7.2.22).

Exercise 7.4.

(1) Construct the Green function $\mathcal{G}_*(\tau, \tau')$ explicitly and express $\mathcal{Q}^{(1)}$ in terms of $\mathcal{Q}^{(0)}$ and $\dot{\mathcal{Q}}^{(0)}$.

(2) Supposing that $V = V^{SBS}$ and the environment is ohmic, compute $\mathcal{Q}^{(1)}$ by use of the kernel (7.2.35).

Exercise 7.5. Once $\mathcal{Q}$ has been obtained, χ may be found by use of Eq. (7.2.21) in the form

$$\chi_\alpha(\tau) = \eta^{1/2}\{\chi_\alpha^{(0)}(\tau) + \eta\chi_\alpha^{(1)}(\tau) + \eta^2\chi_\alpha^{(2)}(\tau) + \cdots\}. \quad (7.4.3)$$

For the bilinear model, in particular,

$$\chi_\alpha^{(n)}(\tau) = \frac{1}{2}\tilde{\gamma}_\alpha\omega_\alpha \int_{-\infty}^{\infty} d\tau' e^{-\omega_\alpha|\tau-\tau'|}\mathcal{Q}^{(n)}(\tau). \quad (7.4.4)$$

Supposing that $V = V^{SBS}$, compute $\chi_\alpha^{(0)}(\tau)$. Also, depict the approximate trajectory of the bounce in the contour diagram drawn in Exercise 7.1 above.

Exercise 7.6. Supposing that $V = V^{SBS}$ and the environment is ohmic, compute $B^{(1)}$ and $B^{(2)}$.

---

[2] The stationary point of $S_\beta[q, x]$ is not unique; if $X^\beta \equiv (\mathcal{Q}^\beta, \chi^\beta)$ is a stationary point, X_c^β defined as $X_c^\beta(\tau) := X^\beta(\tau - c)$ is another for an arbitrary constant c. Since $S_\beta[X_c^\beta]$ is independent of c, however, one may evaluate $S_\beta[X^\beta]$ uniquely by arbitrarily choosing one of the infinite stationary points.

8

General versus harmonic environments

We have introduced the harmonic environment in Chapter 5 and studied its influence on MQC and MQT in Chapters 6 and 7, respectively. The harmonic environment has been shown to reproduce the friction associated with the motion of the macrosystem in the classical situation on one hand, and to exert both the dissipation and the dephasing effects on the macrosystem in the quantum-mechanical situation on the other. In this sense, it is a plausible model for an actual environment. To what extent, then, is it a valid model? Is it possible at all to ascertain its validity without detailed knowledge of actual environments which are complicated in general? This chapter tries to specify the conditions which would allow a general environment to be replaced by the harmonic one.

8.1 Modified Born–Oppenheimer basis

Suppose that the macroscopic degree of freedom R (as well as its conjugate momentum P) consists of D components:

$$R \equiv \{R_a \mid a = 1, 2, \ldots, D\}, \quad P \equiv \{P_a \mid a = 1, 2, \ldots, D\}, \quad (8.1.1)$$

$$[\hat{R}_a, \hat{P}_b] = i\hbar \delta_{ab} . \quad (8.1.2)$$

Suppose, on the other hand, that the environment consists of N degrees of freedom for which the canonical variables are denoted by ξ and π:

$$\xi \equiv \{\xi_k \mid k = 1, 2, \ldots, N\}, \quad \pi \equiv \{\pi_k \mid k = 1, 2, \ldots, N\}, \quad (8.1.3)$$

$$[\hat{\xi}_k, \hat{\pi}_{k'}] = i\hbar \delta_{kk'}. \quad (8.1.4)$$

In order to proceed, we need some information on the interaction between the environment and the macrosystem. Let us assume that the total Hamiltonian $\hat{H}_{\text{tot}}$ for the entire system consisting of the macrosystem and the environment has the following property.

Assumption $\mathcal{A}0$: The interaction between the macrosystem and the environment does not depend on the momentum of the macrosystem, namely

$$\hat{H}_{\text{tot}} = \hat{H}_S^{(0)} + \hat{H}^{(1)}, \tag{8.1.5}$$

$$\hat{H}_S^{(0)} \equiv \frac{1}{2M}\hat{P}^2 + U^{(0)}(\hat{R}), \tag{8.1.6}$$

$$\hat{H}^{(1)} \equiv U^{(1)}(\hat{\xi}, \hat{\pi}, \hat{R}), \tag{8.1.7}$$

where M is a positive constant, $U^{(0)}(R)$ and $U^{(1)}(\xi, \pi, R)$ are arbitrary functions, and

$$\hat{P}^2 \equiv \sum_a \hat{P}_a^2. \tag{8.1.8}$$

The validity of this assumption, which seems fairly plausible, has to be examined, case by case, for a given system.

Let us treat $\hat{H}^{(1)}$ essentially in the spirit of the Born–Oppenheimer approximation without, however, resorting to any approximation. It is convenient to employ the position representation. Denote the position representation of an operator $\hat{A}$ with A, in general. Then,

$$H^{(1)} = U^{(1)}\left(\xi, \frac{\hbar}{i}\nabla_\xi, R\right). \tag{8.1.9}$$

For a fixed R, the quantity $H^{(1)}$ may be regarded as a Hamiltonian for the environment alone. Let the associated orthonormal set of eigenfunctions be denoted by

$$\{\chi_{\nu,R}(\xi) \mid \nu \in \mathbf{N}\}, \tag{8.1.10}$$

and enumerate the corresponding energy eigenvalue $\mathcal{E}_{\nu,R}$ in ascending order:

$$\mathcal{E}_{0,R} < \mathcal{E}_{1,R} < \mathcal{E}_{2,R} < \cdots, \tag{8.1.11}$$

$$H^{(1)}\chi_{\nu,R}(\xi) = \mathcal{E}_{\nu,R}\chi_{\nu,R}(\xi). \tag{8.1.12}$$

For each fixed R, the set (8.1.10) constitutes a complete orthonormal set:[1]

$$\int d\xi \, \chi_{\mu,R}^*(\xi)\chi_{\nu,R}(\xi) = \delta_{\mu\nu}, \tag{8.1.13}$$

$$\sum_\nu \chi_{\nu,R}(\xi)\chi_{\nu,R}^*(\xi') = \delta(\xi - \xi'), \tag{8.1.14}$$

where

$$\int d\xi \equiv \prod_k \int_{-\infty}^{\infty} d\xi_k, \quad \delta(\xi - \xi') \equiv \prod_k \delta(\xi_k - \xi_k'). \tag{8.1.15}$$

---

[1] Hereafter, Kronecker's delta is to be understood as Dirac's delta whenever a continuous spectrum is involved.

The potential $U^{(0)}(R)$ augmented by $\mathcal{E}_{0,R}$ is to be called the macrosystem potential[2] $U(R)$:

$$U(R) := U^{(0)}(R) + \mathcal{E}_{0,R}, \tag{8.1.16}$$

$$\hat{H}_\mathcal{S} := \hat{H}_\mathcal{S}^{(0)} + \mathcal{E}_{0,\hat{R}} = \frac{1}{2M}\hat{P}^2 + U(\hat{R}). \tag{8.1.17}$$

Let the complete set of eigenstates of $\hat{H}_\mathcal{S}$ be denoted by

$$\{|n\rangle \mid n \in \mathbf{N}\}, \quad \langle n|n'\rangle = \delta_{nn'}, \tag{8.1.18}$$

and $\psi_n(R)$ be the wavefunction representing $|n\rangle$:

$$\psi_n(R) \equiv \langle R|n\rangle. \tag{8.1.19}$$

The Hilbert space $\mathcal{H}_\mathcal{S}$ for the macrosystem is spanned by the set (8.1.18):

$$\mathcal{H}_\mathcal{S} = Span\{|n\rangle \mid n \in \mathbf{N}\}, \tag{8.1.20}$$

that is

$$\int dR \ \psi_m^*(R)\psi_n(R) = \delta_{mn}, \tag{8.1.21}$$

$$\sum_n \psi_n(R)\psi_n^*(R') = \delta(R - R'), \tag{8.1.22}$$

where

$$\int dR \equiv \prod_a \int_{-\infty}^{\infty} dR_a, \quad \delta(R - R') \equiv \prod_a \delta(R_a - R_a'). \tag{8.1.23}$$

With the preparation made so far, we define a wavefunction $\Psi_{n\nu}(R, \xi)$ for the entire system:

$$\Psi_{n\nu}(R, \xi) := \psi_n(R)\chi_{\nu,R}(\xi). \tag{8.1.24}$$

The set of these wavefunctions is complete and orthonormal:

$$\int\int dRd\xi \ \Psi_{m\mu}^*(R, \xi)\Psi_{n\nu}(R, \xi) = \delta_{mn}\delta_{\mu\nu}, \tag{8.1.25}$$

$$\sum_{n\nu} \Psi_{n\nu}(R, \xi) \ \Psi_{n\nu}^*(R', \xi') = \delta(R - R')\delta(\xi - \xi'). \tag{8.1.26}$$

The set of functions $\{\Psi_{n\nu}(R, \xi) \mid n, \nu \in \mathbf{N}\}$ is to be called the _modified Born–Oppenheimer basis_, which is slightly different from the usual _Born–Oppenheimer basis_ (hence the qualification _modified_) and is more suitable for the present purpose than the latter.

---

[2] Conventionally, $U(R)$ is called the "_adiabatic potential_ (associated with the environment in its ground state $\chi_{0,R}$)".

8.2 Modified Born–Oppenheimer representation for Hamiltonian

Let us compute matrix elements of $\hat{H}_{\text{tot}}$ with respect to the modified Born–Oppenheimer basis. We begin by noting that

$$H_S^{(0)}\Psi_{n\nu}(R,\xi) = \left\{H_S^{(0)}\psi_n(R)\right\}\chi_{\nu,R}(\xi) + \Phi_{n\nu}^{(1)}(R,\xi) + \Phi_{n\nu}^{(2)}(R,\xi), \quad (8.2.1)$$

$$\Phi_{n\nu}^{(1)}(R,\xi) \equiv -\frac{\hbar^2}{M}(\nabla_R\psi_n(R))\cdot(\nabla_R\chi_{\nu,R}(\xi)), \quad (8.2.2)$$

$$\Phi_{n\nu}^{(2)}(R,\xi) \equiv -\frac{\hbar^2}{2M}\psi_n(R)\nabla_R^2\chi_{\nu,R}(\xi), \quad (8.2.3)$$

$$H^{(1)}\Psi_{n\nu}(R,\xi) = \psi_n(R)H^{(1)}\chi_{\nu,R}(\xi). \quad (8.2.4)$$

Hence, we find by use of Eq. (8.1.12) that

$$H_{\text{tot}}\Psi_{n\nu}(R,\xi) = \left\{\left(H_S^{(0)}+\mathcal{E}_{\nu,R}\right)\psi_n(R)\right\}\chi_{\nu,R}(\xi) + \Phi_{n\nu}^{(1)}(R,\xi) + \Phi_{n\nu}^{(2)}(R,\xi). \quad (8.2.5)$$

The first term on the right-hand side may be rewritten with the aid of Eq. (8.1.17). Thus

$$
\begin{aligned}
H_{m\mu,n\nu} &:= \int\int \mathrm{d}R\mathrm{d}\xi \ \Psi_{m\mu}^*(R,\xi)H_{\text{tot}}\Psi_{n\nu}(R,\xi) \\
&= \delta_{\mu\nu}\int \mathrm{d}R \ \psi_m^*(R)\{H_S+\mathcal{E}_{\nu 0}(R)\}\psi_n(R) + \mathcal{V}_{m\mu,n\nu}^{(1)} + \mathcal{V}_{m\mu,n\nu}^{(2)}, \quad (8.2.6)
\end{aligned}
$$

$$\mathcal{E}_{\nu 0}(R) := \mathcal{E}_{\nu,R} - \mathcal{E}_{0,R}, \quad (8.2.7)$$

$$\mathcal{V}_{m\mu,n\nu}^{(\sigma)} := \int\int \mathrm{d}R\mathrm{d}\xi \ \Psi_{m\mu}^*(R,\xi)\Phi_{n\nu}^{(\sigma)}(R,\xi). \quad (8.2.8)$$

In order to express the above in a more transparent manner, we define the *modified Born–Oppenheimer effective vector potential* $\{\mathbf{A}(R)\}_{\mu\nu}$ as

$$\{\mathbf{A}(R)\}_{\mu\nu} := -\int \mathrm{d}\xi \ \chi_{\mu,R}^*(\xi)\frac{\hbar}{i}\nabla_R\chi_{\nu,R}(\xi), \quad (8.2.9)$$

where ∇_R, which denotes differentiation with respect R, is not to be confused with ∇_ξ. As well as ∇_R, the quantity $\{\mathbf{A}(R)\}_{\mu\nu}$ is a D-component vector, whose a-th component is to be denoted by $\{A_a(R)\}_{\mu\nu}$. Also, let $A_a(R)$ be the matrix whose element at the μ-th row and the ν-th column is $\{A_a(R)\}_{\mu\nu}$, and $\mathbf{A}(R)$ be the vector whose a-th component is $A_a(R)$, then

$$\{\mathbf{A}(R)\}^\dagger = \mathbf{A}(R), \quad (8.2.10)$$

$$\{A_a(R)A_b(R)\}_{\mu\nu} = \hbar^2\int \mathrm{d}\xi (\partial_a\chi_{\mu,R}^*(\xi))(\partial_b\chi_{\nu,R}(\xi)), \quad (8.2.11)$$

where † attached to a matrix denotes Hermitian conjugation, and

$$\partial_a \equiv \frac{\partial}{\partial R_a}. \tag{8.2.12}$$

The following identity turns out to be useful later:

$$-\hbar^2 \int d\xi \, \chi_{\mu,R}^*(\xi) \nabla_R^2 \, \chi_{\nu,R}(\xi) = \{(\mathbf{A}(R))^2\}_{\mu\nu} - \frac{\hbar}{i}\{\nabla_R \cdot \mathbf{A}(R)\}_{\mu\nu}. \tag{8.2.13}$$

To prove this, one has only to integrate both sides of

$$\chi_{\mu,R}^*(\xi)\nabla_R^2 \chi_{\nu,R}(\xi) = -(\nabla_R \chi_{\mu,R}^*(\xi)) \cdot (\nabla_R \chi_{\nu,R}(\xi)) + \nabla_R \{\chi_{\mu,R}^*(\xi)\nabla_R \chi_{\nu,R}(\xi)\} \tag{8.2.14}$$

with respect to ξ and use Eqs. (8.2.9) and (8.2.11).

In what follows, we use the following convention: for an arbitrary function $\psi(R)$,

$$(\nabla_R \cdot \mathbf{A}(R))\psi(R) = \{\nabla_R \cdot \mathbf{A}(R) - \mathbf{A}(R) \cdot \nabla_R\}\psi(R), \tag{8.2.15}$$

where ∇_R on the left-hand side acts only on $\mathbf{A}(R)$, this restriction being stipulated by the round bracket ( ) embracing $\nabla_R \cdot \mathbf{A}(R)$, while the first ∇_R on the right-hand side acts on $\mathbf{A}(R)\psi(R)$ as a whole. We may now express $\mathcal{V}_{m\mu,n\nu}^{(\sigma)}$ in a compact form as

$$\mathcal{V}_{m\mu,n\nu}^{(1)} + \mathcal{V}_{m\mu,n\nu}^{(2)}$$
$$= \frac{1}{2M} \int dR \, \psi_m^*(R) \left\{ -\frac{\hbar}{i}(\nabla_R \cdot \mathbf{A}(R) + \mathbf{A}(R) \cdot \nabla_R) + (\mathbf{A}(R))^2 \right\}_{\mu\nu} \psi_n(R).$$

This combined with Eq. (8.2.6) gives

$$H_{m\mu,n\nu} = \int dR \, \psi_m^*(R) \left\{ \frac{1}{2M} \left(\frac{\hbar}{i}\nabla_R - \mathbf{A}(R) \right)^2 + U(R) + \mathcal{E}_{\nu 0}(R) \right\}_{\mu\nu} \psi_n(R). \tag{8.2.16}$$

Now, introduce a new Hilbert space $\mathcal{H}_\mathcal{E}$:

$$\mathcal{H}_\mathcal{E} := Span\{|\nu\rangle \,|\, \nu \in \mathbf{N}\}, \quad \langle\mu|\nu\rangle = \delta_{\mu\nu}, \tag{8.2.17}$$

where $|\nu\rangle$ has nothing to do with $|n\rangle$ in Eq. (8.1.18). Let $\hat{\mathbf{A}}(R)$ and $\hat{H}_\mathcal{E}(R)$ be operators defined on $\mathcal{H}_\mathcal{E}$ as

$$\hat{\mathbf{A}}(R) := \sum_{\mu\nu} |\mu\rangle\{\mathbf{A}(R)\}_{\mu\nu}\langle\nu|, \tag{8.2.18}$$

$$\hat{H}_\mathcal{E}(R) := \sum_\nu |\nu\rangle\mathcal{E}_{\nu 0}(R)\langle\nu|. \tag{8.2.19}$$

Also, introduce the states

$$|n, \nu\rangle\rangle := |n\rangle|\nu\rangle, \tag{8.2.20}$$

which span the direct-product space $\mathcal{H}_S \otimes \mathcal{H}_\mathcal{E}$ to be denoted by $\mathcal{H}$:

$$\mathcal{H} := \mathcal{H}_S \otimes \mathcal{H}_\mathcal{E} = Span\{|n, \nu\rangle\rangle \mid n, \nu \in \mathbf{N}\}. \tag{8.2.21}$$

With these notations, Eq. (8.2.16) is re-expressed as

$$H_{m\mu,n\nu} = \int dR \ \psi_m^*(R) \left\langle \mu \left| \left| \left\{ \frac{1}{2M} \left(\frac{\hbar}{i} \nabla_R - \hat{\mathbf{A}}(R) \right)^2 + U(R) + \hat{H}_\mathcal{E}(R) \right\} \right| \nu \right\rangle \right| \psi_n(R)$$

$$= \langle\langle m, \mu | \hat{H} | n, \nu \rangle\rangle, \tag{8.2.22}$$

where

$$\hat{H} := \frac{1}{2M}(\hat{P} - \hat{\mathbf{A}}(\hat{R}))^2 + U(\hat{R}) + \hat{H}_\mathcal{E}(\hat{R}). \tag{8.2.23}$$

Note that $\hat{\mathbf{A}}(\hat{R})$ and $\hat{H}_\mathcal{E}(\hat{R})$ are operators on $\mathcal{H}$; they result from the substitution of the operator $\hat{R}$ on $\mathcal{H}_S$ for R in $\hat{\mathbf{A}}(R)$ and $\hat{H}_\mathcal{E}(R)$, respectively. It has thus been shown that the description in terms of $\hat{H}_{\text{tot}}$ (given by Eq. (8.1.5)) is equivalent to that in terms of the operator $\hat{H}$ (given by Eq. (8.2.23)) defined on the Hilbert space (8.2.21). The formal argument so far does not involve any approximation.

In what follows, the ground state of the environment is to be denoted by $|vac\rangle$, namely

$$|vac\rangle \equiv |\nu\rangle_{\nu=0}, \tag{8.2.24}$$

and the environmental states other than $|vac\rangle$ are to be labeled with α:

$$\mathcal{H}_\mathcal{E} = Span\{|vac\rangle, |\alpha\rangle \mid \alpha = 1, 2, \ldots\}. \tag{8.2.25}$$

8.3 Caldeira–Leggett assumption

Let us introduce a series of assumptions to be called the *Caldeira–Leggett assumptions*.

Assumption $\mathcal{A}$1: $\mathcal{E}_{\alpha 0}(R)$ does not depend on R, that is

$$\mathcal{E}_{\alpha 0}(R) = \hbar\omega_\alpha, \text{ with } \omega_\alpha \text{ being a positive constant.} \tag{8.3.1}$$

Assumption $\mathcal{A}$2: $\{\mathbf{A}(R)\}_{00} = 0$.
Assumption $\mathcal{A}$3: $|\{\mathbf{A}(R)\}_{\alpha\beta}| \ll |\{\mathbf{A}(R)\}_{0\alpha}|$.
Assumption $\mathcal{A}$4: The initial state of the environment is the ground state $|vac\rangle$.
Assumption $\mathcal{A}$5: The contribution of $\{\mathbf{A}(R)\}_{\alpha\beta}$ to the time evolution of the entire system is negligible.

The validity of the assumption $\mathcal{A}2$ is automatically assured if the ground state of the environment is non-degenerate, and $\mathcal{A}5$ will be assured also if $\mathcal{A}3 \wedge \mathcal{A}4$ is valid. The validity of $\mathcal{A}1$, $\mathcal{A}3$ and $\mathcal{A}4$, on the other hand, may only be checked case by case by detailed examination of the system in question.

Suppose that the above assumptions are acceptable. Then,

$$\hat{\mathbf{A}}(R) \simeq \sum_\alpha \{\mathbf{A}(R)\}_{\alpha 0} |\alpha\rangle \langle \text{vac}| + \text{h.c.}, \tag{8.3.2}$$

which implies that, among the environmental operators, only the following three kinds are effectively involved in $\hat{H}$:

$$|\alpha\rangle\langle\alpha|, \quad |\alpha\rangle\langle\text{vac}|, \quad |\text{vac}\rangle\langle\alpha|. \tag{8.3.3}$$

Writing formally as

$$\hat{b}_\alpha \equiv |\text{vac}\rangle\langle\alpha|, \quad \hat{b}_\alpha^\dagger \equiv |\alpha\rangle\langle\text{vac}|, \tag{8.3.4}$$

and introducing an arbitrary positive constant m_α of dimension of mass to define[3]

$$\mathbf{A}_\alpha(R) := \frac{1}{i} \left(\frac{2}{\hbar m_\alpha \omega_\alpha} \right)^{1/2} \{\mathbf{A}(R)\}_{\alpha 0}, \tag{8.3.5}$$

one finds

$$\hat{\mathbf{A}}(R) \simeq \sum_\alpha \mathbf{A}_\alpha(R)\, \hat{p}_\alpha, \tag{8.3.6}$$

$$\hat{H}_{\mathcal{E}}(R) \simeq \hat{H}_{\mathcal{E}} \equiv \sum_\alpha \hbar\omega_\alpha \hat{b}_\alpha^\dagger \hat{b}_\alpha, \tag{8.3.7}$$

where

$$\hat{p}_\alpha := \frac{1}{i} \left(\frac{\hbar m_\alpha \omega_\alpha}{2} \right)^{1/2} (\hat{b}_\alpha - \hat{b}_\alpha^\dagger). \tag{8.3.8}$$

Hence,

$$\hat{H} \simeq \hat{H}_{\text{CL1}}$$
$$\equiv \frac{1}{2M} \left\{ \hat{P} - \sum_\alpha \mathbf{A}_\alpha(\hat{R}) \hat{p}_\alpha \right\}^2 + U(\hat{R}) + \sum_\alpha \hbar\omega_\alpha \hat{b}_\alpha^\dagger \hat{b}_\alpha. \tag{8.3.9}$$

Note that the operators $\hat{b}_\alpha$ and $\hat{b}_\alpha^\dagger$ defined by Eq. (8.3.4) do not obey the boson commutation relation.

At this point, we introduce two more assumptions.

---

[3] Without loss of generality, we assume that $\mathbf{A}_\alpha(R)$ is real; this can be assured by an appropriate choice of the phase of $|\alpha\rangle$.

Assumption $\mathcal{A}6$: In Eq.(8.3.9), $\hat{b}_\alpha$ may be regarded as a boson operator, that is,

$$[\hat{b}_\alpha, \hat{b}_\beta^\dagger] \simeq \delta_{\alpha\beta}. \tag{8.3.10}$$

This is a bold assumption; under this assumption there can arise such states as

$$\hat{b}_\alpha^\dagger \hat{b}_\beta^\dagger |\text{vac}\rangle, \tag{8.3.11}$$

namely states with multiply excited bosons, of which there are no counterparts in the space $\mathcal{H}_\mathcal{E}$. The assumption $\mathcal{A}6$ will be justified if $\mathcal{A}4$ is valid and if the environmental effect is so weak that multiply-excited states do not contribute to the final result.

Assumption $\mathcal{A}7$: Modified Born–Oppenheimer effective vector potential is a derivative of a scalar potential,[4] that is, there exists a function $f_\alpha(R)$ such that

$$\mathbf{A}_\alpha(R) = \nabla_R f_\alpha(R). \tag{8.3.12}$$

This assumption, which is necessarily valid if R consists of a single component alone, is not valid in general. Under the assumption $\mathcal{A}6 \wedge \mathcal{A}7$, one may introduce the unitary operator

$$\hat{\mathcal{U}} := \exp\left(\frac{i}{\hbar} \sum_\alpha f_\alpha(\hat{R})\hat{p}_\alpha\right), \tag{8.3.13}$$

and transform the Hamiltonian (8.3.9) as

$$\begin{aligned}
\hat{H}_{\text{CL2}} &:= \hat{\mathcal{U}}^\dagger \hat{H}_{\text{CL1}} \hat{\mathcal{U}} \\
&= \frac{1}{2M} \hat{P}^2 + U(\hat{R}) \\
&\quad + \sum_\alpha \left[\frac{1}{2m_\alpha} \hat{p}_\alpha^2 + \frac{1}{2} m_\alpha \omega_\alpha^2 \{\hat{x}_\alpha - f_\alpha(\hat{R})\}^2 - \frac{1}{2}\hbar\omega_\alpha\right],
\end{aligned} \tag{8.3.14}$$

$$\hat{x}_\alpha := \left(\frac{\hbar}{2m_\alpha\omega_\alpha}\right)^{1/2} (\hat{b}_\alpha + \hat{b}_\alpha^\dagger). \tag{8.3.15}$$

The Hamiltonian $\hat{H}_{\text{CL2}}$, which is unitarily equivalent to $\hat{H}_{\text{CL1}}$, is nothing but the Hamiltonian (5.1.1) with which we started in Chapter 5.

A fair number of assumptions have been introduced in the course of the argument leading from the Hamiltonian (8.1.5) to the final result (8.3.14). It is up to the reader to judge to what extent the harmonic-environment model may be said to be general. An important remark is in order here. In general, the raw Hamiltonian (8.1.5) may appear entirely different from that of the harmonic-environment model. Recall that,

---

[4] In general, a vector potential with this property is said to be "exact".

in (8.1.5), R is the macroscopic degree of freedom selected out of many degrees of freedom of the entire system and ξ stands for the remaining degrees of freedom. Accordingly, the specific nature of ξ varies depending on the system in question; it may stand for electrons or spins, for instance. It is only through the series of transformations (to be called the *Born–Oppenheimer-type processing*) introduced in the present chapter that (8.1.5) may be reduced to (8.3.14) if the reduction is possible at all. One should not be so rash as to reject the harmonic-environment model by a mere appearance of (8.1.7). What has been mentioned above may be summarized schematically as follows:

system of many degrees of freedom

$\downarrow$

rearrangement into the macroscopic degrees of freedom and the environmental ones

$\mid$

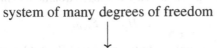

Born–Oppenheimer-type processing

$\downarrow$?

harmonic-environment model

The arguments in Chapters 5 to 7 are based on the premise that the Hamiltonian is given unqualifiedly by (5.1.1). From the point of view of the present chapter, however, the Hamiltonian (8.3.9) may be adopted only with the proviso that the initial state of the environment is $|\mathrm{vac}\rangle$. This circumstance incurs the following issue. If the Schrödinger equation

$$i\hbar \frac{\mathrm{d}}{\mathrm{d}t}|\Psi(t)\rangle\rangle = \hat{H}_{\mathrm{CL1}}|\Psi(t)\rangle\rangle \tag{8.3.16}$$

is unitary-transformed to

$$i\hbar \frac{\mathrm{d}}{\mathrm{d}t}|\widetilde{\Psi(t)}\rangle\rangle = \hat{H}_{\mathrm{CL2}}|\widetilde{\Psi(t)}\rangle\rangle, \quad |\widetilde{\Psi(t)}\rangle\rangle \equiv \hat{\mathcal{U}}^{\dagger}|\Psi(t)\rangle\rangle, \tag{8.3.17}$$

the corresponding transformation must be effected on the initial condition as well, that is, the above-mentioned initial state

$$|\Psi(0)\rangle\rangle = |\psi, \mathrm{vac}\rangle\rangle \equiv |\psi\rangle|\mathrm{vac}\rangle \tag{8.3.18}$$

must be unitary-transformed to

$$|\widetilde{\Psi(0)}\rangle\rangle = \hat{\mathcal{U}}^{\dagger}|\psi, \mathrm{vac}\rangle\rangle, \tag{8.3.19}$$

where the right-hand side is an entangled state of the form of Eq. (6.4.13). By contrast, the argument in chapter 6 adopted

$$|\widetilde{\Psi(0)}\rangle\rangle = |\psi, \mathrm{vac}\rangle\rangle \tag{8.3.20}$$

for simplicity. The choice between the above two initial conditions is related inti-
mately to the status assigned to the harmonic-environment model.

Exercises

Exercise 8.1. The Born–Oppenheimer basis $\{\Psi_{n\nu}^{BO}(R, \xi) \mid n, \nu \in \mathbf{N}\}$ consists of

$$\Psi_{n\nu}^{BO}(R, \xi) := \psi_n^{(\nu)}(R)\chi_{\nu, R}(\xi), \tag{8.3.21}$$

where $\{\psi_n^{(\nu)}(R) \mid n \in \mathbf{N}\}$ is the complete set of orthonormal eigenfunctions associated with
"the macrosystem Hamiltonian in the case of the environment being in its ν-th state", namely

$$H_S^{(0)} + \mathcal{E}_{\nu, R}. \tag{8.3.22}$$

This basis depends on ν in contrast to the modified Born–Oppenheimer basis which does
not. In general, the R-dependence of $\mathcal{E}_{\nu, R}$ varies significantly with ν, and that is why the
Born–Oppenheimer basis is used. By contrast, the modified Born–Oppenheimer basis is
more transparent to work with if the assumption $\mathcal{A}1$ is valid even approximately.

(1) Show that the functions (8.3.21) constitute a complete orthonormal set.
(2) Try to repeat the argument of this chapter by use of the Born–Oppenheimer basis
   (without much success, perhaps).

Exercise 8.2. Study properties of the state (8.3.19).

Exercise 8.3. Discuss quantum resonant oscillation by use of the Hamiltonian (8.3.9),
where $\hat{b}_\alpha$ is not a boson operator but the operator defined by Eq. (8.3.4). Do the same with
the exact Hamiltonian (8.2.23).

9

The cat in the moonlight

The contemporary epistemology of physics is based on quantum mechanics. This framework accommodates even far-reaching theories for those phenomena, ranging from microscopic to ultra-macroscopic, which are totally beyond experimental verification. Nevertheless, one may hesitate to regard it as the final paradigm, for there remains the cat problem. The cat problem, as mentioned in Chapter 1, is a manifestation of the conflict between quantum mechanics and macrorealism. This chapter discusses possible experimental means to decide between the two in the light of Einstein's moon.

9.1 Macrorealism

To pave the way for the ensuing discussion, it is necessary to state precisely what is meant by macrorcalism (to be abbreviated as MR). MR is defined as the epistemological view which accepts the following two assumptions.[1]

$\mathcal{MRA}$1: The assumption of macroscopic definiteness.
If a macrosystem under certain conditions is, whenever observed, found to be in one of two or more macroscopically distinct states $\{S_1, S_2, \ldots\}$, then one can assign to it, at almost all times, the property of "actually being" in a particular one of these states, even when it is not observed.
$\mathcal{MRA}$2: The assumption of non-invasive measurability.
For the macrosystem, it is in principle possible to observe which of the states $\{S_1, S_2, \ldots\}$ it is in without affecting its subsequent behavior.[2]

---

[1] The following statements are quoted from Refs. [2] and [25] with a minor modification; in these references, $\mathcal{MRA}$1 and $\mathcal{MRA}$2 are called the assumptions (or hypotheses) of "macrorealism" and "non-invasive measurability at the macrolevel," respectively. In this book, for brevity, we call $\mathcal{MRA}$1 $\wedge$ $\mathcal{MRA}$2 the macrorealism.
[2] The phrase "without affecting its subsequent behavior" can be replaced by "in such a way that its subsequent behavior is affected only in a precisely predictable manner".

In general, quantum mechanics (to be abbreviated as QM) admits neither $\mathcal{MRA}1$ nor $\mathcal{MRA}2$. Certainly, the macrosystem is found to be in one of the above states when it is observed. This is an empirical fact (or the definition of "observation"), which QM admits as well. It goes without saying that the crux of $\mathcal{MRA}1$ lies in the qualifying phrase "even when it is not observed".

9.2 Leggett–Garg inequality

Suppose that one has observed a phenomenon which appears to be a kind of quantum resonant oscillation. Can one immediately reject MR and conclude in favor of QM that a cat state (QIMDS) has been detected? It is too early to do so. Certainly, a quantum-mechanical description of the observed phenomenon would involve a cat state. In this sense, the observation strongly suggests the detection of a cat state. Nevertheless, it may still be consistent with MR, too. In order to be able to claim detection of a cat state, one has to establish that[3] the experimental results in question are compatible with QM but incompatible with MR. Thus, one needs a quantitative criterion to judge whether or not this statement is warranted.

The famous Bell inequality allows one to sort out phenomena that are compatible with QM but incompatible with LR (i.e. the local realism). It is well known that such phenomena (involving the Einstein–Podolsky–Rosen correlation) have been verified experimentally.[4] In a spirit similar to Bell's, an inequality to decide between QM and MR has been proposed. It is the Leggett–Garg inequality to be derived shortly. One can use this inequality without any specific theories of MR in mind; it serves to confront quantitatively a class of theories satisfying $\mathcal{MRA}1 \wedge \mathcal{MRA}2$ against QM and experiments. Incidentally, although MR can be rejected on the basis of the Leggett–Garg inequality, QM cannot be. These circumstances are analogous to those with the Bell inequality.

Consider a macrosystem which is, whenever observed, found to be in one of the two macroscopically distinct states $\mathcal{S}_\pm$ (e.g. the states localized in the right well or the left well of the symmetric double well). Let $q(t)$ be a variable to indicate which of the two states the macrosystem is in, with $\mathcal{S}_\pm$ corresponding to $q(t) = \pm 1$, respectively. For simplicity, focus attention on four moments of time $\{t_1, t_2, t_3, t_4\}$ such that $t_4 > t_3 > t_2 > t_1$. Suppose that MR is valid, then $\mathcal{MRA}1$ implies[5]

$$q(t_j) = +1 \text{ or } -1 \quad : j \in \{4, 3, 2, 1\}. \tag{9.2.1}$$

This property allows one to conceive of the *joint probability* $\mathcal{P}(\sigma_4 t_4 | \sigma_3 t_3 | \sigma_2 t_2 | \sigma_1 t_1)$ for $\{q(t_4), q(t_3), q(t_2), q(t_1)\} = \{\sigma_4, \sigma_3, \sigma_2, \sigma_1\}$. Being a joint probability, it ought

---

[3] We take it for granted that MR is the only theory competing with QM.

[4] See e.g. N. D. Mermin, *Boojums All The Way Through*, Cambridge University Press (1990), Section 12.

[5] For the time being, we need $\mathcal{MRA}1$ alone; $\mathcal{MRA}2$ will be invoked later when the theory is compared with experiments.

to have the following two properties:

$$P(\sigma_4 t_4 | \sigma_3 t_3 | \sigma_2 t_2 | \sigma_1 t_1) \geq 0, \qquad (9.2.2)$$

$$\sum_{\sigma_1 \sigma_2 \sigma_3 \sigma_4} P(\sigma_4 t_4 | \sigma_3 t_3 | \sigma_2 t_2 | \sigma_1 t_1) = 1, \qquad (9.2.3)$$

where each σ_j ($j = 1 \sim 4$) takes the values $+1$ or -1, of course. With the joint probability in hand, one may define time-correlation functions for the variable $q(t)$ by the following equation; for any pair (j, i),

$$C_{ji}^{\mathrm{MR}} := \sum_{\sigma_1 \sigma_2 \sigma_3 \sigma_4} \sigma_j \sigma_i \, P(\sigma_4 t_4 | \sigma_3 t_3 | \sigma_2 t_2 | \sigma_1 t_1) \quad : 4 \geq j > i \geq 1. \qquad (9.2.4)$$

These correlation functions satisfy the *Leggett–Garg inequalities* $\mathcal{L}\mathcal{G}1 - \mathcal{L}\mathcal{G}3$:

$$\mathcal{L}\mathcal{G}1: \qquad \mathcal{K}_1 \equiv \left| C_{32}^{\mathrm{MR}} - C_{31}^{\mathrm{MR}} \right| + C_{21}^{\mathrm{MR}} \leq 1, \qquad (9.2.5)$$

$$\mathcal{L}\mathcal{G}2: \qquad \mathcal{K}_2 \equiv -\left(C_{32}^{\mathrm{MR}} + C_{21}^{\mathrm{MR}} + C_{31}^{\mathrm{MR}} \right) \leq 1, \qquad (9.2.6)$$

$$\mathcal{L}\mathcal{G}3: \quad -1 \leq \mathcal{K}_3 \equiv \frac{1}{2} \left(C_{43}^{\mathrm{MR}} + C_{32}^{\mathrm{MR}} + C_{21}^{\mathrm{MR}} - C_{41}^{\mathrm{MR}} \right) \leq 1. \qquad (9.2.7)$$

The proof, of which the mathematical structure is in common with the Bell inequality, may run as follows.

[Proof] Let us begin by deriving some auxiliary inequalities. Let σ, σ' and σ'' be binary variables taking the values $+1$ or -1, then

$$\left| \sigma \sigma'' + \sigma' \sigma'' \right| = \left| \sigma \sigma''(1 + \sigma \sigma') \right| = 1 + \sigma \sigma'. \qquad (9.2.8)$$

Putting $(\sigma, \sigma', \sigma'') = (\sigma_1, \pm \sigma_2, \sigma_3)$ in this identity, we find

$$1 \pm \sigma_1 \sigma_2 = \left| \sigma_1 \sigma_3 \pm \sigma_2 \sigma_3 \right|, \qquad (9.2.9)$$

which implies

$$1 \pm \sigma_2 \sigma_1 + \sigma_3 \sigma_1 \pm \sigma_3 \sigma_2 \geq 0, \qquad (9.2.10)$$

$$1 \pm \sigma_2 \sigma_1 - (\sigma_3 \sigma_1 \pm \sigma_3 \sigma_2) \geq 0. \qquad (9.2.11)$$

Likewise, putting $(\sigma, \sigma', \sigma'')$ equal to $(\sigma_1, \sigma_3, \sigma_2)$ or $(\sigma_1, -\sigma_3, \sigma_4)$, we find

$$\left| \sigma_1 \sigma_2 + \sigma_3 \sigma_2 \right| = 1 + \sigma_1 \sigma_3, \qquad (9.2.12)$$

$$\left| \sigma_1 \sigma_4 - \sigma_3 \sigma_4 \right| = 1 - \sigma_1 \sigma_3, \qquad (9.2.13)$$

which implies

$$\left| \sigma_3 \sigma_2 + \sigma_2 \sigma_1 \right| + \left| \sigma_4 \sigma_3 - \sigma_4 \sigma_1 \right| = 2, \qquad (9.2.14)$$

$$-2 \leq \sigma_3 \sigma_2 + \sigma_2 \sigma_1 \pm (\sigma_4 \sigma_3 - \sigma_4 \sigma_1) \leq 2. \qquad (9.2.15)$$

So much for preliminaries. In this proof, we abbreviate C_{ji}^{MR} as C_{ji}. It follows from (9.2.10) that

$$1 \pm C_{21} + C_{31} \pm C_{32}$$
$$= \sum_{\sigma_1\sigma_2\sigma_3\sigma_4} (1 \pm \sigma_2\sigma_1 + \sigma_3\sigma_1 \pm \sigma_3\sigma_2)\mathcal{P}(\sigma_4 t_4|\sigma_3 t_3|\sigma_2 t_2|\sigma_1 t_1) \geq 0, \quad (9.2.16)$$

which gives $\mathcal{LG}2$. Similarly, it follows from (9.2.11) that

$$1 \pm C_{21} - (C_{31} \pm C_{32}) \geq 0. \tag{9.2.17}$$

This combined with (9.2.16) leads to

$$|C_{32} \pm C_{31}| \mp C_{21} \leq 1, \tag{9.2.18}$$

which gives $\mathcal{LG}1$. Incidentally, since the indices may be re-labeled freely, we may exchange indices 2 and 3 in (9.2.18) to find

$$|C_{32} \pm C_{21}| \mp C_{31} \leq 1. \tag{9.2.19}$$

Finally, it follows from (9.2.15) that

$$-2 \leq C_{32} + C_{21} \pm (C_{43} - C_{41}) \leq 2, \tag{9.2.20}$$

which implies

$$C_{32} + C_{21} + |C_{43} - C_{41}| \leq 2, \tag{9.2.21}$$
$$-2 \leq C_{32} + C_{21} - |C_{43} - C_{41}|. \tag{9.2.22}$$

Hence

$$|C_{32} + C_{21}| + |C_{43} - C_{41}| \leq 2. \tag{9.2.23}$$

Noting that the left-hand side is greater than $|2\mathcal{K}_3|$, we obtain $\mathcal{LG}3$.　　Q.E.D.

9.3 Measurable correlation functions

Experimentally measurable quantities are the following.[6]

- $\mathcal{P}^{\text{emp}}(\sigma t_i) \equiv$ "the probability for $q(t_i) = \sigma$",
- $W^{\text{emp}}(\sigma' t_j|\sigma t_i) \equiv$ "the conditional probability for $q(t_j) = \sigma'$ under the condition that $q(t_i) = \sigma$ (namely, the transition probability from the state $\mathcal{S}_\sigma$ to the state $\mathcal{S}_{\sigma'}$)".

---

[6] The superscript "emp" stands for "empirical".

With these data in hand, the experimenter can compute the following correlation function:

$$C_{ji}^{\text{emp}} := \sum_{\sigma'\sigma} \sigma'\sigma \; W^{\text{emp}}(\sigma't_j|\sigma t_i)\mathcal{P}^{\text{emp}}(\sigma t_i). \tag{9.3.1}$$

Experiments are performed on an ensemble of macrosystems.[7] Supposing that the non-invasive measurement of $\mathcal{MRA}\,2$ is realized, decompose the ensemble of the macrosystems into two sub-ensembles, the one on which the non-invasive measurement has been performed and the other on which no measurements at all have been performed yet. The non-invasiveness of the measurement means that the statistical property of the former ensemble is the same as that of the latter, where the statistical property in question is embodied by the joint probability $\mathcal{P}(\sigma_4t_4|\sigma_3t_3|\sigma_2t_2|\sigma_1t_1)$. In other words, once $\mathcal{MRA}\,2$ is accepted, such relations as

$$W^{\text{emp}}(\sigma't_3|\sigma t_2)\mathcal{P}^{\text{emp}}(\sigma t_2) = \sum_{\sigma_4\sigma_1} \mathcal{P}(\sigma_4t_4|\sigma't_3|\sigma t_2|\sigma_1t_1) \tag{9.3.2}$$

hold. If MR is valid, therefore, it follows from Eqs. (9.2.4), (9.3.1) and (9.3.2) that

$$C_{ji}^{\text{emp}} = C_{ji}^{\text{MR}}. \tag{9.3.3}$$

No such relations as (9.3.2) hold in QM, where one faces the problem[8] whether it is possible to introduce the very concept of a joint probability $\mathcal{P}(\sigma_4t_4|\sigma_3t_3|\sigma_2t_2|\sigma_1t_1)$; it is impossible to do so in general.

9.4 Quantum-mechanical correlation function

Suppose for simplicity that the states $\mathcal{S}_{\mp}$ are mutually symmetric; to be explicit, one may consider the ground states $|\mp\rangle$ associated with the left well and the right well, respectively, of the symmetric double well. Also, assume that the Hamiltonian is time-translation invariant.[9] As in Chapter 6, let $\mathcal{P}_+(t)$ be the quantum-mechanical probability for the macrosystem to be found in the state $\mathcal{S}_+$ at time t assuming that it was in the state $\mathcal{S}_-$ at time 0. By symmetry, the probability for the macrosystem to be found in the state $\mathcal{S}_-$ at time t assuming that it was in the state $\mathcal{S}_+$ at time 0 is

---

[7] That is, one begins by preparing a large number of macrosystems with identical properties. Or one may effectively produce such an ensemble by setting up an identical situation repeatedly for a single macrosystem.

[8] That is, the problem whether it is possible to define a joint probability concerning a time-series of events (i.e. a probability concerning the history of a given system) in such a way that the additivity of probability amplitudes, the additivity of probabilities and the normalization of the total probability are mutually compatible. This problem is the subject of the theory of consistent (or non-interfering) histories. See, e.g. R. Omnès, *The Interpretation of Quantum Mechanics*, Princeton University Press (1994), Chapter 4.

[9] In Section 9.2, where we were concerned with MR whose detailed nature was totally unknown, we assumed neither symmetry of the states $\mathcal{S}_{\mp}$ nor time-translation invariance.

also $\mathcal{P}_+(t)$. Therefore, the quantum-mechanical transition probability is given as

$$W^{\text{QM}}(\sigma' t_j | \sigma t_i) = \{1 - \mathcal{P}_+(t_j - t_i)\}\delta_{\sigma'\sigma} + \mathcal{P}_+(t_j - t_i)\delta_{\sigma'-\sigma}. \quad (9.4.1)$$

This may be used to define the quantum-mechanical correlation function as[10]

$$C_{ji}^{\text{QM}} := \sum_{\sigma'\sigma} \sigma'\sigma W^{\text{QM}}(\sigma' t_j | \sigma t_i)\mathcal{P}^{\text{emp}}(\sigma t_i) \quad (9.4.2)$$

$$= \sum_{\sigma} \left[\{1 - \mathcal{P}_+(t_j - t_i)\} - \mathcal{P}_+(t_j - t_i)\right]\mathcal{P}^{\text{emp}}(\sigma t_i). \quad (9.4.3)$$

The quantity embraced by the square bracket [ ] on the right-hand side does not depend on σ. Since

$$\sum_{\sigma} \mathcal{P}^{\text{emp}}(\sigma t_i) = 1 \quad (9.4.4)$$

by definition of probability, it follows that

$$C_{ji}^{\text{QM}} = 1 - 2\mathcal{P}_+(t_j - t_i). \quad (9.4.5)$$

The right-hand side ought to coincide with C_{ji}^{emp} if QM is valid.

Hereafter, let us consider the case of the symmetric double well. By use of the quantum-mechanical prediction (6.3.22), we find[11]

$$C_{ji}^{\text{QM}} = C(t_j - t_i), \quad C(t) \approx e^{-t/T_2}\cos(t\tilde{\Delta}). \quad (9.4.6)$$

Thus, the quantum-mechanical correlation function is given by the interference term in $\mathcal{P}_\pm$.

Let $\mathcal{K}_l^{\text{QM}}$ be the $\mathcal{K}_l$ in Leggett–Garg inequalities with C_{ji}^{MR} replaced by C_{ji}^{QM} (namely, $C(t_j - t_i)$), then the following theorem holds:

Theorem If the correlation function $C(t)$ is concave (or convex), $\mathcal{K}_1^{\text{QM}}$ (or $\mathcal{K}_2^{\text{QM}}$) is maximum for fixed t_3 and t_1 when $t_3 - t_2 = t_2 - t_1$. Also, if $C(t)$ is concave, $\mathcal{K}_3^{\text{QM}}$ is maximum for fixed t_4 and t_1 when t_4, t_3, t_2 and t_1 are equally separated.

In view of this theorem, one may take

$$t_3 - t_2 = t_2 - t_1 = \tau/\tilde{\Delta}, \quad (9.4.7)$$

---

[10] Although the transition probability $W^{\text{QM}}(\sigma' t_j | \sigma t_i)$ can be computed theoretically within QM, the initial probability $\mathcal{P}^{\text{emp}}(\sigma t_i)$ cannot; the latter has to be determined by experiment. It is often the case, however, that experimenters do not measure $\mathcal{P}^{\text{emp}}(\sigma t_i)$. In that case, one infers it by invoking the principle of equal *a priori* probability, the *canonical* distribution, and so on, as the case may be. At any rate, Eq. (9.4.2) is not a result of QM alone but of QM combined with empirical fact (or inference from it).

[11] For simplicity, θ_0 is neglected. Also, $2/\Gamma_1$ is denoted by T_2. Although Eq. (6.3.22) was derived in the temporal domain (6.3.18), we use it for times of $\mathcal{O}(1/\Delta)$ for the purpose of getting a feel about the Leggett–Garg inequality. If a more precise discussion is desired, one has only to use the $\mathcal{P}_+(t)$ obtained from (6.3.14) instead.

where τ is a constant such that $\cos \tau > 0 > \cos 2\tau$, to find

$$\mathcal{K}_1^{\mathrm{QM}} = |e^{-\gamma\tau} \cos \tau - e^{-2\gamma\tau} \cos 2\tau| + e^{-\gamma\tau} \cos \tau$$

$$= 2e^{-\gamma\tau} \cos \tau - e^{-2\gamma\tau} \cos 2\tau, \qquad (9.4.8)$$

$$\gamma \equiv (T_2 \tilde{\Delta})^{-1} = \Gamma_1/2\tilde{\Delta}. \qquad (9.4.9)$$

If decoherence due to the environment is negligible ($\gamma = 0$), then

$$\mathcal{K}_1^{\mathrm{QM}} = -2 \left(\cos \tau - \frac{1}{2} \right)^2 + \frac{3}{2}, \qquad (9.4.10)$$

$$\max_{\tau} \mathcal{K}_1^{\mathrm{QM}} = \mathcal{K}_1^{\mathrm{QM}}\big|_{\tau=\pi/3} = \frac{3}{2} > 1, \qquad (9.4.11)$$

which contradicts $\mathcal{LG}1$. Hence, $\mathcal{LG}1$ decides between QM and MR. On the other hand, if decoherence is so pronounced that $\gamma \gg 1$, then $|\mathcal{K}_1^{\mathrm{QM}}| \ll 1$, which does not contradict $\mathcal{LG}1$. In this case, QM and MR are compatible as far as $\mathcal{LG}1$ is concerned. To make a rough estimate of the maximum value of γ violating $\mathcal{LG}1$, one may put[12] $\tau = \pi/3$ to find

$$\mathcal{K}_1^{\mathrm{QM}} = z_1 + \frac{1}{2}z_1^2, \quad z_1 \equiv e^{-\pi\gamma/3}. \qquad (9.4.12)$$

Accordingly,

$$\mathcal{K}_1^{\mathrm{QM}} > 1 \iff z_1 > \sqrt{3} - 1 \iff \gamma < 0.297 \cdots . \qquad (9.4.13)$$

Similar consideration with $\mathcal{LG}2$ or $\mathcal{LG}3$ gives the estimate 0.15 or 0.31, respectively, for the maximum value of γ. Hence, among the three inequalities, $\mathcal{LG}3$ works over the widest range of γ. MQT-experiments performed so far correspond[13] to the case of large γ and no results have been found to violate Leggett–Garg inequalities.

By use of Eqs. (6.2.37) and (9.4.9), one may estimate that $\gamma \sim \eta/h$. Then, the condition (9.4.13) may be re-expressed as $\eta/h \lesssim 0.3$. For a SQUID, for instance, the present technology is able to realize this condition by setting up a situation of sufficiently high resistance (cf. Eq. (3.1.49)).

9.5 Thought experiment to test Leggett–Garg inequality

MR is rejected if the experimentally-determined correlation function C_{ji}^{emp} violates the Leggett–Garg inequality. It is, of course, necessary that the experiment in question realizes a non-invasive measurement[14] in the sense of $\mathcal{MRA}\,2$. If, furthermore,

---

[12] The optimal value of τ depends on γ. However, its γ-dependence is not very important as seen from the following result.

[13] This is a qualitative correspondence since the potential in the MQT-situation is not a symmetric double well.

[14] Logically speaking, we need to establish a criterion to judge whether or not a given experiment is non-invasive.

the experimental result is such that

$$C_{ji}^{\text{emp}} \simeq C_{ji}^{\text{QM}}, \tag{9.5.1}$$

then not only MR is rejected but also QM is confirmed.

The innocent statement that Eq. (9.5.1) should hold if QM is valid overlooks a grave problem. In order to determine the correlation function C_{ji}^{emp}, the state of the macrosystem has to be measured at the two moments of time $\{t_j, t_i\}$. In general, however, whenever a measurement is performed, even in such a way that all disturbances conceivable in the classical physics are eliminated so as to make it non-invasive, the macrosystem necessarily becomes entangled with the measurement apparatus. This is inevitable if QM is universally valid; the measurement process itself must be treated as a physical process to be described by QM. Thus, the measurement performed at time t_i could modify the subsequent quantum-mechanical time evolution of the macrosystem (or, to be precise, the entire system consisting of the macrosystem and the apparatus). Accordingly, the transition probability $W^{\text{emp}}(\sigma' t_j | \sigma t_i)$ determined as a result of the second measurement at time $t_j (> t_i)$ could be different from the transition probability $W^{\text{QM}}(\sigma' t_j | \sigma t_i)$ predicted for the intrinsic time evolution of the macrosystem. In other words, if QM is valid, the very fact that it is could hinder measurement of the quantum-mechanical transition probability. Are we in a hopeless situation? Fortunately, we are not. For $W^{\text{QM}}(+t_2| - t_1)$ to be extractable from experiments, it is sufficient that the time evolution in the case of[15] the macrosystem being in the state S_- at time t_1 is kept intact, whereas that in the case of the macrosystem being in the state S_+ at time t_1 may be modified drastically. This observation offers the opportunity for a carefully designed experiment to verify Eq. (9.5.1). Such an experiment has been proposed for the quantum resonant oscillation in SQUID. It is *Tesche's thought experiment* to be introduced below.

Hereafter, we assume that QM is valid. The apparatus to be used to measure the flux through $\mathcal{SQUID}$[16] is a dc-SQUID. The latter is a superconducting ring, too. In contrast to the former, it is equipped with terminals allowing external current to flow in and out, and has two Josephson junctions inserted as marked by $\times$ in Fig. 9.1. The current flowing through the terminal depends sensitively on the flux threading through the dc-SQUID. Hence, by monitoring the current, one can make the dc-SQUID function as a flux meter. If the dc-SQUID is placed near $\mathcal{SQUID}$, a part of the flux of the former threads through the latter and vice versa. Its magnitude is determined by the mutual inductance, which can be controlled by varying the

---

[15] See the ensuing discussion for the precise meaning of the phrase in the case of, which is difficult to express without the use of equations.

[16] The SQUID (the rf-SQUID to be precise) of Section 3.1 is to be denoted as $\mathcal{SQUID}$ so as to be distinguished from the dc-SQUID below.

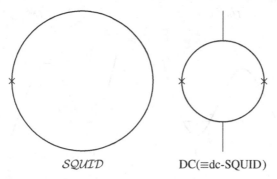

$\mathcal{SQUID}$ DC($\equiv$dc-SQUID)

Fig. 9.1. The dc-SQUID to measure the flux of $\mathcal{SQUID}$.

relative geometrical configuration.[17] Hence, the dc-SQUID may be used to observe the behavior of the flux through $\mathcal{SQUID}$. It is to be noted, however, that the flux potential U ($\equiv U(\Phi)$) for $\mathcal{SQUID}$ suffers a change in general due to the effect of the dc-SQUID, since the presence of the latter modifies the external flux Φ_{ex} through $\mathcal{SQUID}$ (cf. Eq. (3.1.20)). The flux Φ^{dc} through the dc-SQUID may be treated quantum-mechanically in a similar manner as in the case of $\mathcal{SQUID}$. If the value of the flux is restricted to an appropriate range, the flux potential turns out to be a tilted double well U^{dc} ($\equiv U^{dc}(\Phi^{dc})$). The degree of tilt and the height of the barrier can be varied continuously by the current I_{ex}^{dc} injected externally from the terminal and the externally applied flux Φ_{ex}^{dc} through the ring. Hereafter, this dc-SQUID is to be abbreviated as DC.[18] Suppose that U is set up to be the symmetric double well in the absence of the mutual inductance. Let the relative spatial configuration of $\mathcal{SQUID}$ and DC be such that U gets tilted down leftwards or rightwards if DC[19] gets localized in the left well or the right well, respectively, of U^{dc}. Of course, $\mathcal{SQUID}$ affects DC as well; the configuration is supposed to be such that U^{dc} gets tilted down leftwards or rightwards if $\mathcal{SQUID}$ gets localized in the left well or the right well, respectively, of U.

The initial setup of the experiment is as follows. First, Φ_{ex}^{dc} is adjusted to make U^{dc} tilt down leftwards so much that DC settles down in the left-well ground state. Next, Φ_{ex}^{dc} is readjusted to make U^{dc} tilt down rightwards so that the left-well ground state energy lies in between the right-well ground and first-excited states[20] (Fig. 9.2(0)). The barrier is adjusted to be so large as to virtually prohibit tunneling from the left-well ground state to a right-well state, thereby keeping DC in the left-well ground state $|-\rangle_{dc}$. At this stage, U is tilted down leftwards due to the effect of the

---

[17] Fig. 9.1 is not intended to represent an actual configuration.

[18] As mentioned below, DC functions not only as a probe to observe the flux of $\mathcal{SQUID}$, but also as a measurement apparatus (i.e. a detector or a counter).

[19] In this section, the quantum state for the flux of DC is to be abbreviated as DC and similarly for $\mathcal{SQUID}$.

[20] More generally, the right-well nth and $(n + 1)$th excited states will do.

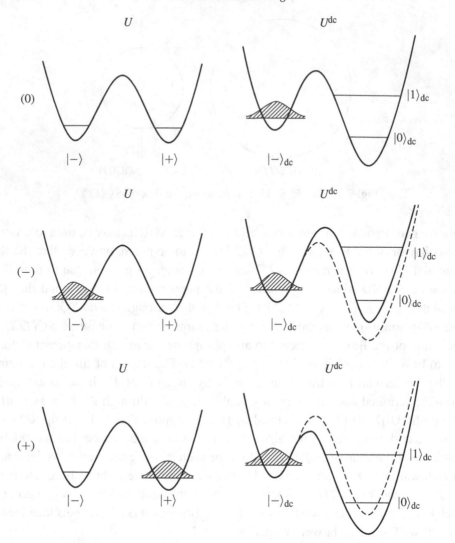

Fig. 9.2. The initial setup of the experiment (the broken curves in Figs. ($\mp$) being identical to the U^{dc} in Fig. (0)).

flux from DC. Then, Φ_{ex} is readjusted so that U reverts to the symmetric situation. The U^{dc} in Fig. 9.2(0) refers to the fictitious case that the effect of $\mathcal{SQUID}$ is negligible. Actually, the effect of $\mathcal{SQUID}$ is such that the barrier of U^{dc} increases or decreases depending on $\mathcal{SQUID}$ being localized in the left well or the right well, respectively (Fig. 9.2($\mp$)). It has to be ensured, however, that the barrier is large enough, not only in the case of Fig. 9.2($-$) but also in the case of Fig. 9.2($+$), to keep DC in the left-well ground state. Although the left-well ground state in the case of ($+$) or ($-$) is slightly different from that in the case of (0), since the potential is distorted slightly, the difference is assumed to be negligible. Hence the

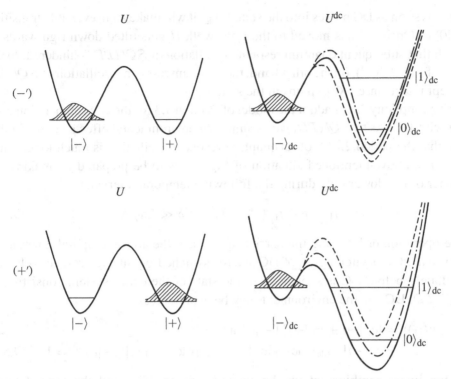

Fig. 9.3. The situation immediately after the dc-SQUID is switched on: the broken curve is identical to the U^{dc} in Fig. 9.2(0), and the dot-dashed curves in ($\mp'$) are identical to those U^{dc} in Figs. 9.2($\mp$), respectively.

common symbol $|-\rangle_{\mathrm{dc}}$ is used for all three cases. Likewise, strictly speaking, the potential U in ($+$) or ($-$) is slightly asymmetric and different from that in (0). But this difference is assumed to be negligible as well.

So much for the initial setup. Now, we lower the barrier of U^{dc} temporarily (cf. (9.5.3) below) so as to produce the following situation (Fig. 9.3):

If and only if $\mathcal{SQUID}$ is localized in the right well, the energy of $|-\rangle_{\mathrm{dc}}$ coincides with that of the right-well first excited state $|1\rangle_{\mathrm{dc}}$.

In the case of Fig. 9.3($+'$), DC finds itself in the resonant-tunneling situation and would start oscillating between $|-\rangle_{\mathrm{dc}}$ and $|1\rangle_{\mathrm{dc}}$ with a period $2\pi/\Delta^{\mathrm{dc}}$ if it were not for environmental effects. Actually, DC interacts with the environment and will decay from the state $|1\rangle_{\mathrm{dc}}$ to the right-well ground state $|0\rangle_{\mathrm{dc}}$. If the environmental effect is so large that

the lifetime T_1^{dc} of the state $|1\rangle_{\mathrm{dc}}$ is much shorter than π/Δ^{dc}, (9.5.2) then, as soon as DC moves into the state $|1\rangle_{dc}$, it will make an irreversible transition to $|0\rangle_{dc}$. Once DC has moved to the right well, U gets tilted down rightwards so much that subsequent quantum resonant oscillation of $\mathcal{SQUID}$ is hindered. In the case of Fig. 9.3($-'$), on the other hand, the quantum resonant oscillation of $\mathcal{SQUID}$ is kept intact since DC remains in the state $|-\rangle_{dc}$.

For simplicity, let us adopt the model of Section 4.3 for the environment interacting with DC. As to $\mathcal{SQUID}$, we assume that environmental effects are negligible and that the period $2\pi/\Delta$ of its quantum resonant oscillation is much longer than T_1^{dc}. The above-mentioned situation of Fig. 9.3 is to be prepared by making the barrier of U^{dc} lower only during the following temporal interval:

$$t_1 < t < t_1 + \delta \ : T_1^{dc} \ll \delta \ll 2\pi/\Delta. \tag{9.5.3}$$

The operation of lowering the barrier of U^{dc} and the above temporal interval are to be called the <u>switching-on of DC</u> and the <u>switched-on interval</u>, respectively. Immediately before DC is switched on, the state of the entire system consisting of $\mathcal{SQUID}$, DC and the environment may be expressed as

$$\begin{aligned}|\Psi(t_1)\rangle &= \{c_-|-\rangle + c_+|+\rangle\}\,|-\rangle_{dc}|\text{vac}\rangle \\ &= c_-|-\rangle|-\rangle_{dc}|\text{vac}\rangle + c_+|+\rangle|-\rangle_{dc}|\text{vac}\rangle, \quad |c_-|^2 + |c_+|^2 = 1. \tag{9.5.4}\end{aligned}$$

In the linear combination on the second line, the first and the second terms represent[21] <u>the case of $(-)$</u> and <u>the case of $(+)$</u>, respectively, of Fig. 9.2. Immediately after DC is switched on, these terms represent the case of $(-')$ and the case of $(+')$, respectively, of Fig. 9.3. During the switched-on interval, time evolution of $\mathcal{SQUID}$ on its own is negligible under the condition (9.5.3). Hence, immediately after the switched-on interval, the first term of (9.5.4) has remained intact, while in the second term DC and $\mathcal{SQUID}$ have evolved to $|0\rangle_{dc}$ and $|\widetilde{+}\rangle$, respectively, where $|\widetilde{+}\rangle$ is the right-well ground state of the rightwards-tilted double well (Fig. 9.4):

$$|\Psi(t_1 + \delta)\rangle \simeq c_-|-\rangle|-\rangle_{dc}|\text{vac}\rangle + c_+|\widetilde{+}\rangle|0\rangle_{dc}\sum_\alpha c_\alpha(\delta)|\alpha\rangle. \tag{9.5.5}$$

This is an entangled state in which the states $|-\rangle$ and $|\widetilde{+}\rangle$ for $\mathcal{SQUID}$ are *correlated* with the states $|-\rangle_{dc}$ and $|0\rangle_{dc}$ for DC, respectively. As a result of the emergence of this correlation, the initial linear combination (see the first line of Eq. (9.5.4)) for $\mathcal{SQUID}$ by itself has been destroyed utterly. It is, however, precisely this sort of correlation that is required of a measurement to determine which of the states $|\pm\rangle$ $\mathcal{SQUID}$ is in. Note that the environment is also involved in the entanglement in the state (9.5.5). Otherwise, the right-hand side of the above equation would be

---

[21] Thus, the word <u>the case</u> in this section should not be endowed with connotation of a underlying classical classification; it is meant to specify one of the terms in a linear combination.

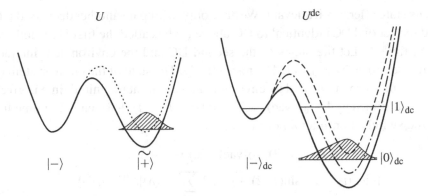

Fig. 9.4. The situation where the dc-SQUID has moved to the right well: the dotted curve in the left figure is identical to the U in Fig. 9.2, and the three curves in the right figure are the same as those in Fig. 9.3($+'$).

replaced by

$$\left\{ c_- |-\rangle |-\rangle_{dc} + c_+ |\widetilde{+}\rangle \left(\cos\left(\frac{\Delta^{dc}}{2}\delta\right) |-\rangle_{dc} + i\sin\left(\frac{\Delta^{dc}}{2}\delta\right) |1\rangle_{dc} \right) \right\} |vac\rangle,$$

(9.5.6)

and the above-mentioned correlation would be lost eventually unless δ is adjusted to be equal to π/Δ^{dc}. By contrast, the entangled state (9.5.5) involving the environment as well conserves the correlation without such adjustment.[22] Thus, the situation (9.5.2) is a desirable one for DC to function as a measurement apparatus.

Let us follow the time evolution further. In the first term of (9.5.5), $\mathcal{SQUID}$ executes the quantum resonant oscillation on its own and DC remains unchanged, whereas in the second term both $\mathcal{SQUID}$ and DC remain unchanged. Hence, the state of the entire system at time $t_2 (> t_1 + \delta)$ is given as[23]

$$|\Psi(t_2)\rangle \simeq c_-(t_2 - t_1)\{\cos(\tau/2)|-\rangle + i\sin(\tau/2)|+\rangle\}|-\rangle_{dc}|vac\rangle$$
$$+ c_+(t_2 - t_1)|\widetilde{+}\rangle|0\rangle_{dc} \sum_\alpha c_\alpha(\delta) e^{-i(t_2-t_1)\omega_\alpha}|\alpha\rangle, \quad \tau \equiv (t_2 - t_1)\Delta,$$

(9.5.7)

$$c_-(t) \equiv c_- \exp(-it E_-^{dc}/h), \quad c_+(t) \equiv c_+ \exp(-it E_{\mp}/h).$$

(9.5.8)

It is seen that the first term retains the intrinsic time evolution of $\mathcal{SQUID}$ in the case of it being initially in the state $|-\rangle$.

What remains to be done is to measure the state of $\mathcal{SQUID}$ at time t_2. This is a relatively easy task; this measurement is allowed to disturb the system to any extent,

---

[22] It is because the Poincaré period (cf. Section 2.5) is virtually infinite.
[23] The energies of $|-\rangle$ and $|0\rangle_{dc}$ are taken as the origins of energy for $\mathcal{SQUID}$ and DC, respectively, and the energies of $|\widetilde{+}\rangle$ and $|-\rangle_{dc}$ are denoted by $E_{\mp}$ and E_-^{dc}, respectively.

since its state after t_2 is irrelevant. We have only to prepare another dc-SQUID (to be called the second DC) identical to DC above (to be called the first DC), and switch it on at time t_2. Let the states of the second DC and the environment interacting with it be denoted by $|\ \rangle_{dc'}$ and $|\ \rangle'$, respectively. The state of the entire system (now including the second DC and its environment as well) at the initial time is given by (9.5.4) multiplied by $|-\rangle_{dc'}|vac\rangle'$, while at time T $(\equiv t_2 + \delta)$, namely immediately after the switched-on interval of the second DC, it is given as

$$|\Psi(T)\rangle \simeq c_-(t_2 - t_1)\cos(\tau/2)|-\rangle|vac\rangle|vac\rangle'|-,-\rangle\rangle$$

$$+ ic_-(t_2 - t_1)\sin(\tau/2)|\widetilde{+}\rangle|vac\rangle \sum_{\alpha'} c_{\alpha'}(\delta)|\alpha'\rangle'|-,0\rangle\rangle$$

$$+ c_+(t_2 - t_1)|\widetilde{+}\rangle \sum_{\alpha} c_{\alpha}(\delta)e^{-i(t_2-t_1)\omega_\alpha}|\alpha\rangle \sum_{\alpha'} c_{\alpha'}(\delta)|\alpha'\rangle'|0,0\rangle\rangle, \quad (9.5.9)$$

where, for brevity, we have put $E_-^{dc'} = 0$ and

$$|*, *'\rangle\rangle \equiv |*\rangle_{dc}|*'\rangle_{dc'}. \quad (9.5.10)$$

At time T, we may measure the frequency of the event that the second DC has emitted energy into the environment whereas the first DC has not done so yet. The probability for this event to be found is given theoretically by $|c_-|^2(\sin(\tau/2))^2$ as is seen from the second term proportional to $|-,0\rangle\rangle$. The coefficient $|c_-|^2$ $(= 1 - |c_+|^2)$ may be determined by measuring the frequency of the event that both DC's have emitted energy; the probability for this event to be found is given theoretically by $|c_+|^2$ as is seen from the third term proportional to $|0,0\rangle\rangle$. In other words, what an experimenter can "see" directly is the *coarse-grained density operator*[24] with respect to the two DC's alone:

$$\hat{\rho}^{dc,dc'} \equiv \text{Tr}_{SQUID} \text{Tr}_{environment} \text{Tr}_{environment'} |\Psi(T)\rangle\langle\Psi(T)|$$

$$= |c_-\cos(\tau/2)|^2 |-,-\rangle\rangle\langle\langle-,-|$$

$$+ |c_-\sin(\tau/2)|^2 |-,0\rangle\rangle\langle\langle-,0| + |c_+|^2|0,0\rangle\rangle\langle\langle0,0|, \quad (9.5.11)$$

which displays the desired information. We may conclude thus: it is in principle possible to measure the quantum-mechanical transition probability $W^{QM}(+t_2|-t_1)$ $(= \mathcal{P}_+(t_2 - t_1) \simeq (\sin(\tau/2))^2)$.

9.6 Concluding remarks

Today, two decades since its announcement, the Leggett program has become more realistic than ever. Young-type interference experiments, which used to be possible only for particles in a near vacuum, have been performed successfully with electrons

---

[24] Or *reduced density operator*.

in a solid. It used to be thought that such an interference effect would be washed out by the effect of the environment, namely the solid in this case, except for a specially ordered system such as superconducting electrons. The advance of so-called mesoscopic technology, however, has made it possible for a fairly macroscopic solid (e.g. of the size of 1 μm) to be maintained in a sufficiently <u>quiet</u> situation, thereby demonstrating an interference effect which is essentially the same as *in vacuo*. These developments suggest that the presence of an environment does not necessarily imply decoherence (not only theoretically but also in practice) and that it is not outrageous to expect a detection of QIMDS.

Indeed, MQT are thought to have been detected in several systems. For example, MQT confirmed in SQUID[25] is a phenomenon in which a magnetic flux through a superconducting ring tunnels out of the ring (Section 3.1). Since the flux is accompanied by a current flowing along the surface of the ring, and the number of electrons carrying this current is estimated typically to be of the order of 10^{15}, the state with the flux threading through the ring is macroscopically distinct from the one with the flux outside the ring. MQT-experiments do not constitute direct evidence for QIMDS, because they do not allow the measurement of the phase difference between macroscopically distinct states in question.[26] Nevertheless, they are remarkable in that they demonstrate the typically-quantum-mechanical phenomena of tunneling in such a macroscopic situation. As mentioned in Chapter 3, active research is also in progress concerning quantum nucleation in liquid He, quantum tunneling of magnetization in a small magnet, and so on. It is desirable to have many more examples of MQT confirmed and various conceivable routes towards the detection of MQC explored.

If the thought experiment in the preceding section is realized, it will be possible to test Eq. (9.5.1). Realization of this experiment, which is expected also to constitute a non-invasive measurement in the sense of MR, is one of the most significant challenges for twenty-first-century physics.

Note added in the English edition: In 1999, an observation of quantum resonant oscillations in a "single-Cooper-pair box" was reported (see Ref. [27]). This remarkable achievement constitutes an important progress towards detection of MQC.

Exercises

Exercise 9.1. Prove the theorem in Section 9.4 proceeding, for instance, as follows:
Let

$$T \equiv t_3 - t_1, \quad x \equiv (t_2 - t_1)/T, \quad f(x) \equiv C(Tx), \tag{9.6.1}$$

---

[25] As well as in current-biased Josephson junction, which is similar to SQUID in essential respects.
[26] MQT-experiments measure only the absolute value of coefficients of each term in Eq. (2.3.4), for instance.

then

$$\mathcal{K}_2^{QM} = -f(1-x) - f(x) - C(T) : \quad 0 < x < 1, \tag{9.6.2}$$

$$d\mathcal{K}_2^{QM}/dx = -f'(x) + f'(1-x). \tag{9.6.3}$$

By assumption, $f''(x) > 0$, that is, $f'(x)$ increases monotonically. Hence, $\mathcal{K}_2^{QM}$ is maximum at $x = 1 - x$. Other cases may be dealt with similarly.

Exercise 9.2. Confirm the following equations and estimate the maximum value of γ violating $\mathcal{LG}2$ or $\mathcal{LG}3$:

$$\mathcal{K}_2^{QM}\big|_{\tau=2\pi/3} = z_2 + \frac{1}{2}z_2^2, \quad z_2 \equiv e^{-2\pi\gamma/3}, \tag{9.6.4}$$

$$\mathcal{K}_3^{QM}\big|_{\tau=\pi/4} = \frac{1}{2\sqrt{2}}(3z_3 + z_3^3), \quad z_3 \equiv e^{-\pi\gamma/4}. \tag{9.6.5}$$

Exercise 9.3. Construct a Hamiltonian which is able to describe Tesche's thought experiment, and study the time evolution of the entire system in detail.

Appendix A. Euclidean space and Hilbert space

This appendix is not meant to give mathematically rigorous definitions but merely to explain notations used in the text.

A.1 Three-dimensional Euclidean space

The three-dimensional space in the laboratory is regarded as the Euclidean space $\mathbf{E}^3$. An appropriately chosen right-handed orthonormal triad (2.2.13) spans $\mathbf{E}^3$, which is a linear space. This circumstance is expressed as

$$\mathbf{E}^3 = \mathcal{S}pan\{\mathbf{e}_j \mid j = x, y, z\} := \left\{ \sum_j x_j \mathbf{e}_j \,\middle|\, x_j \in \mathbf{R} \right\}. \tag{A.1.1}$$

Let $\mathcal{R}$ be a rotation tensor on $\mathbf{E}^3$:

$$\mathcal{R}\mathbf{e}_j = \sum_k \mathbf{e}_k \mathcal{R}_{kj}, \tag{A.1.2}$$

where $\{\mathcal{R}_{kj}\}$ is a 3×3 matrix of unit determinant (i.e. $\{\mathcal{R}_{kj}\} \in SO(3)$). In order to stipulate that the rotation around the unit vector $\mathbf{n}$ by the angle θ is to be considered, the notations $\mathcal{R}(\theta\mathbf{n})$ and $\mathcal{R}_{kj}(\theta\mathbf{n})$ are used; the rotation in question is specified uniquely by the vector $\theta\mathbf{n}$ of length θ.

A.2 Hilbert space

The linear space (Hilbert space) $\mathcal{H}$ spanned by the orthonormal basis[1]

$$\{|n\rangle \mid n \in \mathbf{N} \;:\; \langle n|n'\rangle = \delta_{nn'}\} \tag{A.2.1}$$

is expressed, in analogy to (A.1.1), as

$$\mathcal{H} = \mathcal{S}pan\{|n\rangle \mid n \in \mathbf{N}\} := \left\{ \sum_n c_n|n\rangle \,\middle|\, c_n \in \mathbf{C} \right\}. \tag{A.2.2}$$

---

[1] Hereafter, an orthonormal basis is abbreviated as a basis. In case a non-orthonormal basis is used, it will be so announced.

$\mathcal{H}$ is the set of all possible linear combinations of $\{|n\rangle\}$. Given this Hilbert space $\mathcal{H}$ and another one

$$\tilde{\mathcal{H}} = \mathcal{S}pan\{|\widetilde{\alpha}\rangle \mid \alpha \in \mathbf{N}\}, \tag{A.2.3}$$

the direct-product space $\mathcal{H} \otimes \tilde{\mathcal{H}}$ is defined as

$$\mathcal{H} \otimes \tilde{\mathcal{H}} := \mathcal{S}pan\{|n, \alpha\rangle\rangle \mid n, \alpha \in \mathbf{N} \; : |n, \alpha\rangle\rangle \equiv |n\rangle|\widetilde{\alpha}\rangle\}$$

$$= \left\{ \sum_{n\alpha} c_{n\alpha} |n, \alpha\rangle\rangle \;\middle|\; c_{n\alpha} \in \mathbf{C} \right\}, \tag{A.2.4}$$

which is the set of all possible linear combinations of $\{|n, \alpha\rangle\rangle\}$. Inner products in this space are specified by linearity and the following definition:

$$\langle\langle n, \alpha | n', \alpha'\rangle\rangle := \langle n | n'\rangle\langle\widetilde{\alpha} \mid \widetilde{\alpha'}\rangle. \tag{A.2.5}$$

Accordingly, the order of $\mathcal{H}$ and $\tilde{\mathcal{H}}$ is immaterial ($\mathcal{H} \otimes \tilde{\mathcal{H}} = \tilde{\mathcal{H}} \otimes \mathcal{H}$). The direct product of more than two Hilbert spaces $\{\mathcal{H}^{(\mu)}\}$ is similarly defined and denoted by $\bigotimes_{\mu} \mathcal{H}^{(\mu)}$.

Appendix B. Virtual ground state of a system of a single degree of freedom and its decay

This appendix is an elementary introduction to the <u>quantum theory of decay</u>, namely the theory used to describe quantum-decay phenomena. In what follows, H_S is the Hamiltonian given by (2.1.10) and V is the bumpy slope shown in Fig. 2.1(b).

B.1 Energy eigenfunction

Those readers who might find the following material a little too abstract are advised to confirm it by solving the square-well problem exactly, or the bumpy-slope problem with the naive WKB method.

Define the energy eigenfunction $\mathcal{U}_E$ by the following equation:

$$H_S \mathcal{U}_E(q) = E \mathcal{U}_E(q), \tag{B.1.1}$$

$$\mathcal{U}_E(-\infty) = 0. \tag{B.1.2}$$

For a given E, Eq. (B.1.1), being a second-order differential equation, has two linearly-independent solutions, of which the <u>leftward</u> boundary condition (B.1.2) singles out a particular linear combination. For $E > V_\star$, $\mathcal{U}_E$ is a wave reflected totally by the infinitely high wall on the left. In the <u>rightward region</u> $q > q_\star$ where the potential is constant, $\mathcal{U}_E$ may be expressed as

$$\mathcal{U}_E(q) = \frac{1}{2i}\left\{A_+ e^{ik(E)q} - A_- e^{-ik(E)q}\right\} \quad : q > q_\star, \tag{B.1.3}$$

$$k(E) := \{2(E - V_\star)\}^{1/2}/h, \tag{B.1.4}$$

where $A_\pm$ are complex coefficients to be determined. Since the boundary condition (B.1.2) fixes the ratio A_-/A_+ as a function of E, $\mathcal{U}_E$ is determined uniquely up to an overall multiplicative factor, which is allowed to depend on E. We choose this factor appropriately to cast $\mathcal{U}_E$ into the form

$$\mathcal{U}_E(q) = \begin{cases} b_E(q) & : q < 1, \\ \dfrac{1}{2i}\left\{A_+(E)e^{ik(E)q} - A_-(E)e^{-ik(E)q}\right\} & : q > q_\star, \end{cases} \tag{B.1.5}$$

with b_E and $A_\pm(E)$ being smooth functions of E and $b_E(q)$ being a real normalized function of q localized around $q \sim 0$.

Since E is a mere parameter in the differential equation (B.1.1), what has been mentioned above, except for the physical picture of the reflected wave, is valid[1] for an arbitrary complex E. For a complex E, we adopt the convention of choosing the branch of $k(E)$ as

$$\arg\{|E - V_\star|e^{i\theta}\}^{1/2} = \theta/2 \quad : \quad -\pi < \theta \le \pi. \tag{B.1.6}$$

Hence, in particular,

$$k(E) = i|k(E)| \quad : \quad \theta = \pi. \tag{B.1.7}$$

Since H_S is a real differential operator, it follows from the choice of the overall multiplicative factor as specified by Eq. (B.1.5) that

$$\mathcal{U}_E(q) \in \mathbf{R} \quad : \quad E \in \mathbf{R}. \tag{B.1.8}$$

Accordingly[2]

$$A_-(E) = \{A_+(E)\}^* \quad : \quad E > V_\star. \tag{B.1.9}$$

Furthermore, the energy eigenfunction has the following properties $\mathcal{P}0 - \mathcal{P}3$.

$\mathcal{P}0$: If the overall multiplicative factor is chosen appropriately, b_E and $A_\pm(E)$ are analytic with respect to E except for the branch point at $E = V_\star$.

This property should be fairly obvious. Formally speaking, it is a result of the Poincaré theorem; if the coefficients of a differential equation are analytic with respect to a parameter, so are its solutions. Clearly, the coefficients of Eq. (B.1.1) are analytic with respect to E. Hence, so is $\mathcal{U}_E$. $A_\pm(E)$ have a branch point only because $\mathcal{U}_E$ has been expressed in the form of (B.1.5) where an element (B.1.4) having a branch point at $E = V_\star$ has been introduced; $\mathcal{U}_E$ as a whole does not have any branch points at all.

Let $\mathbf{C}_c$ be the domain in the complex E-plane with a cut along the semi-infinite interval $(-\infty, V_\star)$ on the real axis.

$\mathcal{P}1$: On the domain $\mathbf{C}_c$, the zeros of $A_\pm(E)$ are located symmetrically with respect to the real axis.

This property follows from $\mathcal{P}0$ and (B.1.9).

$\mathcal{P}2$: Let the zeros of $A_-(E)$ be

$$E_n - \frac{1}{2}ih\Gamma_n \quad : \quad n = 0, 1, \ldots, \quad E_n, \quad \Gamma_n \in \mathbf{R}, \quad E_0 < E_1 < \cdots, \tag{B.1.10}$$

then $\Gamma_n > 0$. $\tag{B.1.11}$

This will be proven in the next section.

$\mathcal{P}3$: In the quasi-classical situation,

$$E_n - E_{n-1} = \mathcal{O}(h\omega_0) \quad : \quad E_{-1} \equiv 0, \tag{B.1.12}$$

$$h\Gamma_n/(E_n - E_{n-1}) = \mathcal{O}(e^{-1/h}), \tag{B.1.13}$$

for each n.

---

[1] We somewhat abuse language to call $\mathcal{U}_E$ the energy eigenfunction even if E is a complex number.
[2] In general, $z > 0$ is meant to imply that the complex number z is real and positive.

This property ensures the existence of a positive constant $\bar{\epsilon}$ such that

$$h\Gamma_n \ll \bar{\epsilon} \ll E_n - E_{n-1}, \tag{B.1.14}$$

$$A_\pm(E) \simeq \left\{ E - \left(E_n \pm \frac{1}{2} ih\Gamma_n \right) \right\} \tilde{a}_{n\pm} \quad : \ |E - E_n| < \bar{\epsilon}, \tag{B.1.15}$$

where $\tilde{a}_{n\pm}$ are E-independent constants. Since $\tilde{a}_{n-} = \tilde{a}_{n+}^*$ by virtue of Eq. (B.1.9), we may write

$$\tilde{a}_{n\pm} =: \sqrt{2\pi/h\Gamma_n} \, a_n \exp(\pm i\delta_n), \tag{B.1.16}$$

with a_n and δ_n being real constants. The foremost factor on the right-hand side has been introduced for convenience to ensure that a_n is of $\mathcal{O}(1)$.

B.2 Quasi-stationary wave

We define the *quasi-stationary wave* $\mathcal{U}_{QSS}$ as

$$\mathcal{U}_{QSS}(q) := \mathcal{U}_{\tilde{E}}(q), \quad \tilde{E} \equiv E_0 - \frac{1}{2}ih\Gamma_0 . \tag{B.2.1}$$

This wave satisfies

$$H_S \, \mathcal{U}_{QSS}(q) = \tilde{E} \, \mathcal{U}_{QSS}(q) . \tag{B.2.2}$$

Since $h\Gamma_0$ is much smaller than E_0, $\mathcal{U}_{\tilde{E}}(q)$ should not differ much from $\mathcal{U}_{E_0}(q)$ for $q < 1$. Hence

$$\lim_{q \downarrow -\infty} \mathcal{U}_{QSS}(q) = 0, \tag{B.2.3}$$

$$\Re \, \mathcal{U}_{QSS}(q) > 0 \quad : \ q < 1. \tag{B.2.4}$$

In other words, $\Re \, \mathcal{U}_{QSS}(q)$ does not have nodes at $q < 1$ and looks like a Gaussian function in the neighborhood of $q = 0$. Also, by virtue of Eqs. (B.1.5) and (B.1.10), we have

$$\mathcal{U}_{QSS}(q) = (2i)^{-1} A_+(\tilde{E}) \exp\{ik(\tilde{E})q\} \quad : \ q > q_\star, \tag{B.2.5}$$

where

$$\Re \, k(\tilde{E}) > 0, \tag{B.2.6}$$

since $\Re \, k(\tilde{E})$ should not differ much from $\Re \, k(E_0)$ either. In view of (B.2.5) and (B.2.6), $\mathcal{U}_{QSS}(q)$ is said to obey the outgoing boundary condition. It follows from Eq. (B.2.2) that

$$\mathcal{U}_{QSS}^*(q) H_S \mathcal{U}_{QSS}(q) - \mathcal{U}_{QSS}(q) H_S \mathcal{U}_{QSS}^*(q) = -ih\Gamma_0 |\mathcal{U}_{QSS}(q)|^2 . \tag{B.2.7}$$

If $\mathcal{U}_{QSS}$ were an ordinary quantum state, the right-hand side would vanish and the law of conservation of probability current (i.e. a Wronskian relation) would result. Guided by this analogy, we define the quasi-stationary-wave probability integral $\mathcal{P}(q)$ and the quasi-stationary-wave probability current $\mathcal{J}(q)$ as[3]

$$\mathcal{P}(q) := \int_{-\infty}^{q} dq' \, |\mathcal{U}_{QSS}(q')|^2, \tag{B.2.8}$$

$$\mathcal{J}(q) := \Re \left(\mathcal{U}_{QSS}^*(q) \frac{h}{i} \frac{d}{dq} \mathcal{U}_{QSS}(q) \right). \tag{B.2.9}$$

---

[3] These are mere nicknames which have nothing to do with quantum-mechanical probabilities.

We integrate both sides of (B.2.7) and use the property (B.2.3) to find

$$\Gamma_0 = \mathcal{J}(q)/\mathcal{P}(q). \tag{B.2.10}$$

This important relation is to be called the *Gamow formula*, which was first derived by Gamow in his classic paper dealing with the α-decay. The crux of this formula is the following:

The ratio of $\mathcal{J}(q)$ and $\mathcal{P}(q)$ is independent of q, although each of them depends on q.

Exploiting this property, we choose q to be greater than q_* so that Eq. (B.2.5) may be used and find

$$\mathcal{J}(q) = h|\mathcal{U}_{\text{QSS}}(q)|^2 \, \Re k(\tilde{E}) > 0, \tag{B.2.11}$$

where the last inequality follows from (B.2.6). This proves (B.1.11) for $n = 0$. Cases of general n can be proven similarly.

B.3 Resonance and virtual bound states

In this and the next sections, we suppose that $E \in \mathbf{R}$ and express $A_\pm(E)$ as

$$A_\pm(E) =: A(E) \exp(\pm i\delta(E)), \quad A(E) \geq 0, \quad \delta(E) \in \mathbf{R}. \tag{B.3.1}$$

Accordingly

$$\mathcal{U}_E(q) = A(E) \sin\{k(E)q + \delta(E)\} \quad : q > q_*. \tag{B.3.2}$$

It follows from Eqs. (B.1.15) and (B.1.16) that

$$\{A(E)\}^2 \simeq a_n^2/\Lambda(E - E_n; h\Gamma_n) \quad : |E - E_n| < \bar{\epsilon}, \tag{B.3.3}$$

$$\Lambda(x; \gamma) := \frac{1}{\pi} \frac{\gamma/2}{x^2 + (\gamma/2)^2}. \tag{B.3.4}$$

The property (B.1.13) implies that $A(E)$ has a sharp dip at $E = E_n$. For example,

$$A(E_0)/A(0) \sim h\Gamma_0/E_0 \ll 1. \tag{B.3.5}$$

Hence, $\mathcal{U}_{E_n}$ represents a wave which is almost localized in the metastable well. E_n and $\mathcal{U}_{E_n}$ are to be called the *nth resonance energy* and the *nth resonance wave*, respectively. $b_{E_n}(q)$ has n nodes. In particular, $b_{E_0}(q)$ is nodeless and looks like a Gaussian function for $q \sim 0$.

Although the resonance wave $\mathcal{U}_{E_n}$ may appear to be a bound state, it is extended to $q \sim \infty$ and is not normalizable; it does not represent a quantum state by itself. However, a normalizable wavefunction localized around $q \sim 0$ may be constructed by superposing $\mathcal{U}_E$ with E around $E = E_n$. The wavefunction

$$\psi_n(q) := \int_{E_n - \bar{\epsilon}}^{E_n + \bar{\epsilon}} dE \, \tilde{\Lambda}_n(E) \mathcal{U}_E(q), \tag{B.3.6}$$

$$\tilde{\Lambda}_n(E) \equiv \Lambda(E - E_n; h\Gamma_n) \bigg/ \int_{E_n - \bar{\epsilon}}^{E_n + \bar{\epsilon}} dE \, \Lambda(E - E_n; h\Gamma_n) \tag{B.3.7}$$

is called the *n*-th virtual bound state. In particular, the 0-th virtual bound state is called the virtual ground state. The coefficient ($\propto \{A(E)\}^{-2}$) of superposition has been so chosen that only those $\mathcal{U}_E$ with the energy E in the close neighborhood of E_n

(i.e. $|E - E_n| \lesssim h\Gamma_n$) contribute to the superposition.[4] $\psi_n(q)$ almost coincides with $b_{E_n}(q)$ inside the well and vanishes to a good approximation outside the well.

B.4 Decay of the virtual ground state

Let us focus attention on the virtual ground state. In order to study its decay, we solve the Schrödinger equation (2.1.9) under the initial condition $\psi(q, 0) = \psi_0(q)$. The formal solution is given as

$$\psi(q, t) = \int_{E_0-\bar{\epsilon}}^{E_0+\bar{\epsilon}} dE \, \tilde{\Lambda}_0(E) \mathcal{U}_E(q) \exp(-iEt/h). \qquad (B.4.1)$$

In particular, inside the metastable well, this wavefunction is of the form

$$\psi(q, t) \simeq b_{E_0}(q) e^{-iE_0 t/h} \mathcal{A}(\Gamma_0 t; 2\bar{\epsilon}/h\Gamma_0) \quad : q < 1, \qquad (B.4.2)$$

$$\mathcal{A}(\tau; \Omega) := \frac{1}{c} \int_{-\Omega}^{\Omega} \frac{dv}{\pi} \frac{e^{-iv\tau/2}}{v^2 + 1}, \quad c \equiv \int_{-\Omega}^{\Omega} \frac{dv}{\pi} \frac{1}{v^2 + 1}. \qquad (B.4.3)$$

The persistence probability $\mathcal{P}(t)$ defined by Eq. (2.3.1) in Section 2.3 is found with the aid of Eq. (B.4.2) as

$$\mathcal{P}(t) \simeq |\mathcal{A}(\Gamma_0 t; 2\bar{\epsilon}/h\Gamma_0)|^2. \qquad (B.4.4)$$

In the quasi-classical situation, where $\Omega \gg 1$, the integral (B.4.3) may be evaluated approximately to give

$$\mathcal{A}(\tau; \Omega) \simeq \begin{cases} 1 - (\Omega/4\pi)\tau^2 & : \tau \ll \Omega^{-1}, \\ \exp(-\tau/2) + (4/\pi\Omega^2\tau)\sin(\tau\Omega/2) & : \tau \gg \Omega^{-1}. \end{cases} \qquad (B.4.5)$$

In the temporal domain

$$1/\Omega \ll \tau/2 \ll \log \Omega, \qquad (B.4.6)$$

therefore, the exponential-decay law (2.3.2) is valid:

$$\mathcal{A}(\tau; \Omega) \simeq \mathcal{A}(\tau; \infty) = \exp(-\tau/2), \qquad (B.4.7)$$

$$\Gamma = \Gamma_0 = -2 \, \Im \, \tilde{E}/h. \qquad (B.4.8)$$

It is thus seen[5] that one may obtain the decay rate for the virtual ground state by solving the eigenvalue problem (B.2.2) under the conditions (B.2.3), (B.2.4), and (B.2.5). This problem may be solved by use of the proper WKB method to lead to the formula (2.4.5).

---

[4] One may choose $\{A(E)\}^{-p}$, $p > 1$, instead of $\{A(E)\}^{-2}$.

[5] It is to be emphasized that the quasi-stationary wave itself, being unnormalizable, does not represent a quantum state, and hence no physical meaning can be ascribed to it. It merely provides us with convenient means to calculate the decay rate without requiring the explicit solution of the time-dependent Schrödinger equation or analytical evaluation of the complex zero of $A_\pm(E)$, which are more difficult.

Appendix C. Functional derivative

A function of a function is called a functional. An example is the S defined by Eq. (2.6.12); S is a functional of q and the value of S at q is denoted by $S[q]$. In dealing with a functional, it is desirable to distinguish between a function q and its value $q(\tau)$ at τ.[1] Suppose that the variable q is required to belong to the function space $\mathcal{F}_\infty$ defined by Eq. (2.6.13), then S is said to be a functional defined on the domain $\mathcal{F}_\infty$.

For a preliminary consideration, let us recall partial derivatives of a function $\mathcal{S}(q_1, q_2, \ldots, q_N)$ of N variables. For a given $r_1, r_2, \ldots, r_N$, we may write

$$\mathcal{S}(q_1 + r_1, q_2 + r_2, \ldots, q_N + r_N) - \mathcal{S}(q_1, q_2, \ldots, q_N) = \sum_{n=1}^{\infty} \delta_r^n \mathcal{S}(q_1, q_2, \ldots, q_N),$$

(C.0.1)

where $\delta_r^n \mathcal{S}(\cdots)$ represents the n-th order terms when the left-hand side is expanded in powers of $r_1, \ldots, r_N$. Rewrite r_j as δq_j, and abbreviate $\delta_{\delta q}^n \mathcal{S}(\cdots)$ as $\delta^n \mathcal{S}(\cdots)$, then

$$\delta^1 \mathcal{S}(q_1, q_2, \ldots, q_N) = \sum_{j=1}^{N} \frac{\partial \mathcal{S}(q_1, q_2, \ldots, q_N)}{\partial q_j} \delta q_j,$$ (C.0.2)

$$\delta^2 \mathcal{S}(q_1, q_2, \ldots, q_N) = \frac{1}{2!} \sum_{j=1}^{N} \sum_{k=1}^{N} \frac{\partial^2 \mathcal{S}(q_1, q_2, \ldots, q_N)}{\partial q_j \partial q_k} \delta q_j \delta q_k,$$ (C.0.3)

$$\cdots\cdots\cdots .$$

These equations give $\delta^n \mathcal{S}(\cdots)$ in terms of the n-th order partial derivative of $\mathcal{S}$. Alternatively, with $\delta^n \mathcal{S}(\cdots)$ defined by Eq. (C.0.1), these equations may be regarded as defining partial derivatives.

The latter view may be extended from the case of a N-variable function to that of a functional to define functional derivatives. Let r be an arbitrary function belonging to $\mathcal{F}_\infty$. In analogy to Eq. (C.0.1), we write

$$S[q + r] - S[q] = \sum_{n=1}^{\infty} \delta_r^n S[q],$$

(C.0.4)

---

[1] An often-encountered convention of writing $S[q(\tau)]$ for $S[q]$ is not recommended; $S[q]$ is not a quantity depending on τ.

where $\delta_r^n S[q]$ represents terms of the n-th order with respect to the function r. Thus, it should be possible to express $\delta_r^1 S[q]$ in the form

$$\delta_r^1 S[q] = \int_{-\infty}^{\infty} d\tau \; f(\tau) \, r(\tau), \tag{C.0.5}$$

where $f(\tau)$ is some function of τ. This function is denoted by the symbol $\delta S[q]/\delta q(\tau)$ and is called the functional derivative.[2] Similarly, it should be possible to express $\delta_r^2 S[q]$ in the form

$$\delta_r^2 S[q] = \frac{1}{2!} \int_{-\infty}^{\infty} d\tau \int_{-\infty}^{\infty} d\tau' \; g(\tau, \tau') \; r(\tau) r(\tau'), \tag{C.0.6}$$

where $g(\tau, \tau')$ is some function of τ and τ'. This function is denoted by the symbol $\delta^2 S[q]/\delta q(\tau)\delta q(\tau')$ and is called the second-order functional derivative. One may deal with higher-order ones similarly. Now, rewrite r as δq and define

$$\delta^n S[q] := \delta_{\delta q}^n S[q], \tag{C.0.7}$$

which is called the n-th variation of $S[q]$, then

$$\delta^1 S[q] =: \int_{-\infty}^{\infty} d\tau \; \frac{\delta S[q]}{\delta q(\tau)} \delta q(\tau), \tag{C.0.8}$$

$$\delta^2 S[q] =: \frac{1}{2!} \int_{-\infty}^{\infty} d\tau \int_{-\infty}^{\infty} d\tau' \; \frac{\delta^2 S[q]}{\delta q(\tau)\delta q(\tau')} \delta q(\tau)\delta q(\tau'), \tag{C.0.9}$$

$$\cdots \cdots \cdots .$$

Note that the equality symbol is not := but =: . With $\delta^n S[q]$ defined by Eqs. (C.0.4) and (C.0.7), the above equations (C.0.8), (C.0.9), $\cdots$ define functional derivatives. Equation (C.0.8) may be viewed as a continuous version of Eq. (C.0.2); the former would result if the discrete index j and the summation over it in the latter were replaced by the continuous variable τ and the integral over it, respectively. It is this correspondence that motivates the notation $\delta S[q]/\delta q(\tau)$. This correspondence, however, is purely formal; Eq. (C.0.8) is not derived as a continuum limit of Eq. (C.0.2). Nevertheless, with this formal correspondence in mind, the functional S is said to be differentiated when the functional derivative $\delta S[q]/\delta q(\tau)$ is computed.[3]

$\delta S[q]/\delta q(\tau)$ is also a functional of q (and is a function of τ as well). Hence, its functional derivative is defined according to the above rule. It has the following property:

$$\frac{\delta}{\delta q(\tau)}\left(\frac{\delta S[q]}{\delta q(\tau')}\right) = \frac{\delta^2 S[q]}{\delta q(\tau)\delta q(\tau')}, \tag{C.0.10}$$

where the left-hand side is the functional derivative of the functional $\delta S[q]/\delta q(\tau')$, while the right-hand side is the second-order functional derivative of S as defined by Eq. (C.0.9). Incidentally, given a function q, its value $q(\tau)$ is determined by q (for any τ). For each τ, therefore, $q(\tau)$ is a functional of q:

$$q(\tau) = \int_{-\infty}^{\infty} d\tau' \; \delta(\tau - \tau')q(\tau'). \tag{C.0.11}$$

---

[2] More formally, called the variational derivative.
[3] On the right-hand side of Eq. (C.0.8), the last factor $\delta q(\tau)$, which is independent of $q(\tau)$, is the value of the function δq at τ; to avoid confusion, it is better to be written as $(\delta q)(\tau)$.

Hence

$$\frac{\delta q(\tau')}{\delta q(\tau)} = \delta(\tau' - \tau). \tag{C.0.12}$$

As an example, consider the functional S defined by Eq. (2.6.12), then

$$S[q + r] - S[q]$$

$$= \int_{-\infty}^{\infty} d\tau \left[\frac{1}{2} \{\dot{q}(\tau) + \dot{r}(\tau)\}^2 - \frac{1}{2} \{\dot{q}(\tau)\}^2 + V(q(\tau) + r(\tau)) - V(q(\tau)) \right]$$

$$= \int_{-\infty}^{\infty} d\tau \{\dot{q}(\tau)\dot{r}(\tau) + V'(q(\tau))r(\tau)\} + \frac{1}{2} \int_{-\infty}^{\infty} d\tau [\{\dot{r}(\tau)\}^2 + V''(q(\tau))\{r(\tau)\}^2]$$

$$+ \frac{1}{3!} \int_{-\infty}^{\infty} d\tau \, V'''(q(\tau))\{r(\tau)\}^3 + \cdots, \tag{C.0.13}$$

$$V'(q(\tau)) := \left. \frac{dV(q)}{dq} \right|_{q=q(\tau)}, \quad \cdots. \tag{C.0.14}$$

The first integral on the right-hand side of Eq. (C.0.13) may be cast into the form of (C.0.5) by integrating the term involving $\dot{r}$ by parts.[4] It follows that

$$\frac{\delta S[q]}{\delta q(\tau)} = -\ddot{q}(\tau) + V'(q(\tau)). \tag{C.0.15}$$

Likewise, the second integral may be cast into the form

$$\frac{1}{2} \int_{-\infty}^{\infty} d\tau \, r(\tau)\mathcal{D}_{\tau}^{[q]} r(\tau), \quad \mathcal{D}_{\tau}^{[q]} := -\frac{d^2}{d\tau^2} + V''(q(\tau)), \tag{C.0.16}$$

where the symbol $\mathcal{D}_{\tau}^{[q]}$ stipulates that it is a differential operator with respect to τ and at the same time is a functional of q. Re-expressing the above further in the form of (C.0.6), we find

$$\frac{\delta^2 S[q]}{\delta q(\tau)\delta q(\tau')} = \mathcal{D}_{\tau}^{[q]} \delta(\tau - \tau'). \tag{C.0.17}$$

Similar argument gives

$$\frac{\delta^3 S[q]}{\delta q(\tau)\delta q(\tau')\delta q(\tau'')} = \delta(\tau - \tau')\delta(\tau - \tau'')V'''(q(\tau)). \tag{C.0.18}$$

Next, let us consider the case that the variable of a functional is also a functional. Let $\tilde{S}[x]$ be a functional of x and $x^{[q]}$ be another functional of q. We may define a new functional of q as

$$S[q] := \tilde{S}[x^{[q]}], \tag{C.0.19}$$

which is called a composite functional. Since $x^{[q]}$ is a function of τ as well as a functional of q, its value at τ may be denoted by $x^{[q]}(\tau)$. Then, the following rule holds:

$$\frac{\delta S[q]}{\delta q(\tau)} = \int_{-\infty}^{\infty} d\tau' \left. \frac{\delta \tilde{S}[x]}{\delta x(\tau')} \right|_{x=x^{[q]}} \frac{\delta x^{[q]}(\tau')}{\delta q(\tau)}. \tag{C.0.20}$$

---

[4] Note the fact that $q, r \in \mathcal{F}_{\infty}$.

A function q belonging to the function space $\mathcal{F}_\infty$ is called a point q of the space $\mathcal{F}_\infty$. Suppose that the functional derivative of S vanishes at a point Q:

$$\left.\frac{\delta S[q]}{\delta q(\tau)}\right|_{q=Q} = 0 . \tag{C.0.21}$$

Such a point Q is called a *stationary point* of S, and $S[Q]$ is called the *stationary value* of S at the stationary point.

Appendix D. Miscellanea about spin

D.1 Spin operator

Let $\{\hat{S}_a \mid a = 1, 2, 3\}$ be the spin operator[1] of magnitude S:

$$[\hat{S}_a, \hat{S}_b] = i \sum_c \varepsilon_{abc} \hat{S}_c, \qquad \sum_a \hat{S}_a^2 = S(S+1), \tag{D.1.1}$$

where indices a, b, ... have nothing to do with the three-dimensional Euclidean space $\mathbf{E}^3$ representing the space in the laboratory. With the aid of the rotation matrix (see Appendix A) on $\mathbf{E}^3$, we define the spin-operator vector $\hat{\mathbf{S}}$ as[2]

$$\hat{\mathbf{S}} := \sum_a \sum_j \hat{S}_a \mathcal{R}_{aj} \mathbf{e}_j. \tag{D.1.2}$$

Note that $\hat{\mathbf{S}}$ is a vector of $\mathbf{E}^3$ as well as an operator. Hereafter, unless otherwise stated, we take $\mathcal{R}_{aj} = \delta_{aj}$, which is to be called the standard choice. When effects of anisotropy of a magnet or magnetic field are considered, there appears an operator of the form $\mathbf{e}_j \cdot \hat{\mathbf{S}}$ (i.e. the $\mathbf{e}_j$-component of $\hat{\mathbf{S}}$), which couples the spin space to $\mathbf{E}^3$. For the purpose of emphasizing this circumstance, we use the symbol $\hat{S}_j := \mathbf{e}_j \cdot \hat{\mathbf{S}}$. With the standard choice above, we have $\hat{S}_x = \hat{S}_1, \ldots$. Of course, the operators $\{\hat{S}_a\}$ are specified by the condition (D.1.1) alone, and it is arbitrary which of the three to call $\hat{S}_1$. Likewise, one may argue that it is merely a matter of nomenclature whether to use the symbols $\hat{S}_x, \ldots$ or $\hat{S}_1, \ldots$. However, in view of the fact that the spin operators are defined without any reference to $\mathbf{E}^3$ at all, it is advisable to keep a conceptual distinction between the abstract indices a, b, ... and the $\mathbf{E}^3$-bound ones x, y, ...

For a given unit vector $\mathbf{m}$ and a real number θ, the operator

$$\hat{R}(\theta\mathbf{m}) := \exp(-i\theta\mathbf{m} \cdot \hat{\mathbf{S}}) \tag{D.1.3}$$

is called the *spin-rotation operator* associated with the rotation $\mathcal{R}(\theta\mathbf{m})$ on $\mathbf{E}^3$. Let $\hat{R}^\dagger(\theta\mathbf{m})$ be the Hermitian conjugate of $\hat{R}(\theta\mathbf{m})$, then

$$\hat{R}^\dagger(\theta\mathbf{m})\hat{\mathbf{S}}\hat{R}(\theta\mathbf{m}) = \mathcal{R}(\theta\mathbf{m})\hat{\mathbf{S}}, \tag{D.1.4}$$

---

[1] Taken to be dimensionless.

[2] $\mathcal{R}_{aj}$ is obtained from $\mathcal{R}_{kj}$ by replacing the left index k with a. Although these two quantities carry an index of different nature, their values are the same; $\mathcal{R}_{1j} = \mathcal{R}_{xj}, \ldots$.

where $\mathcal{R}(\theta\mathbf{m})$ on the right-hand side acts on $\{\mathbf{e}_j\}$, of course, in accordance with Eq. (A.1.2).

D.2 Spin-1/2 decomposition of spin S

To deal with the spin operator $\hat{\mathbf{S}}$ of magnitude S, we introduce an auxiliary tool, namely a set $\{\hat{\mathbf{s}}^{(\alpha)} \mid \alpha = 1, 2, \ldots, 2S\}$ of $2S$ independent spin operators with each $\hat{\mathbf{s}}^{(\alpha)}$ being of magnitude 1/2. The total spin $\hat{\mathbf{T}} := \sum_\alpha \hat{\mathbf{s}}^{(\alpha)}$ satisfies the following theorem.

Theorem D.2.1 [spin-1/2 decomposition theorem for operators]
For an arbitrary polynomial $f(\mathbf{x}) \equiv f(x_1, x_2, x_3)$ of three variables,[3]

$$f(\hat{\mathbf{T}}) = 0 \implies f(\hat{\mathbf{S}}) = 0.$$

(The converse is false, of course; for example, $\hat{\mathbf{S}}^2 = S(S+1) \neq \hat{\mathbf{T}}^2$.) Perhaps, the validity of this theorem is obvious intuitively. A particle (or a magnet) of spin S is in reality a collection of elementary particles each of spin 1/2.[4] Hence, the property of the former should be the same as that of the latter, provided that the latter behaves cooperatively as an entity of spin S.[5]

A formal proof may run as follows. Let $\mathcal{H}_S$ be the $(2S + 1)$-dimensional Hilbert space on which $\hat{\mathbf{S}}$ acts (cf. Eq. (3.3.20)), and

$$\mathcal{H}_{1/2}^{(\alpha)} := \mathcal{S}pan\left\{\left|\pm\frac{1}{2}\right\rangle^{(\alpha)}\right\} \tag{D.2.1}$$

be the two-dimensional Hilbert space on which $\hat{\mathbf{s}}^{(\alpha)}$ acts. The total spin $\hat{\mathbf{T}}$ acts on the direct-product space $\tilde{\mathcal{H}}$:

$$\tilde{\mathcal{H}} := \bigotimes_\alpha \mathcal{H}_{1/2}^{(\alpha)} = \tilde{\mathcal{H}}_S \oplus \tilde{\mathcal{H}}_{S-1} \oplus \cdots, \tag{D.2.2}$$

where the right-hand side is the decomposition of $\tilde{\mathcal{H}}$ into its irreducible subspaces with $\tilde{\mathcal{H}}_S$ being the spin-S subspace. Hence

$$\begin{aligned} f(\hat{\mathbf{T}}) = 0 &\iff f(\hat{\mathbf{T}})\tilde{\mathcal{H}} = 0 \\ &\implies f(\hat{\mathbf{T}})\tilde{\mathcal{H}}_S = 0 \\ &\iff f(\hat{\mathbf{S}})\mathcal{H}_S = 0 \iff f(\hat{\mathbf{S}}) = 0, \end{aligned} \tag{D.2.3}$$

where $\hat{A}\tilde{\mathcal{H}} = 0$ is the abbreviation of $\hat{A}|\psi\rangle = 0$ for $^\forall|\psi\rangle \in \tilde{\mathcal{H}}$.[6] In the above chain of reasoning, the last but one equivalence follows from the fact that the action of $\hat{\mathbf{T}}$ on $\tilde{\mathcal{H}}_S$ is

---

[3] The variables are not assumed to be commutative; for instance, $x_1x_2 \neq x_2x_1$.

[4] An anecdote about Eugene Wigner tells that he always tested a theorem on matrices with 2×2 matrices. One might suspect that S was always decomposed as $(1/2) \times 2S$ in his brain.

[5] Equations in which some of the individual spins play distinguished roles are out of the scope of theorem D.2.1; for example, although the equation $\hat{\mathbf{T}}^2 - \frac{3}{2} - 2\hat{\mathbf{s}}^{(1)} \cdot \hat{\mathbf{s}}^{(2)} = 0$ holds for $S = 1$, the theorem does not apply to this equation since it can not be cast into the form of $f(\hat{\mathbf{T}}) = 0$.

[6] A state belonging to $\tilde{\mathcal{H}}$ is to be denoted by a round ket $|\cdots)$.

isomorphic to that of $\hat{\mathbf{S}}$ on $\mathcal{H}_S$. To see this isomorphism, define

$$|S) := \prod_\alpha \left|\frac{1}{2}\right)^{(\alpha)},$$

$$|m-1) := \{S(S+1) - m(m-1)\}^{-1/2}\,\hat{T}_-|m), \tag{D.2.4}$$

and note that these states span $\tilde{\mathcal{H}}_S$:

$$\tilde{\mathcal{H}}_S = \mathcal{S}pan\{|m) \mid m = -S, -S+1, \ldots, S\}. \tag{D.2.5}$$

Hence, there is a one-to-one correspondence between $\tilde{\mathcal{H}}_S$ and $\mathcal{H}_S$:

$$|\psi) \equiv \sum_m c_m|m) \longleftrightarrow \sum_m c_m|m\rangle \equiv |\psi\rangle, \tag{D.2.6}$$

$$f(\hat{\mathbf{T}})|\psi) \longleftrightarrow f(\hat{\mathbf{S}})|\psi\rangle. \tag{D.2.7}$$

(The above correspondence respects phase factors as well.) [Q.E.D.]

As a corollary of (D.2.6) and (D.2.7), we obtain the following theorem.

Theorem D.2.2 [spin-1/2 decomposition theorem for matrix elements]
Given a pair of states $|\psi_n\rangle$ $(n = 1, 2)$ belonging to $\mathcal{H}_S$, let $|\psi_n)$ $(\in \tilde{\mathcal{H}}_S)$ be the pair of states corresponding to the former in accordance with (D.2.6), then

$$\langle\psi_2|f(\hat{\mathbf{S}})|\psi_1\rangle = (\psi_2|f(\hat{\mathbf{T}})|\psi_1).$$

Note that this is not a mere correspondence but an equality. Thus, in order to evaluate the left-hand side, which is complicated in general, one has only to compute matrix elements of products of at most p spin-1/2 operators if the degree of the polynomial $f(\mathbf{x})$ is p.

Theorem D.2.1 may be applied typically to the proof of a relation of the form

$$f_1(\hat{\mathbf{S}}) = f_2(\hat{\mathbf{S}}), \tag{D.2.8}$$

where each $f_\nu(\mathbf{x})$ $(\nu = 1, 2)$ is a product of exponential functions with the exponent of each of them being linear in $\mathbf{x}$. In this case, $f_\nu(\hat{\mathbf{T}})$ is factorizable as

$$f_\nu(\hat{\mathbf{T}}) = \prod_\alpha f_\nu(\hat{\mathbf{s}}^{(\alpha)}) \quad : \nu = 1, 2. \tag{D.2.9}$$

Accordingly, one may prove $f_1(\hat{\mathbf{T}}) = f_2(\hat{\mathbf{T}})$ simply by demonstrating (D.2.8) for spin 1/2. Once this is proven, theorem D.2.1 may be invoked with $f \equiv f_1 - f_2$ to prove (D.2.8) for an arbitrary S.

D.3 Spin disentanglement theorem

Consider operating the rotation operator $\hat{R} \equiv \exp(-i\theta\hat{S}_2)$ on the state $|S\rangle$. A direct calculation by expanding $\hat{R}$ in the Taylor series of the exponent, for instance, is fairly cumbersome. Instead, pay attention to the relation $2i\hat{S}_2 = \hat{S}_+ - \hat{S}_-$. If $\hat{S}_\pm$ were mutually commutative, $\hat{R}$ would be equal to $\exp(\theta\hat{S}_-/2)\exp(-\theta\hat{S}_+/2)$ and $\hat{R}|S\rangle$ would reduce to $\exp(\theta\hat{S}_-/2)|S\rangle$ since $\hat{S}_+|S\rangle = 0$, thereby simplifying the calculation tremendously. This suggests the following conjecture:

$$\exp(-i\theta\hat{S}_2) = \exp(\lambda\hat{S}_-)\hat{\mathcal{M}}\exp(-\lambda\hat{S}_+), \tag{D.3.1}$$

where λ is a function of θ and $\hat{\mathcal{M}}$ is a function of $\hat{S}_3$ and θ. It should not be difficult to guess that a $\hat{S}_3$-dependent factor $\hat{\mathcal{M}}$ arises from the non-commutativity of $\hat{S}_\pm$. The right-hand side is a kind of normal ordering; the operators $\hat{S}_+$ and $\hat{S}_-$, which are entangled on the left-hand side, are disentangled and sorted out towards the right and the left, respectively. The above relation is equivalent to

$$\hat{\mathcal{M}} = \exp(-\lambda \hat{S}_-)\exp(-i\theta \hat{S}_2)\exp(\lambda \hat{S}_+). \tag{D.3.2}$$

If we succeed in choosing λ in such a way that the right-hand side is a function of $\hat{S}_3$ alone, then the conjecture (D.3.1) turns out to be correct and we are done.

If one tried to compute the right-hand side of (D.3.2) for a general S from scratch, one would soon be at a loss unless one is exceptionally energetic. Fortunately, by virtue of the spin disentanglement theorem proven in the preceding section, it is sufficient to do the computation only for $S = 1/2$. Let $\hat{s}$ be the spin-1/2 operator. For $S = 1/2$, one may use the identities $(2\hat{s}_2)^2 = 1$ and $\hat{s}_\pm^2 = 0$ to reduce (D.3.2) to

$$\hat{\mathcal{M}} = (1 - \lambda \hat{s}_-)\{\cos(\theta/2) - (\hat{s}_+ - \hat{s}_-)\sin(\theta/2)\}(1 + \lambda \hat{s}_+)$$

$$= \{1 + (\lambda - t)(\hat{s}_+ - \hat{s}_-) - \lambda(\lambda - 2t)\hat{s}_-\hat{s}_+\}c, \tag{D.3.3}$$

$$t \equiv \tan(\theta/2), \quad c \equiv \cos(\theta/2). \tag{D.3.4}$$

Noting that $\hat{s}_-\hat{s}_+ = 1/2 - \hat{s}_3$, one finds that the right-hand side depends only on $\hat{s}_3$ if λ is chosen to be t. With this choice,

$$\hat{\mathcal{M}} = \left\{1 + t^2\left(\frac{1}{2} - \hat{s}_3\right)\right\}c = \left(\frac{1}{2} + \hat{s}_3\right)c + \left(\frac{1}{2} - \hat{s}_3\right)\frac{1}{c}. \tag{D.3.5}$$

The operators $1/2 \pm \hat{s}_3$ on the right-hand side are the projectors onto the subspaces corresponding to the eigenvalues $\pm 1/2$ of $\hat{s}_3$, respectively. Hence, $\hat{\mathcal{M}}$ is equal to c or $1/c$ depending on $2\hat{s}_3$ being equal to 1 or -1. Accordingly, one finds

$$\hat{\mathcal{M}} = \exp(2\hat{s}_3 \log c). \tag{D.3.6}$$

Thus, the following lemma has been proven.

Theorem D.3.1 [lemma for spin disentanglement theorem]

$$\exp(-i\theta \hat{S}_2) = \exp(\lambda \hat{S}_-)\exp\left\{2\hat{S}_3 \log\left(\cos\frac{\theta}{2}\right)\right\}\exp(-\lambda \hat{S}_+), \tag{D.3.7}$$

$$\lambda \equiv \tan(\theta/2). \tag{D.3.8}$$

The spin operator appearing on the left-hand side is $\hat{S}_2$, namely $\mathbf{e}_y \cdot \hat{\mathbf{S}}$. The direction $\mathbf{e}_y$ may be generalized to an arbitrary direction

$$\mathbf{m} \equiv -\mathbf{e}_x \sin\phi + \mathbf{e}_y \cos\phi \tag{D.3.9}$$

in the xy-plane by ϕ-rotation of the spin around the $\mathbf{e}_z$-axis. Let $\hat{R} \equiv \hat{R}(\phi \mathbf{e}_3)$, then

$$\hat{R}\hat{S}_2\hat{R}^\dagger = \mathbf{m} \cdot \hat{\mathbf{S}}, \quad \hat{R}\hat{S}_+\hat{R}^\dagger = e^{-i\phi}\hat{S}_+, \quad \hat{R}\hat{S}_3\hat{R}^\dagger = \hat{S}_3. \tag{D.3.10}$$

Multiplying Eq. (D.3.7) by $\hat{R}$ and $\hat{R}^\dagger$ from the left and the right, respectively, one obtains with the help of the above relations the following theorem.[7]

---

[7] Those readers who are fond of highbrow techniques may prove this theorem by use of the representation of $SL(2, \mathbf{C})$, for instance. However, as has been explained above, the essence of the theorem is exhausted by $SU(2)$, namely spin-1/2; highbrow methods of proof tend to make simple matters appear difficult.

Theorem D.3.2 [spin disentanglement theorem]

$$\hat{R}(\theta\mathbf{m}) \equiv \exp\{i\theta(\hat{S}_1 \sin\phi - \hat{S}_2 \cos\phi)\}$$

$$= \exp(\mu\hat{S}_-)\exp\left\{2\hat{S}_3 \log\left(\cos\frac{\theta}{2}\right)\right\}\exp(-\mu^*\hat{S}_+), \qquad (D.3.11)$$

$$\mu \equiv e^{i\phi}\tan(\theta/2). \qquad (D.3.12)$$

The constant μ is a quantity familiar in the *stereographic projection* from the south pole of the Riemann sphere. In the case of letting $\hat{R}(\theta\mathbf{m})$ act on $|-S\rangle$ in place of $|S\rangle$, it is more convenient to arrange the operators in the reverse order; rotating the spin by π around the $\mathbf{e}_x$-axis in (D.3.11), one obtains

Theorem D.3.3 [reverse-ordered spin disentanglement theorem]

$$\hat{R}(\theta\mathbf{m}) = \exp(-\mu^*\hat{S}_+)\exp\left\{-2\hat{S}_3 \log\left(\cos\frac{\theta}{2}\right)\right\}\exp(\mu\hat{S}_-). \qquad (D.3.13)$$

D.4 Spin coherent state

For the purpose of understanding spin intuitively, let us try to construct a state in which the spin is in **n**-direction for a given unit vector **n**. We start from the relation

$$\mathbf{e}_z \cdot \hat{\mathbf{S}}\,|S\rangle = S\,|S\rangle. \qquad (D.4.1)$$

Now, any unit vector **n** can be obtained by rotating $\mathbf{e}_z$ by $\theta_\mathbf{n}$ around

$$\mathbf{m} := \frac{\mathbf{e}_z \times \mathbf{n}}{|\mathbf{e}_z \times \mathbf{n}|} = -\mathbf{e}_x \sin\phi_\mathbf{n} + \mathbf{e}_y \cos\phi_\mathbf{n}, \qquad (D.4.2)$$

where $\theta_\mathbf{n}$ and $\phi_\mathbf{n}$ are the azimuth angles for **n** with respect to the triad $\{\mathbf{e}_x, \mathbf{e}_y, \mathbf{e}_z\}$. It follows from Eq. (D.1.4) that

$$\mathbf{e}_z \cdot \hat{\mathbf{S}} = \hat{R}^\dagger(\theta_\mathbf{n}\mathbf{m})\,\mathbf{n} \cdot \hat{\mathbf{S}}\,\hat{R}(\theta_\mathbf{n}\mathbf{m}). \qquad (D.4.3)$$

This relation combined with Eq. (D.4.1) gives Eq. (3.3.21), where

$$|\mathbf{n}\rangle := \hat{R}(\theta_\mathbf{n}\mathbf{m})|S\rangle \qquad (D.4.4)$$

$$= \exp(-\zeta_\mathbf{n}^*\hat{S}_+ + \zeta_\mathbf{n}\hat{S}_-)|S\rangle, \qquad \zeta_\mathbf{n} \equiv \frac{1}{2}\theta_\mathbf{n}\exp(i\phi_\mathbf{n}). \qquad (D.4.5)$$

For $S = 1/2$, in particular, $|\mathbf{n}\rangle$ may be expressed explicitly by expanding $\hat{R}(\theta_\mathbf{n}\mathbf{m})$ in the Taylor series as

$$|\mathbf{n}\rangle = \cos\frac{\theta_\mathbf{n}}{2}\left|\frac{1}{2}\right\rangle + e^{i\phi_\mathbf{n}}\sin\frac{\theta_\mathbf{n}}{2}\left|-\frac{1}{2}\right\rangle \quad : S = 1/2. \qquad (D.4.6)$$

If **n** is parallel to $\mathbf{e}_z$, the vector **m** (or the angle $\phi_\mathbf{n}$) is not defined and hence (D.4.4) is meaningless. However, (D.4.5) is well-defined even for $\mathbf{n} = \mathbf{e}_z$ (i.e. $\theta_\mathbf{n} = 0$). Hence, we define $|\mathbf{n}\rangle$ with (D.4.5) for $\mathbf{n} = \mathbf{e}_z$ as well:

$$|\mathbf{e}_z\rangle := |S\rangle. \qquad (D.4.7)$$

Thus, $|\mathbf{n}\rangle$ has been defined to be a continuous function of **n** for all **n** except for the south pole ($\mathbf{n} = -\mathbf{e}_z$); the exclusion of the south pole is an inevitable consequence of the topology of the sphere. The states $|\mathbf{n}\rangle$ are called *spin coherent states*. To be precise, they are called spin coherent states with north-pole convention, since they are generated by rotating $|S\rangle$ namely $|\mathbf{e}_z\rangle$.[8]

It follows immediately from Eq. (3.3.21) that $\langle \mathbf{n}| \, \mathbf{n} \cdot \hat{\mathbf{S}} \, |\mathbf{n}\rangle = S$. On the other hand, for any $\mathbf{l}(\perp\mathbf{n})$, the expectation value of $\mathbf{l} \cdot \hat{\mathbf{S}}$ with respect to the state $|\mathbf{n}\rangle$ vanishes. Therefore, the state $|\mathbf{n}\rangle$ reproduces the classical picture for spin in the following sense:

$$\langle \mathbf{n}| \, \hat{\mathbf{S}} \, |\mathbf{n}\rangle = S\mathbf{n}. \tag{D.4.8}$$

Furthermore, theorem D.3.2 applied to (D.4.4) gives

$$|\mathbf{n}\rangle = (1 + |\mu_{\mathbf{n}}|^2)^{-S} \exp(\mu_{\mathbf{n}} \hat{S}_-)|S\rangle$$
$$= \left(\frac{2}{1 + |\mu_{\mathbf{n}}|^2} \right)^S \sum_m b_{Sm}^{1/2} \, (\mu_{\mathbf{n}})^{S-m} \, |m\rangle, \tag{D.4.9}$$

where $\mu_{\mathbf{n}}$ is defined by (D.3.12) with $(\theta, \phi) = (\theta_{\mathbf{n}}, \phi_{\mathbf{n}})$, and b_{Sm} is the (normalized) binomial coefficient (3.3.38). By use of the above equation, it is easy to show that

$$\int d^2\mathbf{n} \, |\mathbf{n}\rangle\langle \mathbf{n}| = 1, \tag{D.4.10}$$

$$d^2\mathbf{n} := (2S + 1)\frac{d\Omega_{\mathbf{n}}}{4\pi} = \frac{\hbar d\Omega_{\mathbf{n}}}{2\pi(S + 1/2)^{-1}\hbar}, \tag{D.4.11}$$

where $d\Omega_{\mathbf{n}}$ is the area element of the unit sphere.[9] The set $\{|\mathbf{n}\rangle \mid \mathbf{n} \in S^2\}$ thus constitutes a basis. The last expression in Eq. (D.4.11) emphasizes a correspondence with the area element $dp\,dq/2\pi\hbar$ of the phase space for a particle. The Planck constant is put in the numerator in view of the fact that the dimension of $dp\,dq$ is the same as that of $\hbar$. This expression suggests that the role of $\hbar$ for a particle is played by $(S + 1/2)^{-1}\hbar$ in the case of spin.

Equation (D.4.9) gives

$$\langle \mathbf{n}'|\mathbf{n}\rangle = (\Delta_{\mathbf{n}'\mathbf{n}})^{2S}, \tag{D.4.12}$$

$$\Delta_{\mathbf{n}'\mathbf{n}} := \frac{1 + \mu'^*\mu}{(1 + |\mu'|^2)^{1/2}(1 + |\mu|^2)^{1/2}} = c'c + \tilde{s}'^*\tilde{s}, \tag{D.4.13}$$

where

$$\mu \equiv \mu_{\mathbf{n}}, \quad c \equiv \cos(\theta_{\mathbf{n}}/2), \quad \tilde{s} \equiv e^{i\phi_{\mathbf{n}}} \sin(\theta_{\mathbf{n}}/2), \tag{D.4.14}$$

and the corresponding quantities for $\mathbf{n}'$ are stipulated by the prime $'$. Hence,

$$|\Delta_{\mathbf{n}'\mathbf{n}}|^2 = 1 - \frac{|\mu' - \mu|^2}{(1 + |\mu'|^2)(1 + |\mu|^2)}$$
$$= 1 - \frac{1}{4}|\mathbf{n}' - \mathbf{n}|^2 = \frac{1}{2}(1 + \mathbf{n}' \cdot \mathbf{n}). \tag{D.4.15}$$

Therefore, $\{|\mathbf{n}\rangle\}$ is not an orthogonal set but a kind of over-complete basis. Note that it is only $|-\mathbf{n}\rangle$ that is orthogonal to $|\mathbf{n}\rangle$. Incidentally, the result (D.4.12) suggests that it may be derivable much more simply without resorting to such a heavy machinery as (D.4.9). Indeed, the spin-1/2 decomposition discussed in the preceding section may be employed conveniently; it is only necessary to perform the computation for the case of $S = 1/2$,

---

[8] Analogous states may be generated by rotating $|-S\rangle$. They are called spin coherent states with south-pole convention.

[9] Usually, this area element is expressed in terms of azimuth angles as $\sin\theta_{\mathbf{n}} d\theta_{\mathbf{n}} d\phi_{\mathbf{n}}$, which however is not suitable to treat points of the sphere <u>democratically</u>.

which can be done easily by use of Eq. (D.4.6), and raise the result to the power of $2S$. For $\mathbf{n}' \approx \mathbf{n}$, in particular, one finds the approximate formula

$$\langle \mathbf{n}'|\mathbf{n}\rangle \simeq \exp\left\{-\frac{S}{4}|\mathbf{n}' - \mathbf{n}|^2\right\}$$
$$\times \exp\left[-iS\{(\phi_{\mathbf{n}'} - \phi_{\mathbf{n}})(1 - \cos\theta_{\mathbf{n}})\right.$$
$$\left. + \frac{1}{2}(\theta_{\mathbf{n}'} - \theta_{\mathbf{n}})(\phi_{\mathbf{n}'} - \phi_{\mathbf{n}})\sin\theta_{\mathbf{n}}\}\right]. \tag{D.4.16}$$

The most important factor on the right-hand side is the foremost Gaussian one, which shows that the value of $\langle \mathbf{n}'|\mathbf{n}\rangle$ is appreciable only for those $(\mathbf{n}', \mathbf{n})$ which are so close to each other that $|\mathbf{n}' - \mathbf{n}| \lesssim S^{-1/2}$.

Let us generalize the formula (D.4.8) to off-diagonal elements. For brevity, we use the symbol[10]

$$\ll \hat{A} \gg \equiv \langle \mathbf{n}'|\hat{A}|\mathbf{n}\rangle / \langle \mathbf{n}'|\mathbf{n}\rangle. \tag{D.4.17}$$

Calculations are facilitated again by the idea of spin-1/2 decomposition. Paying attention to the correspondence

$$\prod_\alpha |\mathbf{n}\rangle^{(\alpha)} \longleftrightarrow |\mathbf{n}\rangle \quad : \quad |\mathbf{n}\rangle^{(\alpha)} \equiv c\left|\frac{1}{2}\right\rangle^{(\alpha)} + \tilde{s}\left|-\frac{1}{2}\right\rangle^{(\alpha)} \tag{D.4.18}$$

in the sense of (D.2.6), one may employ theorem D.2.2 to find

$$\langle \mathbf{n}'|\hat{S}_3|\mathbf{n}\rangle = \sum_{\alpha''}\left\{\prod_{\alpha'} {}^{(\alpha')}\langle\mathbf{n}'|\right\}\hat{s}_3^{(\alpha'')}\left\{\prod_\alpha |\mathbf{n}\rangle^{(\alpha)}\right\}$$
$$= (\Delta_{\mathbf{n}'\mathbf{n}})^{2S-1}\sum_\alpha {}^{(\alpha)}\langle\mathbf{n}'|\hat{s}_3^{(\alpha)}|\mathbf{n}\rangle^{(\alpha)}. \tag{D.4.19}$$

Since all α's are on equal footing, the sum in the last expression is equal to $2S$ times the contribution of a single α. Hence

$$\ll \hat{S}_3 \gg = S(c'c - \tilde{s}'^*\tilde{s})/\Delta_{\mathbf{n}'\mathbf{n}}. \tag{D.4.20}$$

Likewise,

$$\ll \hat{S}_+ \gg = 2Sc'\tilde{s}/\Delta_{\mathbf{n}'\mathbf{n}}, \quad \ll \hat{S}_- \gg = 2S\tilde{s}'^*c/\Delta_{\mathbf{n}'\mathbf{n}}. \tag{D.4.21}$$

Accordingly, for $\mathbf{n}' \approx \mathbf{n}$ in particular,

$$\ll \hat{\mathbf{S}} \gg = S\mathbf{n}\{1 + \mathcal{O}(|\mathbf{n}' - \mathbf{n}|)\}. \tag{D.4.22}$$

Equivalently, this may be combined with (D.4.16) to give

$$\langle \mathbf{n}'|\hat{\mathbf{S}}|\mathbf{n}\rangle = S\mathbf{n}\{1 + \mathcal{O}(S^{-1/2})\}\langle\mathbf{n}'|\mathbf{n}\rangle. \tag{D.4.23}$$

Matrix elements of a product of $\hat{\mathbf{S}}$ may be computed similarly. For example, those of a quadratic polynomial of $\hat{\mathbf{S}}$ may be summarized with the aid of a pair of complex vectors $\mathbf{a}$ and $\mathbf{b}$ as

$$\ll (\mathbf{a} \cdot \hat{\mathbf{S}})(\mathbf{b} \cdot \hat{\mathbf{S}}) \gg = \left(1 - \frac{1}{2S}\right)\ll \mathbf{a} \cdot \hat{\mathbf{S}} \gg\ll \mathbf{b} \cdot \hat{\mathbf{S}} \gg + \frac{i}{2}\ll (\mathbf{a} \times \mathbf{b}) \cdot \hat{\mathbf{S}} \gg + \frac{S}{2}\mathbf{a} \cdot \mathbf{b}, \tag{D.4.24}$$

$$\mathbf{a} \equiv \sum_j a_j\mathbf{e}_j, \quad \mathbf{b} \equiv \sum_j b_j\mathbf{e}_j \quad : \quad a_j, b_j \in \mathbf{C}. \tag{D.4.25}$$

---

[10] The left-hand side should carry indices $\mathbf{n}'$ and $\mathbf{n}$, which are omitted for brevity.

Bibliography

The following short list cites only some of those works that are either most fundamental or complementary to the present book, which is of the pedagogical nature. (The original Japanese edition cited a few books written in Japanese as well, which are omitted here. Some of the works that have appeared since 1997 are included in this English edition.)

1. Leggett, A. J., Macroscopic quantum systems and the quantum theory of measurement. *Prog. Theor. Phys. Suppl.* **69** (1980), 80.
   The seminal paper in which the Leggett program is announced.
2. Leggett, A. J., Quantum mechanics at the macroscopic level, in *Chance and Matter*, edited by J. Souletie, J. Vannimenus, and R. Stora. Elsevier Science Publishers, 1987, p. 395; in *Directions in Condensed Matter Physics*, edited by G. Grinstein and G. Mazenko. World Scientific, 1986, p. 187.
   An excellent and readable review of Ref. [1] and the subsequent development.
3. Kagan, Yu and Leggett, A. J. (eds), *Quantum Tunnelling in Condensed Media*. North-Holland, 1992.
   A collection of advanced masterly review articles on both experiments and theories of tunneling in systems of many degrees of freedom.
4. Weiss, U., *Quantum Dissipative Systems*. World Scientific, 1993.
   A detailed account of environmental effects in general.
5. Namiki, M., Pascazio, S. and Nakazato, H., *Decoherence and Quantum Measurement*. World Scientific, 1997.
   A detailed discussion on quantum Zeno effect as well as decoherence in general.
6. Eckern, U., Schön, G. and Ambegaokar, V., *Phys. Rev.* **B30** (1984), 6419.
   The frequently-cited standard paper on the microscopic derivation of the quantum mechanics of a magnetic flux (cf. Section 3.1).
7. Gunther, L. and Barbara, B. (eds.), *Quantum Tunneling of Magnetization – QTM '94*. Kluwer Academic Publishers, 1995.
   A collection of early works on magnetization tunneling.
8. Chudnovsky, E. M. and Tejada, J., *Macroscopic Quantum Tunneling of the Magnetic Moment*. Cambridge University Press, 1998.
   The more recent account of magnetization tunneling.
9. Kondo, J., Quantum diffusion of positive muons in solids, in *Perspectives of Meson Science*, edited by T. Yamazaki, K. Nakai, and K. Nagamine. Elsevier Science Publishers, 1992, p. 137.

A readable review on the quantum diffusion of a muon in a solid, which is closely related to the subject of the present book (see also Ref. [3], p. 37).

Since experiments are in progress even while this book is being written, the author has decided, with some hesitation, not to quote experimental data in the text. The following are only a few of the more recent experimental papers which may not be cited in the above references:

10. Arndt, M., Nairz, O., Vos-Andreae, J., Keller, C., van der Zouw, G. and Zeilinger, A., *Nature* **401** (1999), 680. (C_{60})
11. Rouse, R., Han, S. and Lukens, J. E., *Phys. Rev. Lett.* **75** (1995), 1614; **76** (1996), 3404. (SQUID)
12. Friedman, J., Patel, V., Chen, W., Tolpygo, S. K. and Lukens, J. E., Nature **406** (2000), 43. (SQUID)
13. van der Wal, C. H., ter Haar, A. C. J., Wilhelm, F. K., Schouten, R. N., Harmans, C. J. P. M., Orlando, T. P., Lloyd, S. and Mooij, J. E., *Science* **290** (2000), 773. (SQUID)
14. Satoh, T., Morishita, M., Ogata, M., Katoh, S., Hatakeyama, K. and Takashima, M., *Phys. Rev. Lett.* **69** (1992), 335; *Physica* **B197** (1994), 397. (^{3}He–^{4}He liquid mixture)
15. Balibar, S., Caupin, F., Guthmann, C., Lambare, H., Roche, P., Rolley, E. and Maris, H. J., *J. Low Temp. Phys.* **101** (1995), 271; **113** (1998), 459. (liquid ^{4}He and liquid ^{3}He)
16. Ruutu, J. P., Hakonen, P. J., Penttilä, J. S., Babkin, A. V., Saramäki, J. P. and Sonin, E. B., *Phys. Rev. Lett.* **77** (1996), 2514. (solid ^{4}He)
17. Friedman, J. R., Sarachik, M. P., Tejada, J. and Ziolo, R., *Phys. Rev. Lett.* **76** (1996), 3830. (magnetic molecule Mn_{12})
18. Wernsdorfer, W. and Sessoli, R., *Science* **284** (1999), 133. (magnetic molecule Fe_8)

The following is an additional list of papers which the author has consulted or made an explicit quotation from in writing the chapter indicated at the end of each of them.

19. Lifshitz, I. M. and Kagan, Yu, *Sov. Phys.* JETP**35** (1972), 206. (Chapter 3)
20. Nakamura, T., Kanno, Y. and Takagi, S., *Phys. Rev.* **B51** (1995), 8446; *Prog. Theor. Phys.* **93** (1995), 1013. (Chapter 3)
21. Leggett, A. J., Chakravarty, S., Dorsey, A. T., Fisher, M. P. A., Garg, A. and Zwerger, W., *Rev. Mod. Phys.* **59** (1987), 1. (Chapters 4 and 6)
22. Fujikawa, K., Iso, S., Sasaki, M. and Suzuki, H., *Phys. Rev.* **B46** (1992), 10295. (Chapter 6)
23. Caldeira, A. O. and Leggett, A. J., *Ann. Phys.* **149** (1983), 374; Erratum **153** (1984), 445(E). (Chapters 5, 7 and 8)
24. Schmid, A., *Ann. Phys.* **170** (1986), 333.
25. Leggett, A. J. and Garg, A., *Phys. Rev. Lett.* **54** (1985), 857. (Chapter 9)
26. Tesche, C. D., *Phys. Rev. Lett.* **64** (1990), 2358. (Chapter 9)
27. Nakamura, Y., Peshkin Yu. A. and Tsai J. S., *Nature* **398** (1999), 786.

It goes without saying, of course, that a vast quantity of textbooks, literature, conferences and conversations with students as well as colleagues, which are too numerous to cite here, have contributed to the author's knowledge, and hence to the writing of this book.

Index

adiabatic condition, 29
adiabatic environment, 150
adiabatic potential, 163
advanced solution, 130
Aharonov–Bohm effect, 87
Arrhenius' formula, 28
arrow of time, 34
attempt frequency, 55

bilinear model, 119
Bogoliubov quasiparticle, 70
Born–Oppenheimer approximation, 162
Born–Oppenheimer basis, 163
Born–Oppenheimer-type processing, 169
Bose–Einstein condensation, 8
bounce, 35, 153
bounce action, 36, 152
Brillouin–Wigner perturbation theory, 147
bubble, 54, 65
bubble potential, 59
bumpy slope, 14, 19, 51

Caldeira–Leggett assumption, 166
Caldeira–Leggett effective action, 156
Caldeira–Leggett formula, 52
Caldeira–Leggett trick, 155
cavitation, 62
classical decay rate, 28
classical nucleation, 54
closure, 79
coarse-grained density operator, 184
coherence, 95
coherent, 95
coherent state, 95, 132, 149
Coleman–Hepp environment, 106
Coleman–Hepp model, 106, 115
collapse of wavepacket, 110
collective degrees of freedom, 7, 13, 119
conditional probability, 174
configuration space, 78
Cooper pair, 44
correlation function, 132, 173, 175
correlation time, 99

counter term, 131
critical bubble, 54
critical excess concentration, 61
critical nucleus, 54
crossover temperature, 28
current-biased Josephson junction, 45

dc-SQUID, 48, 178
DDD, 96
decay rate, 26, 30, 159
decaying state, 26, 32
decoherence, 5, 95
density operator, 11
dephasing, 96, 101, 110, 144, 145
dephasing effect, 101, 144, 145
dephasing factor, 101
direct-product space, 188
disconnectivity, 9
disentanglement, 202
dislocation, 65
dissipation, 96
double bumpy slope, 75
double slits, 1

effective bubble mass, 56
effective Euclidean action functional, 153
entangled state, 4, 169, 183
environment, 5, 13, 99
environmental frequency distribution function, 123
environmental oscillators, 117
Euclidean action functional, 36, 152
Euclidean energy, 36
Euclidean Lagrangian, 37
Euclidean space, 187
Euclidean time, 36, 152

false vacuum, 61
Fermi's golden rule, 135
Feynman democracy, 90
field operator, 71
field potential, 64
first-order phase transition, 61
fluctuating force, 46, 128

fluctuation-dissipation theorem, 43
flux Hamiltonian, 47
flux potential, 47
flux quantum, 45
fluxon, 45
Franck–Condon factor, 150
free tunneling, 15
frequency shift, 102
functional, 195
functional derivative, 196

Gamow formula, 192
gauge invariance, 44
gauge transformation, 44
Gaussian process, 98
generalized Langevin equation, 128
generalized non-linear Schrödinger field, 68
Ginzburg–Landau order-parameter field, 42
ground-state energy splitting, 22, 30

harmonic environment, 117
Hilbert space, 187
homogeneous nucleation, 54

imaginary time, 36
incoherent, 95
instanton, 37
intrinsic nucleation, 53

joint probability, 172
Josephson current, 41
Josephson junction, 41
Josephson-coupling energy, 47
Josephson's acceleration equation, 45
Josephson's current–phase relation, 42

kink, 66
Kramers degeneracy, 88

Langevin equation, 46, 128
Laplace transformation, 103
Leggett program, 6
Leggett–Garg inequality, 5, 173
lifetime, 26, 55, 135
Lifshitz–Kagan formula, 61
linearly-coupled harmonic-environment model, 117
liquid ^{3}He–^{4}He mixture, 53
local realism, 172
London's penetration depth, 45
longitude basis, 79
longitude representation, 79, 82
longitudinal relaxation time, 102
low-energy effective theory, 70

macrorealism, 3, 171
macroscopic definiteness, 171
macroscopic degrees of freedm, 13
macroscopic linear combination, 3
macroscopic quantum decay, 15
macroscopic quantum nucleation, 61
macroscopic quantum phenomenon, 7

macroscopic quantum phenomenon of the first kind, 7
macroscopic quantum phenomenon of the second kind, 7
macroscopic quantum resonant oscillation, 14, 146
macroscopic resonant quantum tunneling, 29
macroscopic resonant tunneling, 29
macroscopic superposition, 3
macroscopically distinct states, 3, 172
macrosystem, 13
macrosystem energy, 100
macrosystem Hamiltonian, 13
magnetic flux, 44
magnetic relaxation, 102
magnetic resonance, 102
Markov approximation, 129
mass conservation law, 57, 69
measurement problem, 4
Meissner effect, 45
memory term, 127
metastable state, 15, 25, 54
metastable well, 15, 25
microscopic degrees of freedom, 7
mixture, 11, 95
MMM, 3
Möbius boundary condition, 80
modified Born–Oppenheimer basis, 163
modified Born–Oppenheimer effective vector potential, 164
modified Born–Oppenheimer representation, 164
modified Lifshitz–Kagan formula, 61
MQC, 13
MQC-situation, 49, 75, 76
MQP, 8
MQT, 13
MQT-situation, 49, 75, 77
MR, 5, 171

naive WKB method, 30
negative-result measurement, 114
noise, 132
non-invasive measurability, 171
non-linear Klein–Gordon field, 64
non-linear Schrödinger field, 68
north-pole convention, 204
nucleation, 54
Nyquist current, 43

ohmic current, 43
ohmic environment, 124
operator ordering, 62
outgoing boundary condition, 191
over-complete basis, 79, 204

path integral, 90
periodic boundary condition, 80
periodic well, 75
persistence probability, 26
phase separation, 53
phase space, 78, 204
Poincaré cycle, 33
Poincaré period, 33, 183

Poincaré recurrence theorem, 34
Poincaré recurrence time, 33
position representation, 18
position-represented wavefunction, 18
potential renormalization, 131
pre-exponential factor, 31
projector, 10
proper WKB method, 31

QIMDS, 4, 172
QM, 5, 172
quantum decay, 15, 27
quantum decay rate, 28
quantum indeterminacy time, 113
quantum Langevin equation, 128
quantum nucleation, 55
quantum nucleation rate, 60
quantum resonance frequency, 147
quantum resonant oscillation, 13, 24, 146
quantum theory of decay, 189
quantum Zeno effect, 110
quantum Zeno paradox, 110
quantum Zeno regime, 33
quasi-classical situation, 18
quasi-stationary wave, 32, 191

random force, 131
Rayleigh–Plesset model, 56
Rayleigh–Schrödinger perturbation theory,
    147
reduced density operator, 184
relaxation time, 102
renormalized perturbation theory, 147
resonance energy, 192
resonance wave, 192
resonant tunneling, 29, 51, 77
retarded Green function, 121
retarded reaction, 122
retarded resistance function, 124, 128
retarded solution, 130
rf-SQUID, 41, 178
rotating-wave approximation, 102
rotation symmetry, 89

SBS, 19
S-Cat, 3
S-Cattiness, 9, 50, 61
Schrödinger's cat, 3
Schrödinger's linear theater, 2
separable model, 118
single-domain magnet, 72
special bumpy slope, 19, 51
special symmetric double well, 19, 51

special tilted double well, 38
spectral function, 123
spin coherent state, 78, 203
spin potential, 75
spin-1/2 decomposition, 200
spin-disentanglement theorem, 202
spin-rotation operator, 199
squeezed state, 132
SQUID, 41, 178
SSDW, 19
stationary point, 37, 153, 198
stationary process, 99
stationary value, 198
stereographic projection, 203
stochastic process, 98
sub-barrier action, 31
sub-ohmic environment, 124
successive measurements, 114
supercurrent density, 44
super-ohmic environment, 124
supersaturated liquid mixture, 53
supersaturation, 53
survival probability, 26
symmetric double well, 14, 19

TAO, 3
Tesche's thought experiment, 178
thermal nucleation, 54
thin-wall approximation, 67, 70
tilted double well, 20
time reversal, 34
time-correlation function, 132, 173
time-reversal symmetry, 88, 130
transverse relaxation time, 102
true vaccum, 61
tunnel splitting, 22
two-state approximation, 21
two-state system, 21

van Hove limit, 106
variational derivative, 196
velocity potential, 68
virtual bound state, 192
virtual ground state, 25, 192
von Klitzing constant, 52

washboard, 49
watched-pot effect, 110
white noise, 114

Young-type interference, 1, 14

Zeno's paradox, 110

Printed in the United States
By Bookmasters

Building Intelligent
Systems

A Guide to Machine Learning
Engineering

Geoff Hulten

Apress®

Building Intelligent Systems: A Guide to Machine Learning Engineering

Geoff Hulten
Lynnwood, Washington, USA

ISBN-13 (pbk): 978-1-4842-3431-0
https://doi.org/10.1007/978-1-4842-3432-7

ISBN-13 (electronic): 978-1-4842-3432-7

Library of Congress Control Number: 2018934680

Managing Director, Apress Media LLC: Welmoed Spahr
Acquisitions Editor: Susan McDermott
Development Editor: Laura Berendson
Coordinating Editor: Rita Fernando

Cover designed by eStudioCalamar

Cover image designed by Freepik (www.freepik.com)

Distributed to the book trade worldwide by Springer Science+Business Media New York, 233 Spring Street, 6th Floor, New York, NY 10013. Phone 1-800-SPRINGER, fax (201) 348-4505, e-mail orders-ny@springer-sbm.com, or visit www.springeronline.com. Apress Media, LLC is a California LLC and the sole member (owner) is Springer Science + Business Media Finance Inc (SSBM Finance Inc). SSBM Finance Inc is a **Delaware** corporation.

For information on translations, please e-mail rights@apress.com, or visit http://www.apress.com/rights-permissions.

Apress titles may be purchased in bulk for academic, corporate, or promotional use. eBook versions and licenses are also available for most titles. For more information, reference our Print and eBook Bulk Sales web page at http://www.apress.com/bulk-sales.

Any source code or other supplementary material referenced by the author in this book is available to readers on GitHub via the book's product page, located at www.apress.com/9781484234310. For more detailed information, please visit http://www.apress.com/source-code.

Printed on acid-free paper

To Dad, for telling me what I needed to hear.
To Mom, for pretty much just telling me what I wanted to hear.
And to Nicole.

Table of Contents

About the Author xvii

About the Technical Reviewer xix

Acknowledgments xxi

Introduction xxiii

Part I: Approaching an Intelligent Systems Project 1

Chapter 1: Introducing Intelligent Systems 3

Elements of an Intelligent System 4

An Example Intelligent System 6

The Internet Toaster 6

Using Data to Toast 7

Sensors and Heuristic Intelligence 8

Toasting with Machine Learning 10

Making an Intelligent System 11

Summary 12

For Thought... 13

Chapter 2: Knowing When to Use Intelligent Systems 15

Types of Problems That Need Intelligent Systems 15

Big Problems 16

Open-Ended Problems 16

Time-Changing Problems 17

Intrinsically Hard Problems 18

Situations When Intelligent Systems Work 18

When a Partial System Is Viable and Interesting 19

When You Can Use Data from the System to Improve 19

When the System Can Interface with the Objective 20

When it is Cost Effective... 21

When You Aren't Sure You Need an Intelligent System.. 22

Summary.. 23

For Thought.. 23

Chapter 3: A Brief Refresher on Working with Data 25

Structured Data.. 25

Asking Simple Questions of Data... 27

Working with Data Models... 29

Conceptual Machine Learning 30

Common Pitfalls of Working with Data 31

Summary.. 33

For Thought.. 34

Chapter 4: Defining the Intelligent System's Goals 35

Criteria for a Good Goal.. 36

An Example of Why Choosing Goals Is Hard 36

Types of Goals.. 38

Organizational Objectives... 38

Leading Indicators 39

User Outcomes 41

Model Properties 42

Layering Goals 43

Ways to Measure Goals... 44

Waiting for More Information.. 44

A/B Testing.. 45

Hand Labeling... 45

Asking Users... 46

Decoupling Goals.. 46

Keeping Goals Healthy 47

Summary.. 48

For Thought 48

Part II: Intelligent Experiences 51

Chapter 5: The Components of Intelligent Experiences 53

Presenting Intelligence to Users 54

    An Example of Presenting Intelligence 55

Achieve the System's Objectives 57

    An Example of Achieving Objectives 58

Minimize Intelligence Flaws... 58

Create Data to Grow the System.. 59

    An Example of Collecting Data 60

Summary.. 61

For Thought.. 62

Chapter 6: Why Creating Intelligent Experiences Is Hard 63

Intelligence Make Mistakes 63

Intelligence Makes Crazy Mistakes.. 65

Intelligence Makes Different Types of Mistakes... 66

Intelligence Changes.. 68

The Human Factor... 70

Summary.. 72

For Thought.. 72

Chapter 7: Balancing Intelligent Experiences... 75

Forcefulness 76

Frequency 78

Value of Success... 79

Cost of Mistakes 81

    Knowing There Is a Mistake 81

    Recovering from a Mistake.. 82

Intelligence Quality 83

Summary 85

For Thought... 86

Chapter 8: Modes of Intelligent Interaction 87

Automate 87

Prompt 89

Organize 90

Annotate 92

Hybrid Experiences 93

Summary 94

For Thought... 95

Chapter 9: Getting Data from Experience 97

An Example: TeamMaker 98

Simple Interactions 98

Making It Fun 99

Connecting to Outcomes 100

Properties of Good Data 100

Context, Actions, and Outcomes 101

Good Coverage 102

Real Usage 103

Unbiased 103

Does Not Contain Feedback Loops 104

Scale 105

Ways to Understand Outcomes 106

Implicit Outcomes 106

Ratings 107

Reports 107

Escalations 108

User Classifications 108

Summary 109

For Thought... 110

Chapter 10: Verifying Intelligent Experiences **111**

Getting Intended Experiences 112

Working with Context 112

Working with Intelligence 114

Bringing it Together 115

Achieving Goals 116

Continual Verification 117

Summary 117

For Thought... 118

Part III: Implementing Intelligence **121**

Chapter 11: The Components of an Intelligence Implementation **123**

An Example of Intelligence Implementation 124

Components of an Intelligence Implementation 127

The Intelligence Runtime 127

Intelligence Management 127

Intelligence Telemetry Pipeline 128

The Intelligence Creation Environment 128

Intelligence Orchestration 129

Summary 130

For Thought... 130

Chapter 12: The Intelligence Runtime **133**

Context 134

Feature Extraction 135

Models 137

Execution 138

Results 139

Instability in Intelligence 139

Intelligence APIs 140

Summary 140

For Thought... 141

Chapter 13: Where Intelligence Lives 143

Considerations for Positioning Intelligence 143

Latency in Updating 144

Latency in Execution 146

Cost of Operation 148

Offline Operation 149

Places to Put Intelligence 150

Static Intelligence in the Product 150

Client-Side Intelligence 151

Server-Centric Intelligence 152

Back-End (Cached) Intelligence 153

Hybrid Intelligence 154

Summary 155

For Thought... 156

Chapter 14: Intelligence Management 157

Overview of Intelligence Management 157

Complexity in Intelligent Management 158

Frequency in Intelligence Management 159

Human Systems 159

Sanity-Checking Intelligence 160

Checking for Compatibility 160

Checking for Runtime Constraints 161

Checking for Obvious Mistakes 162

Lighting Up Intelligence 162

Single Deployment 163

Silent Intelligence 164

Controlled Rollout 165

Flighting 166

Turning Off Intelligence 167

Summary 168

For Thought... 168

Chapter 15: Intelligent Telemetry 171

Why Telemetry Is Needed 171

    Make Sure Things Are Working 172

    Understand Outcomes 173

    Gather Data to Grow Intelligence 174

Properties of an Effective Telemetry System 175

    Sampling 175

    Summarizing 176

    Flexible Targeting 177

Common Challenges 178

    Bias 178

    Rare Events 179

    Indirect Value 180

    Privacy 180

Summary 181

For Thought... 182

Part IV: Creating Intelligence 183

Chapter 16: Overview of Intelligence 185

An Example Intelligence 185

Contexts 187

    Implemented at Runtime 187

    Available for Intelligence Creation 189

Things Intelligence Can Predict 190

    Classifications 190

    Probability Estimates 191

    Regressions 193

    Rankings 194

    Hybrids and Combinations 194

Summary 194

For Thought... 195

Chapter 17: Representing Intelligence 197

Criteria for Representing Intelligence 197

Representing Intelligence with Code 198

Representing Intelligence with Lookup Tables.. 199

Representing Intelligence with Models... 201

    Linear Models.. 202

    Decision Trees 203

    Neural Networks.. 205

Summary.. 207

For Thought.. 207

Chapter 18: The Intelligence Creation Process..................................... 209

An Example of Intelligence Creation: Blinker 210

Understanding the Environment 210

Define Success 212

Get Data 213

    Bootstrap Data.. 214

    Data from Usage... 215

Get Ready to Evaluate.. 216

Simple Heuristics.. 217

Machine Learning 218

Understanding the Tradeoffs.. 219

Assess and Iterate 219

Maturity in Intelligence Creation.. 220

Being Excellent at Intelligence Creation 221

    Data Debugging.. 221

    Verification-Based Approach 222

    Intuition with the Toolbox 222

    Math (?) 223

Summary.. 223

For Thought.. 224

Chapter 19: Evaluating Intelligence **225**

Evaluating Accuracy 226

    Generalization 226

    Types of Mistakes 227

    Distribution of Mistakes 230

Evaluating Other Types of Predictions 230

    Evaluating Regressions 230

    Evaluating Probabilities 231

    Evaluating Rankings 231

Using Data for Evaluation 232

    Independent Evaluation Data 232

    Independence in Practice 233

    Evaluating for Sub-Populations 235

    The Right Amount of Data 237

Comparing Intelligences 238

    Operating Points 238

    Curves 239

Subjective Evaluations 240

    Exploring the Mistakes 241

    Imagining the User Experience 242

    Finding the Worst Thing 242

Summary 243

For Thought. 244

Chapter 20: Machine Learning Intelligence **245**

How Machine Learning Works 245

The Pros and Cons of Complexity 247

    Underfitting 248

    Overfitting 249

    Balancing Complexity 249

Feature Engineering 250

    Converting Data to Useable Format 251

Helping your Model Use the Data 253

Normalizing 254

Exposing Hidden Information 255

Expanding the Context 256

Eliminating Misleading Things 256

Modeling 257

Complexity Parameters 258

Identifying Overfitting 259

Summary 260

For Thought.... 261

Chapter 21: Organizing Intelligence **263**

Reasons to Organize Intelligence 263

Properties of a Well-Organized Intelligence 264

Ways to Organize Intelligence 265

Decouple Feature Engineering 266

Multiple Model Searches 268

Chase Mistakes 269

Meta-Models 270

Model Sequencing 272

Partition Contexts 274

Overrides 275

Summary 277

For Thought.... 278

Part V: Orchestrating Intelligent Systems **279**

Chapter 22: Overview of Intelligence Orchestration **281**

Properties of a Well-Orchestrated Intelligence 282

Why Orchestration Is Needed 282

Objective Changes 283

Users Change 284

Problem Changes 285

Intelligence Changes 286

Costs Change.. 287

Abuse.. 287

The Orchestration Team 288

Summary.. 288

For Thought. 289

Chapter 23: The Intelligence Orchestration Environment........................... 291

Monitor the Success Criteria... 292

Inspect Interactions 293

Balance the Experience 295

Override Intelligence... 296

Create Intelligence 298

Summary.. 299

For Thought. 300

Chapter 24: Dealing with Mistakes... 301

The Worst Thing That Could Happen 301

Ways Intelligence Can Break.. 303

System Outage 303

Model Outage 304

Intelligence Errors 304

Intelligence Degradation 305

Mitigating Mistakes 306

Invest in Intelligence 306

Balance the Experience... 307

Adjust Intelligence Management Parameters.. 307

Implement Guardrails 308

Override Errors 308

Summary.. 309

For Thought. 310

Chapter 25: Adversaries and Abuse 311

  Abuse Is a Business 312

  Abuse Scales 313

    Estimating Your Risk 313

  What an Abuse Problem Looks Like 314

  Ways to Combat Abuse 315

    Add Costs 315

    Becoming Less Interesting to Abusers 315

    Machine Learning with an Adversary 316

    Get the Abuser out of the Loop 316

  Summary 316

  For Thought. 317

Chapter 26: Approaching Your Own Intelligent System 319

  An Intelligent System Checklist 319

    Approach the Intelligent System Project 320

    Plan for the Intelligent Experience 321

    Plan the Intelligent System Implementation 323

    Get Ready to Create Intelligence 325

    Orchestrate Your Intelligent System 327

  Summary 329

  For Thought. 329

Index 331

About the Author

Geoff Hulten is a machine learning scientist and PhD in machine learning. He has managed applied machine learning teams for over a decade, building dozens of Internet-scale Intelligent Systems that have hundreds of millions of interactions with users every day. His research has appeared in top international conferences, received thousands of citations, and won a SIGKDD Test of Time award for influential contributions to the data mining research community that have stood the test of time.

About the Technical Reviewer

 **Jeb Haber** has a BS in Computer Science from Willamette University. He spent nearly two decades at Microsoft working on a variety of projects across Windows, Internet Explorer, Office, and MSN. For the last decade-plus of his Microsoft career, Jeb led the program management team responsible for the safety and security services provided by Microsoft SmartScreen (anti-phishing, anti-malware, and so on.) Jeb's team developed and managed global-scale Intelligent Systems with hundreds of millions of users. His role included product vision/planning/strategy, project management, metrics definition and people/team development. Jeb helped organize a culture along with the systems and processes required to repeatedly build and run global scale, 24×7 intelligence and reputation systems. Jeb is currently serving as the president of two non-profit boards for organizations dedicated to individuals and families dealing with the rare genetic disorder phenylketonuria (PKU).

Acknowledgments

There are so many people who were part of the Intelligent Systems I worked on over the years. These people helped me learn, helped me understand. In particular, I'd like to thank:

Jeb Haber and John Scarrow for being two of the key minds in developing the concepts described in this book and for being great collaborators over the years. None of this would have happened without their leadership and dedication.

Also: Anthony P., Tomasz K., Rob S., Rob M., Dave D., Kyle K., Eric R., Ameya B., Kris I., Jeff M., Mike C., Shankar S., Robert R., Chris J., Susan H., Ivan O., Chad M. and many others...

Introduction

Building Intelligent Systems is a book about leveraging machine learning in practice.

It covers everything you need to produce a fully functioning Intelligent System, one that leverages machine learning and data from user interactions to improve over time and achieve success.

After reading this book you'll be able to design an Intelligent System end-to-end. You'll know:

- When to use an Intelligent System and how to make it achieve your goals.

- How to design effective interactions between users and Intelligent Systems.

- How to implement an Intelligent System across client, service, and back end.

- How to build the intelligence that powers an Intelligent System and grow it over time.

- How to orchestrate an Intelligent System over its life-cycle.

You'll also understand how to apply your existing skills, whether in software engineering, data science, machine learning, management or program management to the effort.

There are many great books that teach data and machine-learning skills. Those books are similar to books on programming languages; they teach valuable skills in great detail. This book is more like a book on software engineering; it teaches how to take those base skills and produce working systems.

This book is based on more than a decade of experience building Internet-scale Intelligent Systems that have hundreds of millions of user interactions per day in some of the largest and most important software systems in the world. I hope this book helps accelerate the proliferation of systems that turn data into impact and helps readers develop practical skills in this important area.

Who This Book Is For

This book is for anyone with a computer science degree who wants to understand what it takes to build effective Intelligent Systems.

Imagine a typical software engineer who is assigned to a machine learning project. They want to learn more about it so they pick up a book, and it is technical, full of statistics and math and modeling methods. These are important skills, but they are the wrong information to help the software engineer contribute to the effort. *Building Intelligent Systems* is the right book for them.

Imagine a machine learning practitioner who needs to understand how the end-to-end system will interact with the models they produce, what they can count on, and what they need to look out for in practice. *Building Intelligent Systems* is the right book for them.

Imagine a technical manager who wants to begin benefiting from machine learning. Maybe they hire a machine learning PhD and let them work for a while. The machine learning practitioner comes back with charts, precision/recall curves, and training data requests, but no framework for how they should be applied. *Building Intelligent Systems* is the right book for that manager.

Data and Machine Learning Practitioners

Data and machine learning are at the core of many Intelligent Systems, but there is an incredible amount of work to be done between the development of a working model (created with machine learning) and the eventual sustainable customer impact. Understanding this supporting work will help you be better at modeling in a number of ways.

First, it's important to **understand the constraints** these systems put on your modeling. For example, where will the model run? What data will it have access to? How fast does it need to be? What is the business impact of a false positive? A false negative? How should the model be tuned to maximize business results?

Second, it's important to be able to **influence the other participants**. Understanding the pressures on the engineers and business owners will help you come to good solutions and maximize your chance for success. For example, you may not be getting all the training data you'd like because of telemetry sampling. Should you double down on modeling around the problem, or would an engineering solution make more sense? Or maybe you are being pushed to optimize for a difficult extremely-high precision, when your models are already performing at a very good (but slightly lower) precision. Should you keep chasing that super-high precision or should you work to influence the user experience in ways that reduce the customer impact of mistakes?

Third, it's important to understand how the **supporting systems can benefit you**. The escalation paths, the manual over-rides, the telemetry, the guardrails that prevent against major mistakes—these are all tools you can leverage. You need to understand when to use them and how to integrate them with your modeling process. Should you discard a model that works acceptably for 99% of users but really, really badly for 1% of users? Or maybe you can count on other parts of the system to address the problem.

Software Engineers

Building software that delights customers is a lot of work. No way around it, behind every successful software product and service there is some serious engineering. Intelligent Systems have some unique properties which present interesting challenges. This book describes the associated concepts so you can design and build Intelligent Systems that are efficient, reliable, and that best-unlock the power of machine learning and data science.

First, this book will identify the **entities and abstractions** that need to exist within a successful Intelligent System. You will learn the concepts behind the intelligence runtime, context and features, models, telemetry, training data, intelligence management, orchestration, and more.

Second, the book will give you a **conceptual understanding of machine learning and data sciences**. These will prepare you to have good discussions about tradeoffs between engineering investments and modeling investments. Where can a little bit of your work really enable a solution? And where are you being asked to boil the ocean to save a little bit of modeling time?

Third, the book will explore **patterns for Intelligent Systems** that my colleagues and I have developed over a decade and through implementing many working systems. What are the pros and cons or running intelligence in a client or in a service? How do you bound and verify components that are probabilistic? What do you need to include in telemetry so the system can evolve?

Program Managers

Machine learning and Data Sciences are hot topics. They are fantastic tools, but they are tools; they are not solutions. This book will give you enough conceptual understanding so you know what these tools are good at and how to deploy them to solve your business problems.

The first thing you'll learn is to develop an **intuition for when machine learning and data science are appropriate**. There is nothing worse than trying to hammer a square peg into a round hole. You need to understand what types of problems can be solved by machine learning. But just as importantly, you need to understand what types of problems can't be—or at least not easily. There are so many participants in a successful endeavor, and they speak such different, highly-technical, languages, that this is particularly difficult. This book will help you understand enough so you can ask the right questions and understand what you need from the answers.

The second is to get an intuition on return on investment so you can determine **how much Intelligent System to use**. By understanding the real costs of building and maintaining a system that turns data into impact you can make better choices about when to do it. You can also go into it with open eyes, and have the investment level scoped for success. Sometimes you need all the elements described in this book, but sometimes the right choice for your business is something simpler. This book will help you make good decisions and communicate them with confidence and credibility.

Finally, the third thing a program manager will learn here is to understand how to **plan, staff, and manage an Intelligent System project**. You will get the benefit of our experience building many large-scale Intelligent Systems: the life cycle of an Intelligent System; the day-to-day process of running it; the team and skills you need to succeed.

PART I

Approaching an Intelligent Systems Project

Chapters 1-4 set the groundwork for a successful Intelligent Systems project. This part describes what Intelligent Systems are and what they are good for. It refreshes some important background. It explains how to ensure that an Intelligent System has a useful, achievable goal. And it gives an overview of what to expect when taking on an Intelligent Systems project.

CHAPTER 1

Introducing Intelligent Systems

Intelligent Systems are all around us. In our light bulbs. In our cars. In our watches. In our thermostats. In our computers. How do we make them do the things that make our lives better? That delight us?

When should a light bulb turn on? When should an e-commerce site show us a particular product? When should a search engine take us to a particular site? When should a speaker play some music?

Answering questions like these, and answering them extremely well, is fundamental to unlocking the value of Intelligent Systems. And it is really hard.

Some of the biggest, most valuable companies in the world have their core business built around answering simple questions, like these:

- What web page should I display based on a short query?

- What ad should I present in a particular context?

- What product should I show to this shopper?

- What movie would this user enjoy right now?

- What book would this person like to read?

- Which news stories will generate the most interest?

- Which programs should I block from running to keep a machine safe?

Answering each of these questions "well enough" has made companies worth billions—in a few cases hundreds of billions—of dollars. And it has done so by making lots of people smarter, more productive, happier, and safer. But this is just the tip of the iceberg.

3

© Geoff Hulten 2018
G. Hulten, *Building Intelligent Systems*, https://doi.org/10.1007/978-1-4842-3432-7_1

There are tens of thousands of similar questions we could try to answer: When should my front door unlock? What exercise should a fitness app suggest next? What type of song should an artist write next? How should a game evolve to maximize player engagement?

This book is about reliably and efficiently unlocking the potential of Intelligent Systems.

Elements of an Intelligent System

Intelligent Systems connect users to artificial intelligence (machine learning) to achieve meaningful objectives. An Intelligent System is one in which the intelligence evolves and improves over time, particularly when the intelligence improves by watching how users interact with the system.

Successful Intelligent Systems have all of the following:

- **A meaningful objective**. An Intelligent System must have a reason for being, one that is meaningful to users and accomplishes your goals, and one that is achievable by the Intelligent System you will be able to build and run. Selecting an objective is a critical part of achieving success, but it isn't easy to do. The first part of this book will help you understand what Intelligent Systems do, so you'll know when you should use one and what kinds of objectives you should set for it.

- **The intelligent experience**. An intelligent experience must take the output of the system's intelligence (such as the predictions its machine learning makes) and present it to users to achieve the desired outcomes.

  To do this it must have a user interface that adapts based on the predictions and that puts the intelligence in a position to shine when it is right—while minimizing the cost of mistakes it makes when it is wrong. The intelligent experience must also elicit both implicit and explicit feedback from users to help the system improve its intelligence over time. The second part of this book will explore intelligent experiences, the options and the pitfalls for connecting users with intelligence.

- **The implementation of the intelligence**. The Intelligent System implementation includes everything it takes to execute intelligence, to move the intelligence where it needs to be, to manage it, to light up the intelligent experiences based on it, to collect telemetry to verify the system is functioning, and to gather the user feedback that will improve the intelligence over time. The third part of this book describes all the components of an intelligence implementation. It will prepare you to design and implement an Intelligent System of your own.

- **Intelligence creation**. Intelligent Systems are about setting intelligence up for success. This intelligence can come from many different places, ranging from simple heuristics to complex machine learning. Intelligence must be organized so that the right types of intelligence address the right parts of the problem, and so it can be effectively created by a team of people over an extended time. The fourth part of this book discusses the act of creating and growing intelligence for Internet-scale Intelligent Systems. It will prepare you to leverage all available approaches and achieve success.

- **The orchestration**. An Intelligent System lives over time, and all its elements must be kept in balance to achieve its objectives. This orchestration includes controlling how the system changes, keeping the experience in sync with the quality of the intelligence, deciding what telemetry to gather to track down and eliminate problems, and how much money to spend building and deploying new intelligence. It also involves dealing with mistakes, controlling risk, and defusing abuse. The fifth part of this book explains everything it takes to orchestrate an Intelligent System and achieve its goals through all phases of its life-cycle.

An Intelligent System is one way to apply machine learning in practice. An Intelligent System takes *intelligence* (produced via machine learning and other approaches) and leverages and supports it to achieve your objectives and to improve over time.

An Example Intelligent System

Intelligent Systems might be used to implement search engines, e-commerce sites, self-driving cars, and computer vision systems that track the human body and know who a person is and when they are smiling. But these are big and complicated systems.

Let's look at a much simpler example to see how a solution might evolve from a traditional system into an Intelligent System.

The Internet Toaster

Let's consider an Internet-connected smart-toaster. A good idea? Maybe, maybe not. But let's consider it. Our toaster has two controls: a slider that controls the intensity of the toasting and a lever that starts the toast.

It seems simple enough. The toaster's intelligence simply needs to map settings of the intensity slider to toast times. At a low setting, the toaster runs for, say, 30 seconds. At high settings the toaster runs for two minutes. That kind of thing.

So sit down, think for a while, come up with some toast-time settings and send the toaster to customers. What could go wrong?

Well, if you choose a maximum intensity that toasts for too long it could burn the things it toasts. Most customers who use that setting will be unhappy, throw away their cinder-toast, and start again.

You can imagine other failure cases, all the little irritations of toast-making that result in customers standing over their toasters, hands on the levers, ready to cut the toast short. Or customers repeatedly toasting the same piece of bread, a bit at a time, to get it the way they like it.

That's not good. If we are going to build a toaster, we want to build a really good one.

So maybe we do some testing, tweaking the toaster until both the high and low settings make toast that we think is desirable. Not too crispy, not too cool.

Great.

Did we get it right? Will this toaster do what our customers want?

It's hard to know. No matter how much toast we've eaten in our lives, it's really impossible to prove we've gotten the settings right for all the types of toast all our customers might want to make.

And so we realize we need to incorporate the opinions and experiences of others into our toaster-building process. But how?

Maybe we start with a focus group. Bring in dozens of members of the toast-making public, put them in a toasting lab, and take notes as they toast.

Then we tweak the toast-time settings again to reflect the way these volunteers toast. Now do we have the perfect toaster? Will this focus-group–tuned toaster make all the right toast that hundreds of thousands of people all around the world will want?

What if someone puts something frozen into the toaster? Or something from the fridge? Or what if someone has a tradition of making toaster s'mores? Or what if someone invents a new toaster-product, unlike anything humanity has ever toasted before? Is our toaster right for all of these situations? Probably not.

Using Data to Toast

So maybe making a perfect toasting machine is a bit harder than just asking a few people what they like.

There are just too many use cases to optimize by hand if we want to get the perfect toast in every conceivable situation. We could run focus groups every day for the rest of our lives and we still wouldn't see all the types of toast a toaster might make.

We need to do better. It's time for some serious data science.

The toaster is Internet connected, so we can program it to send telemetry back to our service. Every time someone toasts something we can know what setting they used and how long the toasting went before it stopped.

We ship version 1 of the toaster (perhaps to a controlled set of users), and the toast telemetry starts flooding in to our servers.

Now we know exactly which intensity settings people are using in their real lives (not in some contrived lab setting). We know how many times people push the lever down to start a toast job and how many times they pop the lever up to stop one early.

Can we use this data to make a better toaster?

Of course!

We could set the maximum intensity to something that at least a few users are actually using. Then we could set up metrics to make sure we don't have the toaster biased to over-toasting. For example, we could monitor the percentage of toasts that are early-stopped by their user (presumably because they were about to burn something) and tweak and tweak till we get those under control.

We could set the minimum intensity to something reasonable too. Something that users seem to use. We could track the double-toast rate (where someone toasts something and immediately re-toasts it) and tweak to make sure the toaster isn't biased to under-toasting.

Heck, we can even set the default intensity, the one in the middle of the range, to the most commonly used toast time.

Since our toasters are Internet-connected, we can update them with the new settings by having them pull the data from our servers. Heck, we could tweak the toaster's settings every single day, twice on Sunday—this is the age of miracles and wonders!

There are some seams with this approach, some things we had to assume. For example, we had to assume that toasting multiple times in quick succession is a sign of failure, that the customer is re-toasting the same bread instead of making several pieces of toast rapid-fire.

We had to assume that an early-stop is a sign the bread was starting to burn and not a sign that the customer was late for work and rushing out the door.

Also when we deploy the new settings to toasters, how do we ensure that users are going to like them? We are pretty sure (based on data science) that the new settings better-match what our overall user population is doing, so that's good.

But what about the user who was getting their own brand of perfectly toasted bagel yesterday and today they get... something different?

Despite these problems, we've got a pretty decent toaster. We've got telemetry to know the toaster is roughly doing its job. We've got a way to service it and improve it over time. Now let's really blow the doors off this thing.

Sensors and Heuristic Intelligence

If we want to make the best toaster, we're going to need more than a single slider and a toast lever. Let's add some sensors:

- A weight sensor to know how much toast is in the toaster and to determine when a customer places something in the toaster and when they take something out of it.

- A temperature sensor to know if the item placed in the toaster is chilled, frozen, or room temperature.

- A location sensor to know what region of the world the toaster is in so it can adapt to different tastes in different locales.

- A proximity sensor to know if someone is near the toaster and a camera to identify who it is.

- A clock to know if a toast is a breakfast toast or a dinner one.

- A little memory to know what's been toasted recently and to monitor the pattern of setting changes and toastings.

- A smoke sensor to know when the toaster has made a bad mistake and is about to burn something.

Now when a customer walks up to the toaster and puts something in it, the toaster can look at who the user is, try to guess what they are trying to toast, and automatically suggest a setting.

Heck, if the toaster is good enough there is no longer any need for the intensity setting or the toasting lever at all. We could update the toasting experience to be totally automatic. We could ship toasters with no buttons or knobs or anything. The customer drops something in, walks away, and comes back to delightfully toasted—anything!

If we could just figure out a way to turn all these sensor readings into the correct toast times for... anything.

To do that we need intelligence—the program or rules or machine-learned model that makes these types of decisions.

Let's start simply, by hand-crafting some intelligence.

Let's write a set of rules that consider the sensor readings and output intensity suggestions. For example: If something cold and heavy is put into the toaster, then toast for 5 minutes at high intensity. But for every degree above freezing, reduce the toast time by 2 seconds. But in Britain add 15 seconds to the toast time (because they like it that way). But if the weight and size match a known toaster product (some big brand toaster meal), then toast it the "right amount of time" based on the product's directions.

And that kind of stuff.

Every time a user complains, because the toaster toasted something wrong, you can add a new rule.

Whenever the telemetry shows a spike of double-toastings you can tweak the rules to improve.

Every time you add a new rule or update an old one you can update all the toasters you shipped all around the world by having them download the new settings from your server.

With this approach, you'll probably need to write and maintain a lot of rules to deal with all the possibilities. It's going to be a lot of work. You could employ a dozen people and give them months of rule-writing-time, and you might not be happy with the outcome.

You might never be.

Toasting with Machine Learning

In situations like this, where there is an optimization problem that is too hard or too expensive to do by hand, people turn to machine learning.

At a high level, machine learning can observe examples of users using their Internet toasters and automatically produce a set of rules to control the toaster just the way users would want it. Kind of like the hand-crafted heuristic ones we just talked about, only machine-crafted.

And machines can make lot more rules than humans can. Machines can balance inputs from dozens of sensors; they can optimize for thousands of different types of users simultaneously. They can incorporate new data and reoptimize everything every day, every hour—sometimes even faster. They can personalize rules to each user.

In order to work, machine learning needs data that shows the situation the user faced (the sensor readings) the action they took (how they set the toaster) and the outcome they got (if they liked the toast or had to tweak something and try again). Machine learning needs examples of when things are going right as well as examples of when things are going wrong. Lots and lots of examples of them.

To find examples of when things are going right, we can look through the telemetry for times when the user put something in the toaster, pressed the start lever, waited to completion, took the item out, and walked away. These are cases where the user probably got what they wanted. We have a record of all the sensor settings, the toasting time. We can use this as training for the machine learning system. Perfect.

And for examples of when things went wrong we can look through the telemetry again. This time we find all the places where users had to fiddle with the toaster, every time they stopped the toaster early or retoasted the same item multiple times. We have records of the sensor readings and intensity settings that gave users bad outcomes, too.

So combining the good outcomes and the bad outcomes and some machine learning, we can automatically train the perfect toaster-controlling algorithm.

Put in data, get out a program that looks at sensor readings and determines the best intensity and time settings to use.

We push this learned program to our customers' toasters.

We can do it every day, every minute.

We can learn from hundreds of thousands of customer interactions—from hundreds of millions of them.

Magic!

Making an Intelligent System

And this is what Intelligent Systems do. Building an effective Intelligent System requires balancing five major components: the objective, the experience, the implementation, the intelligence, and the orchestration.

We need a problem that is suitable to solve with an Intelligent System, and worth the effort. We need to control the toaster based on what the intelligence thinks. We need to do it in a way the user finds helpful, and we also need to give the user the right controls to interact with the toaster and to give feedback we can use to learn. We need to build all the services and tools and code that gathers telemetry, produces intelligence, moves it where it needs to be, and hooks it up with users.

We need to create the intelligence every day, over and over, in a way that is predictable. And we need to keep everything running over time as new toastable products come into the market and tastes change.

We need to decide how much telemetry we should collect to trade off operational costs with the potential value. We need to decide how much to change the intelligence on any particular day so that the toaster improves, but without confusing or upsetting users.

We need to monitor our key metrics and react if they start to degrade. Maybe tune the experience? Maybe call an emergency intelligence intervention? And we need to deal with mistakes. And mistakes happen.

Intelligence (particularly machine-learned intelligence) can make bad mistakes—spectacular, counter-intuitive, highly customer-damaging mistakes.

For example, our toaster might learn that people in a specific zip code love cinder-toast. No matter what you put into the thing, if you live in that zip code—cinder. So we need ways to identify and manage these mistakes. We need ways to place guardrails on machine learning.

And unfortunately... people are people. Any time we use human feedback to optimize something—like using machine learning to build a toaster—we have to think about all the ways a human might benefit from things going wrong.

Imagine a major bread-maker paying a sweatshop of people to toast their competitor's products using crazy settings. They could carry out millions of toasts per week, flooding our telemetry system with misleading data. Our machine learning might pick up on this and "learn" what this toast-hacker is trying to teach it—our toaster might start regularly undercooking the competitor's product, leading to food safety issues.

Not good.

A successful Intelligent System will have to consider all of these issues and more.

Summary

This chapter introduced Intelligent Systems along with five key conceptual challenges that every Intelligent System must address—the objective, the experience, the implementation, the intelligence, and the orchestration.

It should now be clear that there are many ways to approach these challenges and that they are highly interrelated. To have a successful Intelligent System you must balance them. If one of the conceptual challenges is difficult in your context, the others will need to work harder to make up for it.

For example, if you are trying to retrofit an Intelligent System into an existing system where the experience has already been defined (and can't be changed), the Intelligent System might need to accept a less aggressive objective, to invest more in intelligence, or to have a fuller mistake-mitigation strategy.

But here's another way to look at it: there are a lot of ways to succeed.

Deploy an Intelligent System and discover that the intelligence problem is harder than you thought? No need to panic. There are lots of ways to compensate and make a system that delights customers and helps your business while the intelligence takes time to grow.

The rest of this book will give you the tools to approach Intelligent Systems projects with confidence.

For Thought...

After reading this chapter you should be able to:

- Identify Intelligent Systems in the world around you.

- See the potential that Intelligent Systems can unlock.

- Understand the difference between intelligence (which makes predictions about the world) and an Intelligent System (which combines objective, experience, implementation, intelligence, and orchestration to achieve results).

- Articulate all the conceptually hard things you will need to address to build a successful Intelligent System.

- Understand how these difficult things interact, including some of the tradeoffs and ways they can support one another.

You should be able to answer questions like these:

- What services do you use that (you suspect) are built by turning customer data into intelligence?

- What is the most magical experience you've had with one of these services?

- What is the worst experience you've had?

- Can you identify how the user experience supports the intelligence?

- Can you find any information on how its intelligence is produced? Maybe in a news story or publication?

- Can you determine any of the ways it detects and mitigates intelligence mistakes?

CHAPTER 2

Knowing When to Use Intelligent Systems

So you have a problem you want to solve. An existing system you need to optimize. A new idea to create a business. Something you think your customers will love. A machine-learning–based solution that isn't producing the value you'd hoped. Is an Intelligent System right for you? Sometimes yes. Sometimes no.

This chapter discusses when Intelligent Systems might be the right approach, and provides guidance on when other approaches might be better. It begins by describing the types of problems that can benefit from Intelligent Systems. It then discusses some of the properties required to make Intelligent Systems work.

Types of Problems That Need Intelligent Systems

It is always best to do things the easy way. If you can solve your problem without an Intelligent System, maybe you should. One key factor in knowing whether you'll need an Intelligent System is how often you think you'll need to update the system before you have it right. If the number is small, then an Intelligent System is probably not right.

For example, imagine implementing intelligence for part of a banking system. Your "intelligence" needs to update account balances as people withdraw money. The solution is pretty simple:

```
NewBalance = OldBalance - WithdrawalAmount
```

Something like that (and a bunch of error checking and logging and transaction management), not rocket science. Maybe you'll have a bug. Maybe you'll miss an error case (for example, if the balance goes negative). Then you might need to update the "intelligence" to get it right. Maybe a couple times? Or maybe you're super sloppy and you need to update it a dozen times before you're done.

© Geoff Hulten 2018
G. Hulten, *Building Intelligent Systems*, https://doi.org/10.1007/978-1-4842-3432-7_2

Intelligent Systems are not for problems like this. They are for problems where you think you're going to have to update the system much, much more. Thousands of times, tens of thousands of times, every hour for as long as the system exists. That kind of thing.

There are four situations that clearly require that level of iteration:

- Big problems, that require a lot of work to solve.

- Open-ended problems, which continue to grow over time.

- Time-changing problems, where the right answer changes over time.

- Intrinsically hard problems, which push the boundaries of what we think is possible.

The rest of this section will explore these each in turn.

Big Problems

Some problems are big. They have so many variables and conditions that need to be addressed that they can't really be completed in a single shot.

For example, there are more web pages than a single person could read in their lifetime—more than a hundred people could read. There are so many books, television programs, songs, video games, live event streams, tweets, news stories, and e-commerce products that it would take thousands of person-years just to experience them all.

These problems and others like them require massive scale. If you wanted to build a system to reason about one of these, and wanted to completely finish it before deploying a first version... Well, you'd probably go broke trying.

When you have a big problem that you don't think you can finish in one go, an Intelligent System might be a great way to get started, and an efficient way to make progress on achieving your vision, by giving users something they find valuable and something they are willing to help you improve.

Open-Ended Problems

Some problems are more than big. Some problems are open-ended. That is, they don't have a single fixed solution at all. They go on and on, requiring more work, without end. Web pages, books, television programs, songs, video games, live event streams—more and more of them are being created every day.

Trying to build a system to reason about and organize things that haven't even been created yet is hard.

In these cases, a static solution—one where you build it, deploy it, and walk away—is unlikely to work. Instead, these situations require services that live over long periods of time and grow throughout their lifetimes.

If your problem has a bounded solution, an Intelligent System might not be right. But if your problem is big and on-going, an Intelligent System might be the right solution.

Time-Changing Problems

Things change. Sometimes the right answer today is wrong tomorrow. For example:

- Imagine a system for identifying human faces—and then facial tattoos become super popular.

- Imagine a system for predicting stock prices—and then an airplane crashes into a building.

- Imagine a system for moving spam email to a junk folder—and then a new genius-savant decides to get in the spam business and changes the game.

- Or Imagine a UX that users struggle to use—and then they begin to learn how to work with it.

One thing's for certain—things are going to change.

Change means that the intelligence you implemented yesterday—which was totally right for what was happening, which was making a lot of users happy, maybe even making your business a lot of money—might be totally wrong for what is going to happen tomorrow.

Addressing problems that change over time requires the ability to detect that something has changed and to adapt quickly enough to be meaningful.

If your domain changes slowly or in predictable ways, an Intelligent System might not be needed. On the other hand, if change in your domain is unpredictable, drastic, or frequent, an Intelligent System might be the right solution for you.

Intrinsically Hard Problems

Some problems are just hard. So hard that humans can't quite figure out how to solve them. At least not all at once, not perfectly. Here are some examples of hard problems:

- Understanding human speech.

- Identifying objects in pictures.

- Predicting the weather more than a few minutes in the future (apparently).

- Competing with humans in complex, open-ended games.

- Understanding human expressions of emotion in text and video.

In these situations, machine learning has had great success, but this success has come on the back of years (or decades) of effort, gathering training data, understanding the problems, and developing intelligence. These types of systems are still improving and will continue to improve for the foreseeable future.

There are many ways to make progress on such hard problems. One way is to close the loop between users and intelligence creation in a meaningful application using an Intelligent System as described in this book.

Situations When Intelligent Systems Work

In addition to having a problem that is difficult enough to need an Intelligent System—one that is big, open-ended, time-changing, intrinsically hard, or some combination—an Intelligent System needs a few more things to succeed:

- A problem where a partial solution is viable and interesting.

- A way to get data from usage of the system to improve intelligence.

- An ability to influence a meaningful objective.

- A problem that justifies the effort of building an Intelligent System.

The rest of this section explores these requirements in turn.

When a Partial System Is Viable and Interesting

Intelligent Systems are essentially always incomplete, incorrect in important ways, and likely to make all sorts of mistakes. It isn't important that an Intelligent System is perfect. What matters is that an Intelligent System must be good enough to be interesting to users (and valuable to you).

An Intelligent System is viable when the value of the things it does right is (much) higher than the cost of the things it does wrong.

Consider an Intelligent System that makes "cheap" mistakes. Maybe the system is supposed to show a shopping list in an order that optimizes the user's time in the grocery store. Head to the right, pick up eggs, then milk, then loop to the back of the store for the steak, then into the next aisle for pickles... and on and on. If this system gets a few things backwards it might waste time by sending users in the wrong direction, say 15–20 seconds per mistake. Irritating, but not the end of the world. If the mistakes aren't too frequent—say just one every other shopping trip—it's easy to imagine someone using the system while it is learning the layouts of all the grocery stores in the world and adapting to changes.

Consider another Intelligent System, one that makes "expensive" mistakes. Maybe it is controlling surgical robots, and mistakes put human life at risk. This system would have to meet a much higher quality bar before it is viable. That is, you'd need to produce a much better partial system before deploying it to users. But even this system doesn't need to be perfect to be viable. Remember—surgeons make mistakes, too. An Intelligent System doesn't have to be perfect (and most never will be), it just has to be better than the other guy.

The point is that to be viable, an Intelligent System must provide a good deal for users (and for you) by producing more value with positive outcomes than it produces irritation (and cost) with negative outcomes.

When You Can Use Data from the System to Improve

Intelligent Systems work by closing the loop between usage and intelligence. When users use the system, the intelligence gets better; and when the intelligence gets better, the system produces more value for users.

To make intelligence better with usage, you need to record the interactions between users and the systems in a careful way (which we will discuss in detail later). At a high level, this involves capturing what the user saw, what they did, and whether the outcome was positive or negative. And you need to observe many, many user interactions (the harder the problem, the more you'll need). This type of data can be used to improve intelligence in many ways, including with machine learning.

When the System Can Interface with the Objective

The Intelligent System must be able to change things that influence the objective. This could be through automating a part of the system, or presenting information or options to a user that help them achieve their goals. Intelligent Systems are most effective when the following conditions are met:

- The actions the Intelligent System can take *directly* affect the objective. When there is a positive outcome, it should be largely because the Intelligent System did well; and when there is a negative outcome, it should be largely because the Intelligent System did poorly. The more external factors that affect the outcome, the harder it is to make an effective Intelligent System.

- The actions the Intelligent System can take *quickly* affect the objective. Because the more time between the action and the outcome, the more chance there is for external factors to interfere.

- The actions the Intelligent System can take are in balance with the objectives. That is, the actions the Intelligent System can take are meaningful enough to the outcome that they are usually measurable.

In practice, figuring out where to put Intelligent Systems and how much of your problem to try to solve with them will be key challenges. Too big an objective and you won't be able to connect the Intelligent System to it; too small and it won't be worth all the work.

In fact, many large, complex systems have space for more than one Intelligent System inside them. For example, an operating system might have an Intelligent System to predict what to put into a RAM cache, one to predict what files to show the user when they search, one to recommend apps to users based on usage, and another to manage the tradeoffs between power and performance when the computer is not plugged in.

You can imagine using a single Intelligent System to manage an entire OS. But it probably wouldn't work, because:

- There are too many things to control.

- The controls are too abstract from the goal.

- The types of feedback the system gets would not relate directly enough to the decisions the Intelligent System needs to make.

When it is Cost Effective

Intelligent Systems have different costs than other approaches to building computer systems and services. There are three main components of cost in most similar systems: The Intelligence, the Implementation, and the Orchestration. We will discuss these concepts in Parts 3, 4, and 5 of this book. But roughly *Intelligence* is the ways the system makes decisions about what is right to do and when, the "logic" of the system; *Implementation* is all the services, back-end systems, and client code that implements the system and interfaces intelligence outcomes with users; and *Orchestration* is the managing of the system throughout its life-cycle, such as making sure it is achieving its objectives, and dealing with mistakes.

The intelligence is usually cheaper in Intelligent Systems than other approaches, for a couple of reasons:

1. In general, Intelligent Systems produce the complex logic of a system automatically (instead of using humans). When addressing a big, hard, or open-ended task, Intelligent Systems can lead to substantial savings on intelligence production.

2. With the proper intelligent experience, Intelligent Systems produce the training data (which machine learning requires) automatically as users interact with it. Gathering training data can be a major cost driver for intelligence creation, often requiring substantial human effort, and so getting data from users can drastically shift the economics of a problem.

The implementation of an Intelligent System is similar to or a bit more expensive than the implementation other service-based systems (such as a web service or a chat service). Most of the components are similar between an Intelligent System and a nonintelligent system, but Intelligent Systems have additional requirements related to orchestrating them over their lifecycles, including creating intelligence, changing intelligent experiences, managing mistakes, organizing intelligence, and so on. We will discuss all of these in detail later in the book.

The orchestration costs of an Intelligent System are similar but a bit more expensive compared to running a non-intelligent service (like a web service or a chat service). Intelligent Systems require ongoing work, including tuning the way the intelligence is exposed to users, growing the intelligence to be better and deal with new situations, and identifying and mitigating mistakes.

By the time you've finished this book you'll be able to prepare a detailed analysis of the efficiency of using an Intelligent System for your application.

When You Aren't Sure You Need an Intelligent System

Sometimes you have a place where users, data, and intelligence interact, but you aren't sure you need all the elements of an Intelligent System. This might be for any of these reasons:

- You aren't sure your problem is actually hard, so you don't want to build a whole bunch of fancy intelligence you might not need.

- You realize the problem is hard, but aren't sure the extra work of building an Intelligent System will justify the effort for you.

- You want to take an incremental approach and solve problems as you encounter them, rather than stopping to build an Intelligent System.

That's fine. Intelligent Systems (as described in this book) aren't right for everyone—there are plenty of ways to build systems that solve hard problems. But if you do try to solve a big, open-ended, time-changing, or hard problem, you'll eventually run into many of the challenges described in this book. Even if you choose not to build an Intelligent System as described here, this book will arm you to identify common issues quickly, understand what is going on, and will prepare you to respond with proven approaches.

Summary

Intelligent Systems are most useful when solving big, open-ended, time-changing, or intrinsically hard problems. These are problems where a single solution won't work, and where the system needs to improve over time—months or years.

Intelligent Systems work well when:

- A partial solution is interesting to users and will support a viable business, so that you can start simply and grow the system's capability over time.

- You can get data to improve the system from users as they use the system. Data is usually one of the main costs in building intelligence, so crafting a system that produces it automatically is very valuable.

- When the things the system can control affect the objectives: directly, quickly, and meaningfully. That is, the outcomes the system is getting can be tied to things the Intelligent System is doing.

- When other approaches are too expensive to scale.

When looking for places to use Intelligent Systems, break down your problem into pieces. Each piece should have a clear objective that is critical to overall success, feedback that is quick and direct, and controls (or user experiences your system can control) that directly impact the objective. Each such piece is a candidate for an Intelligent System.

And even if you choose not to use an Intelligent System at all, the patterns in this book about practical machine learning, interfacing intelligence with user experience, and more will benefit just about any machine learning endeavor.

For Thought

After reading this chapter, you should know:

- Whether your system could benefit from an Intelligent System.

- How to find potential Intelligent Systems in the world around you.

You should be able to answer questions like these:

- What is your favorite hobby that would benefit from an Intelligent System?

- What is an example of a common activity you do every day where an Intelligent System might not be right? Why not?

Consider your favorite piece of software:

- Identify three places the software could use an Intelligent System.

- Which one is the most likely to succeed? Which the least? Why?

CHAPTER 3

A Brief Refresher on Working with Data

Data is central to every Intelligent System. This chapter gives a conceptual overview of working with data, introducing key concepts from data science, statistics, and machine learning. The goal is to establish a baseline of understanding to facilitate discussions and decision making between all participants of an Intelligent System project.

This chapter will cover:

- Structured data.

- Asking questions of data.

- Data models.

- Conceptual machine learning.

- Some common pitfalls of working with data.

Structured Data

Data is changing the world, that's for sure.

But what is it? A bunch of pictures on a disk? All the term papers you wrote in high school? The batting averages of every baseball player who ever played?

Yeah, all of that is data. There is so much data—oceans of numbers, filling hard disk drives all over the planet with ones and zeros. In raw form, data is often referred to as unstructured, and it can be quite difficult to work with.

When we turn data into intelligence, we usually work with data that has some structure to it. That is, data that is broken up into units that describe entities or events.

© Geoff Hulten 2018
G. Hulten, *Building Intelligent Systems*, https://doi.org/10.1007/978-1-4842-3432-7_3

For example, imagine working with data about people. Each unit of the data describes a single person by: their weight, their height, their gender, their eye color, that sort of stuff.

One convenient way to think about it is as a table in a spreadsheet (Figure 3-1). There is one row for each person and one column for each property of the person. Maybe the first column of each row contains a number, which is the corresponding person's weight. And maybe the third column of each row contains a number, which is the person's height. On and on.

Weight	Gender	Height (Inches)	Eye Color	...
170	Male	70	Hazel	...
140	Female	60	Brown	...
60	Male	50	Blue	...
...	...	...	...	...

Figure 3-1. *An example of structured data*

The columns will usually contain numbers (like height and weight), or be chosen from a small number of categories (like brown or blue for eye color), or be short text strings like names. There are more advanced options, of course, but this simple scheme is quite powerful and can be used to represent all sorts of things (and unlock our ability to do statistics and machine learning on them).

For example:

- You can represent a web page by: the number of words it contains, the number of images it has in it, the number of links it contains, and the number of times it contains each of 1,000 common keywords.

- You can represent a visit to a search engine by: the query the user typed, the amount of time the user remained on the results page, the number of search results the user clicked, and the URL of the final result they clicked.

- You can represent a toasting event on our Internet toaster by: the temperature of the item placed in the toaster, the intensity setting the user selected, the number of times the user stopped and started the toaster before removing the item, and the total amount of time between when the item was placed in the toaster and when it was taken out.

These are simple examples. In practice, even simple concepts tend to have dozens or hundreds of elements in their representation (dozens or hundreds of columns per row). Choosing the right data sets to collect and the right representations to give them is critical to getting good results from data. And you'll probably get it wrong a few times before you get it right—be prepared to evolve.

Asking Simple Questions of Data

So what can you do with data? You can *ask it questions*. For example, in a data set of people you might want to know:

- What is the average height?

- Who is the tallest person?

- How many people are shorter than 5'5"?

These are pretty easy to answer; simply look through the data and calculate the values (sum up the numbers, or count the rows that meet the criteria).

In an Intelligent System you'll be asking plenty of similar questions. Things like:

- How many times per day do users take a particular action?

- What percentage of users click an ad after they engage with the intelligent experience?

- What's the average revenue per customer per month?

Answering questions like these is important to understanding if a system is meeting its goals (and making sure it isn't going haywire).

Another thing you can do with data is to *make projections*. Imagine a data set with a hundred people in it. Want to know the average height? No problem; just calculate it. But what if you want to know the height of the next person who will be added to the data set? Or whether the next person added to the set will be shorter or taller than 5'?

Can you do these types of things? Sure! Well, sort of, using basic statistics.

With a few simple assumptions, statistics can estimate the most likely height for the next person added to the data set. But statistics can do more. It can express exactly how accurate the estimate is; for example:

> *The most likely height of the next person is 5'10", and with 95% confidence the next person will be between 5'8" and 6' tall.*

This is called a confidence interval. The width of the confidence interval depends on how much data you have, and how "well behaved" the data is. More data results in a narrower window (for example, that the next person is 95% likely to be between 5'9" and 5'11"). Less data results in a wider one (like 5'5" to 6'3"). Why does this matter?

Let's imagine optimizing an Internet business by reducing customer support capacity to save money. Say the current system has capacity for 200 customer support calls per week. So looking into historical telemetry, you can calculate that the average week had 75 support calls and that the maximum number of calls in any week was 98. Intuitively, 200 is much higher than 75—the system must be wasting a lot of money. Cutting capacity in half (down to 100) seems safe, particularly because you don't have any week on record with a higher call volume than that.

But check the confidence interval. What if it came back that the most likely call volume is 75 per week, and with 95% confidence the next week will have between 40 and 110 calls? Well, then 100 doesn't seem like such an obviously good answer. In a data-intensive project, you should always ask for answers. But you should also ask how sure the answers are. And you should make sure the decisions you make take both the answer and the degree of certainty into account.

Working with Data Models

Models capture the answers from data, converting them into *more convenient* or *more useful* formats.

Technically, the projection we discussed previously was a model. Recall that the data set on call center volumes was converted into an average weekly volume of 75 with a 95% confidence interval of 40 - 110. The raw data might have been huge. It might have contained ten terabytes of telemetry going back a decade. But the model contained just four numbers: 75, 95%, 40, and 110.

And this simple model was more useful for making decisions about capacity planning, because it contained exactly the relevant information, and captured it in an intuitive way for the task at hand.

Models can also *fill in gaps in data* and estimate answers to questions that the data doesn't contain.

Consider the data set of people and their heights and weights. Imagine the data set contains a person who is 5'8" and a person who is 5'10", but no one who is 5'9".

What if you need to predict how much a 5'9" person will weigh?

Well, you could build a simple model. Looking into the data, you might notice that the people in the data set tend to weigh about 2.5 pounds per inch of height. Using that model, a 5'9" person would weigh 172.5 pounds. Great. Or is it?

One common way to evaluate the quality of models is called *generalization error*. This is done by reserving a part of the data set, hiding it from the person doing the modeling, and then testing the model on the holdout data to see how well the model does—how well it generalizes to data it has not seen.

For example, maybe there really was a 5'9" person in the data set, but someone hid them from the modeling process. Maybe the hidden 5'9"er weighed 150 pounds. But the model had predicted 172.5 pounds. That's 22.5 pounds off.

Good? Bad? Depends on what you need the model for.

If you sum up these types of errors over dozens or hundreds of people who were held out from the modeling process, you can get some sense of how good the model is. There are lots of technical ways to sum up errors and communicate the quality of a model, two important ones are:

- **Regression Errors** are the difference between a numeric value and the predicted value, as in the weight example. In the previous example, the model's regression error was 22.5.

29

- **Classification Errors** are for models that predict categories, for example one that tries to predict if a person is a male or female based on height and weight (and, yeah, I agree that model wouldn't have many friends). One way to measure classification errors involves accuracy: the model is 85% accurate at telling males from females.

Conceptual Machine Learning

Simple models can be useful. The weight-at-height model is very simple, and it's easy to find flaws with it. For example, it doesn't take gender into account, or body fat percent, or waist circumference, or where the person is from. Statistics are useful for making projections. Create a simple model, use it to answer questions. But models can be complex too—very, very, very complex.

Machine learning uses computers to improve the process of producing (complex) models from data. And these models can make projections from the data, sometimes surprisingly accurate—and sometimes surprisingly inaccurate.

A machine-learning algorithm explores the data to determine the best way to combine the information contained in the representation (the columns in the data set) into a model that generalizes accurately to data you haven't already seen.

For example, a machine-learning algorithm might predict any of the following:

- Gender by combining height, weight, age, and name.

- Weight using gender, height, age, and name.

- Height from the gender and weight.

- And so on...

And the way the model works can be very complicated, multiplying and scaling the various inputs in ways that make no sense to humans but produce accurate results.

There are probably thousands of machine-learning algorithms that can produce models of data. Some are fast; others run for days or weeks before producing a model. Some produce very simple models; some produce models that take megabytes or gigabytes of disk space. Some produce models that can make predictions very quickly on new data, a millisecond or less; some produce models that are computationally intensive to execute. And some do well on certain types of data, but fail at other types of problems.

There doesn't seem to be one universally best machine-learning algorithm to use for all situations—most work well in some situations, but poorly in others. Every few years a new technique emerges that does surprisingly well at solving important problems, and gets a lot of attention.

And some machine-learning people really like the algorithm they are most familiar with and will fight to the death to defend it despite any evidence to the contrary. You have been warned.

The machine-learning process generally breaks down into the following phases, all of which are difficult and require considerable effort from experts:

- **Getting the data to model.** This involves dealing with noise, and figuring out exactly what will be predicted and how to get the training examples for the modeling algorithm.

- **Feature engineering.** This involves processing the data into a data set with the right representation (columns in the spreadsheet) to expose to machine learning algorithms.

- **The modeling.** This involves running one or more machine-learning algorithms on the data set, looking at the mistakes they make, tweaking and tuning things, and repeating until the model is accurate enough.

- **The deployment.** This involves choosing which model to run, how to connect it into the system to create positive impact and minimal damage from mistakes.

- **The maintenance.** This involves monitoring that the model is behaving as expected, and fixing it if it starts to go out of control. For example, by rerunning the training algorithm on new data.

This book is about how to do these steps for large, complex systems.

Common Pitfalls of Working with Data

Data can be complicated, and there are a lot of things that can go wrong. Here are a few common pitfalls to keep in mind when working with data intensive systems, like Intelligent Systems.

Confidence intervals can be broken. Confidence intervals are very useful. Knowing that something is 95% likely to be between two values is great. But a 95% chance of being within an interval means there is a 5% chance of being outside the interval, too. And if it's out, it can be way out. What does that mean? 5% is one in twenty. So if you are estimating something weekly, for example to adjust capacity of a call center, one out of every twenty weeks will be out of interval—that's 2.6 times per year that you'll have too much capacity or not enough. 5% sounds small, like something that might never happen, but keep in mind that even low probability outcomes will happen eventually, if you push your luck long enough.

There is noise in your data. Noise is another word for errors. Like maybe there is a person in the dataset who has a height of 1", or 15'7". Is this real? Probably not. Where does the noise come from? All sorts of places. For example, software has bugs; data has bugs too. Telemetry systems log incorrect values. Data gets corrupted while being processed. The code that implements statistical queries has mistakes. Users turn off their computers at odd times, resulting in odd client states and crazy telemetry. Sometimes when a client has an error, it won't generate a data element. Sometimes it will generate a partial one. Sometimes this noise is caused by computers behaving in ways they shouldn't. Sometimes it is caused by miscommunication between people. Every large data set will have noise, which will inject some error into the things created from the data.

Your data has bias. Bias happens when data is collected in ways that are systematically different from the way the data is used. For example, maybe the people dataset we used to estimate height was created by interviewing random people on the streets of New York. Would a model of this data be accurate for estimating the heights of people in Japan? In Guatemala? In Timbuktu?

Bias can make data less useful and bias can come from all sorts of innocent places, like simple oversights about where the data is collected, or the context of when the data was collected. For example, users who "sorta-liked" something are less likely to respond to a survey than users who loved it or who hated it. Make sure data is being used to do the thing it was meant to do, or make sure you spend time to understand the implications of the bias your data has.

Your data is out of date. Most (simple) statistical and machine learning techniques have a big underlying assumption—that is, that things don't change. But things do change. Imagine building a model of support call volume, then using it to try to estimate support call volume for the week after a major new feature is released. Or the week after a major storm. One of the main reasons to implement all the parts of an Intelligent System is to make the system robust to change.

You want the data to say things it doesn't. Sometimes data is inconclusive, but people like to have answers. It's human nature. People might downplay the degree of uncertainty by saying things like "the answer is 42" instead of saying "we can't answer that question" or "the answer is between 12 and 72." It makes people feel smarter to be precise. It can almost seem more polite to give people answers they can work with (instead of giving them a long, partial answer). It is fun to find stories in the data, like "our best product is the toothpaste, because we redid the display-case last month." These stories are so seductive that people will find them even where they don't exist.

Tip   When working with data, always ask a few questions: Is this right? How sure are we? Is there another interpretation? How can we know which is correct?

Your models of data will make mistakes. Intelligent Systems, especially ones built by modeling data, are wrong. A lot. Sometimes these mistakes make sense. But sometimes they are totally counter-intuitive gibberish. And sometimes it is very hard to fix one mistake without introducing new ones. But this is fine. Don't be afraid of mistakes. Just keep in mind: any system based on models is going to need a plan for dealing with mistakes.

Summary

Data is changing the world. Data is most useful when it is structured in a way that lines up with what it needs to be used for. Structured data is like a spreadsheet, with one row per thing (person, event, web page, and so on), and one column per property of that thing (height, weight, gender).

Data can be used to answer questions and make projections. Statistics can answer simple questions and give you guidance on how accurate the answer is. Machine learning can create very complicated models to make very sophisticated projections. Machine learning has many, many useful tools that can help make accurate models of data. More are being developed all the time.

And data can be misused badly—you have to be careful.

For Thought

After reading this chapter, you should:

- Understand the types of things data is used for.

- Have some intuition about how to apply the things data can do.

- Have some intuition about how data can go wrong and lead to bad outcomes.

You should be able to answer questions like this:

- What is the biggest mistake you know of that was probably made because someone misused data?

CHAPTER 4

Defining the Intelligent System's Goals

An Intelligent System connects intelligence with experience to achieve a desired outcome. Success comes when all of these elements are aligned: the outcome is achievable; the intelligence is targeted at the right problem; and the experience encourages the correct user behavior.

A good success criterion connects these elements. It expresses the desired outcome in plain language. It indicates what sub-problems the intelligence and experience need to solve, and it ties those solutions to the desired larger scale (organizational) outcome. Implicit in good success criteria is a framework that allows all participants to see how their work contributes to the overall goal. This helps prevent them from heading in the wrong direction, even when data and their experiences are conspiring to mislead.

This chapter discusses setting goals for Intelligent Systems, including:

- What makes good goals.

- Why finding good goals is hard.

- The various types of goals a system can have.

- Some ways to measure goals.

© Geoff Hulten 2018

G. Hulten, *Building Intelligent Systems*, https://doi.org/10.1007/978-1-4842-3432-7_4

Criteria for a Good Goal

A successful goal will do all of the following:

1. Clearly **communicate the desired outcome** to all participants.
   Everyone should be able to understand what success looks like
   and why it is important, no matter what their background or
   technical experience.

2. **Be achievable**. Everyone on the team should believe they are set
   up to succeed. The goal can be difficult, but team members should
   be able to explain roughly how they are going to approach success
   and why there is a good chance it will work.

3. **Be measurable**. Intelligent Systems are about optimizing, and
   so Intelligent Systems are about measuring. Because if you can't
   measure something you really aren't going to be able to optimize it.

It isn't easy to know if a goal is correct. In fact, bridging the gap between high-level
objectives and detailed properties of the implementation is often the key challenge
to creating a successful Intelligent System. Some goals will seem perfect to some
participants but make no sense to others. Some will clearly align with positive impact but
be impossible to measure or achieve. There will always be trade-offs, and it is common
to spend a great deal of time refining the definition of success.

But that's OK. It's absolutely worth the effort because failing to define success before
starting a project is the easiest, most certain way to waste time and money.

An Example of Why Choosing Goals Is Hard

Consider an anti-phishing feature backed by an Intelligent System.

One form of phishing involves web sites that look like legitimate banking sites but
are actually fake sites, controlled by abusers. Users are lured to these phishing sites and
tricked into giving their banking passwords to criminals. Not good.

So what should an Intelligent System do?

Talk to a machine-learning person and it won't take long to get them excited. They'll
quickly see how to build a model that examines web pages and predicts whether they are
phishing sites or not. These models will consider things like the text and images on the

CHAPTER 4   DEFINING THE INTELLIGENT SYSTEM'S GOALS

web pages to make their predictions. If the model thinks a page is a phish, block it. If a page is blocked, a user won't browse to it, won't type their banking password into it. No more problem. Easy. Everyone knows what to do.

So the number of blocks seems like a great thing to measure—block more sites and the system is doing a better job.

Or is it?

What if the system is so effective that phishers quit? Every single phisher in the world gives up and finds something better to do with their time?

Perfect!

But then there wouldn't be any more phishing sites and the number of blocks would drop to zero. The system has achieved total success, but the metric indicates total failure.

Not great.

Or what if the system blocks one million phishing sites per day, every day, but the phishers just don't care? Every time the system blocks a site, the phishers simply make another site. The Intelligent System is blocking millions of things, everyone on the team is happy, and everyone feels like they are helping people—but the same number of users are losing their credentials to abusers after the system was built as were losing their credentials before it was built.

Not great.

One pitfall with defining success in an Intelligent System is that there are so many things that can be measured and optimized. It's very easy to find something that is familiar to work with, choose it as an objective, and get distracted from true success.

Recall the three properties of a good success criterion:

1.   Communicate the desired outcome

2.   Be achievable

3.   Be measurable

Using the number of blocked phishing pages as a success metric hits #2 and #3 out of the park, but fails on #1.

The desired outcome of this system isn't to block phishing sites—it is to stop abusers from getting users' banking passwords.

37

Types of Goals

There are many types of things a system can try to optimize, ranging from very concrete to very abstract.

A system's true objective tends to be very abstract (like making money next quarter), but the things it can directly affect tend to be very concrete (like deciding whether a toaster should run for 45 or 55 seconds). Finding a clear connection between the abstract and concrete is a key source of tension in setting effective goals. And it is really, really hard.

One reason it is hard is that different participants will care about different types of goals. For example:

- Some participants will care about making money and attracting and engaging customers.

- Some participants will care about helping users achieve what they are trying to do.

- Some participants will care that the intelligence of the system is accurate.

These are all important goals, and they are related, but the connection between them is indirect. For example, you won't make much money if the system is always doing the wrong thing; but making the intelligence 1% better will not translate into 1% more profit.

This section discusses different ways to consider the success of an Intelligent System, including:

- Organizational objectives

- Leading indicators

- User outcomes

- Model properties

Most Intelligent Systems use several of these on a regular basis but focus primarily on user outcomes and model properties for day-to-day optimization.

Organizational Objectives

Organizational objectives are the real reason for the Intelligent System. In a business these might be things like revenue, profit, or number of units sold. In a nonprofit organization these might be trees saved, lives improved, or other benefits to society.

Organizational objectives are clearly important to optimize. But they are problematic as direct objectives for Intelligent Systems for at least three reasons:

1. They are very distant from what the technology can affect. For example, a person working on an Internet toaster can change the amount of time a cold piece of bread is toasted—how does that relate to number of units sold?

2. They are affected by many things out of the system's control. For example, market conditions, marketing strategies, competitive forces, changes to user behavior over time, and so on.

3. They are very slow indicators. It may take weeks or months to know if any particular action has impacted an organizational objective. This makes them difficult to optimize directly.

Every Intelligent System should contribute to an organizational objective, but the day-to-day orchestration of an Intelligent System will usually focus on more direct measures—like the ones we'll discuss in the next few sections (particularly user outcomes and model properties).

Leading Indicators

Leading indicators are measures that correlate with future success. For example:

- You are more likely to make a profit when your customers like your product than when they hate it.

- You are more likely to grow your customer base when your customers are recommending your product to their friends than when your customers are telling their friends to stay away.

- You are more likely to retain customers when they use your product every day than when they use your product once every couple of months.

Leading indicators are a way to bridge between organizational objectives and the more concrete properties of an Intelligent System (like user outcomes and model properties). If an Intelligent System gets better, customers will probably like it more. That may lead to more sales or it might not, because other factors—like competitors, marketing activities, trends, and so on—can affect sales. Leading indicators factor some of these external forces out and can help you get quicker feedback as you change your Intelligent System.

There are two main types of leading indicators: customer sentiment and customer engagement.

Customer sentiment is a measure of how your customers feel about your product. Do they like using it? Does it make them happy? Would they recommend it to a friend (or would they rather recommend it to an enemy)?

If everyone who uses your product loves it, it is a sign that you are on the right track. Keep going, keep expanding your user base, and eventually you will have business success (make revenue, sell a lot of units, and so on).

On the other hand, if everyone who uses your product hates it you might be in for some trouble. You might have some customers, they might use your product, but they aren't happy with it. They are looking for a way out. If you get a strong competitor your customers are ready to jump ship.

Sentiment is a fuzzy measure, because users' feelings can be fickle. It can also be very hard to measure sentiment accurately—users don't always want to tell you exactly what you ask them to tell you. Still, swings in sentiment can be useful indicators of future business outcomes, and Intelligent Systems can certainly affect the sentiment of users who encounter them.

Customer engagement is a measure of how much your customers use your product. This could mean the frequency of usage. It could also mean the depth of usage, as in using all the various features your product has to offer.

Customers with high engagement are demonstrating that they find value in your product. They've made a habit of your product, and they come back again and again. They will be valuable to you and your business over time.

Customers with low engagement use the product infrequently. These customers may be getting value from your offering, but they have other things on their minds. They might drift away and never think about you or your product again.

Leading indicators have some disadvantages as goals for Intelligent Systems, similar to those that organizational outcomes suffer from:

- They are indirect.

- They are affected by factors out of control of the Intelligent System.

- They aren't good at detecting small changes.

- They provide slow feedback, so they are difficult to optimize directly.

- And they are often harder to measure than organizational objectives (how many surveys do you like to answer?).

Still, leading indicators can be useful, particularly as early indicators of problems—no matter what you think your Intelligent System should be doing, if customers have much worse sentiment after an update than they had before the update, you are probably doing something wrong.

User Outcomes

Another approach for setting goals for Intelligent Systems is to look at the outcomes your users are getting. For example:

- If your system is about helping users find information, are they finding useful information efficiently?

- If your system is about helping users make better decisions, are they making better decisions?

- If your system is about helping users find content they will enjoy, are they finding content that they end up liking?

- If your system is about optimizing the settings on a computer, are the computers it is optimizing faster?

- And if your system is about helping users avoid scams, are they avoiding scams?

Intelligent Systems can set goals around questions and decisions like these and try to optimize the outcomes users get.

This is particularly useful because outcomes rely on a combination of the intelligence and the experience of the Intelligent System. In order for a user to get a good outcome, the intelligence must be correct, and the experience must help the user benefit.

For example, in the anti-phishing example, imagine intelligence that is 100% accurate at identifying scams. If the experience blocks scam pages based on this intelligence, users will get good outcomes. But what if the experience is more subtle? Maybe it puts a little warning on the user's browser when they visit a scam site—a little red X in the address bar. Some users won't notice the warning. Some won't interpret it correctly. Some will ignore it. In this case some users will get bad outcomes (and give their passwords to scammers) even though the intelligence had correctly identified the scam.

User outcomes can make very good targets for Intelligent Systems because they measure how well the intelligence and the experience work together to influence user behavior.

Model Properties

Within every Intelligent System there are concrete, direct things to optimize, for example:

- The error rate of the model that identifies scams.

- The probability a user will have to re-toast their bread.

- The fraction of times a user will accept the first recommendation of what content to use.

- The click-through rate of the ads the system decides to show.

These types of properties don't always line up exactly with user outcomes, leading indicators, or organizational objectives, but they do make very good goals for the people who are working to improve Intelligent Systems.

For example, a model that is right 85% of the time (on test data in the lab) is clearly better than one that is right 75% of the time. Clear and concrete. Easy to get fast feedback. Easy to make progress.

But model properties have some disadvantages as goals for Intelligent Systems:

1. *They are not connected to actual user reality.* For example, if the Internet toaster always gets within 10 seconds of the optimal toast time will users like it? Would 5 seconds of error be better? Sure, obviously, of course. But how much better? Will that error reduction make tastier toast? What should the goal be? If we could get to a model with 4 seconds of error is that enough? Should we stop or press on for 3 seconds of error? How much investment is each additional second worth?

2. *They don't leverage the full system.* A model might make a mistake, but the mistake will be perceived by users in the context of a full system. Maybe the user experience makes the mistake seem so

minor that no one cares. Or maybe there is a really good way for the user to give feedback, which quickly corrects the mistake— way more cheaply than investing in optimizing the last few points of the model's performance.

3. *They are too familiar to machine-learning people.* It is easy to build Intelligent Systems to optimize model properties—it is precisely what machine-learning people spend their lives doing, so it will naturally come up in any conversation about objectives. Be careful with them. They are so powerful and familiar that they may stifle and hijack the system's actual objective.

Optimizing model properties is what intelligence is about, but it is seldom the goal. A good goal will show how improving model properties contributes to having the desired impact on users and the business. A good goal will give guidance on how much model property optimization is worth.

Layering Goals

Success in an Intelligent System project is hard to define with a single metric, and the metrics that define it are often hard to measure. One good practice is to define success on different levels of abstraction and have some story about how success at one layer contributes to the others. This doesn't have to be a precise technical endeavor, like a mathematical equation, but it should be an honest attempt at telling a story that all participants can get behind.

For example, participants in an Intelligent System might:

- On an hourly or daily basis optimize model properties.

- On a weekly basis review the user outcomes and make sure changes in model properties are affecting user outcomes as expected.

- On a monthly basis review the leading indicators and make sure nothing has gone off the rails.

- On a quarterly basis look at the organizational objectives and make sure the Intelligent System is moving in the right direction to affect them.

Revisit the goals of the Intelligent System often during the course of the project. Because things change.

Ways to Measure Goals

One reason defining success is so hard is that measuring success is harder still.

How the heck are we supposed to know how many passwords abusers got with their phishing pages?

When we discuss intelligent experiences in Part II of this book we will discuss ways to design intelligent experiences to help measure goals and get data to make the intelligence better. This section introduces some basic approaches. Using techniques like these should allow more flexibility in defining success.

Waiting for More Information

Sometimes it is impossible to tell if an action is right or wrong at the time it happens, but a few hours or days or weeks later it becomes much easier. As time passes you'll usually have more information to interpret the interaction. Here are some examples of how waiting might help:

- The system recommends content to the user, and the user consumes it completely—by waiting to see if the user consumes the content, you can get some evidence of whether the recommendation was good or bad.

- The system allows a user to type their password into a web page—by waiting to see if the user logs in from eastern Europe and tries to get all their friends to install malware, you can get some evidence if the password was stolen or not.

Waiting can be a very cheap and effective way to make a success criterion easier to measure, particularly when the user's behavior implicitly indicates success or failure.

There are a couple of downsides.

First, waiting adds latency. This means that waiting might not help with optimizing, or making fine-grained measurements.

Second, waiting adds uncertainty. There are lots of reasons a user might change their behavior. Waiting gives more time for other factors to affect the measurement.

A/B Testing

Showing different versions of the feature/intelligence to different users can be a very powerful way to quantify the effect of the feature.

Imagine giving half the users an intelligent experience and the other half a stubbed-out (simple, default) experience. Maybe half the users of the Internet toaster get a toast time of one minute no matter what settings they use or what they put into the toaster. The other half get a toast time that is determined by all the fanciest intelligence you can find.

If users who got the stubbed experience are just as happy/engaged/effective at toasting as the ones who got the full experience—you've got a problem.

A/B testing can be difficult to manage, because it involves maintaining multiple versions of the product simultaneously.

It can also have trouble distinguishing small effects. Imagine testing two versions of the intelligent toaster, one that toasts 1 minute no matter what, and one that toasts 61 seconds no matter what. Is one of them better than the other? Maybe, but it will probably take a long time (and a lot of observations of user interactions) to figure out which.

A/B testing is a great way to make sure that large changes to your system are positive, but it is troublesome with day-to-day optimization.

Hand Labeling

Sometimes a computer can't tell if an action was aligned with success, but a human can. You can hire some humans to periodically examine a small number of events/ interactions and tell you if they were successful or not. In many cases, this hand labeling is easy to do, and doesn't require any particular skill or training.

In order to hand-label interactions, the Intelligent System needs to have enough telemetry to capture and replay interactions. This telemetry must contain enough detail so a human can reliably tell what happened and whether the outcome was good or not (while preserving user privacy). This isn't always possible, particular when it involves having to guess what the user was trying to do, or how they were feeling while they were doing it.

But it never hurts to look at what your system is doing and how it is affecting users. The scale can be small. It can be cheap. But it can also be very useful.

Asking Users

Perhaps the most direct way to figure out if something is succeeding or not is to ask the user. For example, by building feedback mechanisms right into the product:

- The user is shown several pieces of content, selects one. The system pops up a dialog box asking if the user was happy with the choices.

- A self-driving car takes a user to their destination, and as the user is getting out it asks if the user felt safe and comfortable during the trip.

- The toaster periodically asks, "is that toasted just the way you like it?" (And yeah, that could be pretty darn creepy, especially if it does it when you are home alone after dark and the lights aren't on.)

A couple of things to keep in mind:

Users don't always have the answer. For example, asking someone "did you just give your password to a criminal?" might not be very effective—and it might scare a lot of people for no good reason.

Users might not always feel like giving an answer to all of your questions. This will introduce bias. For example, users who are very engaged in their task might not pause to consider a survey, even though they are getting a good outcome.

Users will get sick of being asked questions. This type of feedback should be used sparingly.

But sometimes, asking just .1% of users a simple question once per month can unlock a lot of potential in helping you know if your Intelligent System is succeeding.

Decoupling Goals

Some things are hard to measure directly but can be broken down into simpler pieces, and the pieces can be measured (perhaps using some of the techniques from this section) and then stitched together into an estimate of the whole.

For example, consider the phishing example. The number of credentials lost to phishers can be decoupled into the following:

- The number of your users who visit phish sites (which is estimated by waiting for user reports to identify phishing sites and combining with traffic telemetry to estimate the number of visits).

- The percent of users who type their passwords into phish sites when the system doesn't provide a warning (which is estimated by doing user studies).

- The percent of users who type their passwords into phish sites when the system does provide a warning (which is estimated using telemetry on how many users dismiss the warning and proceed to type into the password box on known phishing sites).

Multiply these together and you have an estimate of a pretty good potential goal for the Intelligent System.

Decoupling is particularly useful when it identifies critical, measurable sub-problems and shows how they chain together with other clear sub-problems to reach overall success.

Keeping Goals Healthy

Sometimes a goal should change. But changing goals is difficult, particularly when people have organized around them. There might be a lot of inertia, many opinions, and no good answers. Having a process for reviewing goals and adapting them is a very good idea. Goals might change if any of the following happen:

- A new data source comes on-line and shows some assumption was wrong.

- A part of the system is functioning very well, and goals around further improvement to it should be changed (before they take on a life of their own, resulting in investing in the wrong things).

- Someone comes up with a better idea about how to connect the Intelligent System's work to actual customer impact.

- The world changes and previous goal no longer reflects success.

And even if the goals don't have to change, it's good to get everyone together once and a while and remind yourselves what you all are trying to accomplish together.

Summary

Having goals is crucial to success in an Intelligent System. But goals are hard to get right. Effective goals should:

1.   Communicate the desired outcome

2.   Be achievable

3.   Be measurable

Goals can be very abstract (like organizational objectives). They can be less abstract (like leading indicators). They can be sort of concrete (like user outcomes). Or they can be super concrete (like model properties).

An effective set of goals will usually tie these various types of goals together into a story that clearly heads toward success.

Most Intelligent Systems will contribute to organizational objectives and leading indicators, but the core work of day-to-day improvement will be focused on user outcomes and model properties.

Goals can be measured through telemetry, through waiting for outcomes to become clear, by using human judgment, and by asking users about their experiences.

And did I mention—goals are hard to get right. They will probably take iteration.

But without effective goals, an Intelligent System is almost certainly doomed to waste time and money—to fail.

For Thought

After reading this chapter, you should:

- Understand the ways you can define success for an Intelligent System, and how to measure whether success is being achieved.

- Be able to define success on several levels of abstraction and tell the story of how the different types of success contribute to each other.

You should also be able to answer questions like these:

Consider your favorite hobby that might benefit from an Intelligent System.

- What organizational objective would the Intelligent System contribute to for its developers?

- What leading outcomes would make the most sense for it?

- What are the specific user outcomes that the Intelligent System would be tracked on?

- Which way would you measure these? Why?

PART II

Intelligent Experiences

Chapters 5-10 explain how to connect intelligence with users to achieve an Intelligent System's objectives. This part discusses the pitfalls and challenges of creating user experiences based on models and machine learning, along with the properties of a successful intelligent experience. It explains how to adapt experiences to get data to grow the system. And it gives a framework for verifying that an intelligent experience is behaving as intended.

CHAPTER 5

The Components of Intelligent Experiences

At the core of every Intelligent System is a connection between the intelligence and the user. This connection is called the *intelligent experience*. An effective intelligent experience will:

- **Present the intelligence to the user** by choosing how they perceive it, balancing the quality of the intelligence with the forcefulness of the experience to engage users.

- **Achieve the system's objectives** by creating an environment where users behave in the ways that achieve the system's objectives—without getting upset.

- **Minimize any intelligence flaws** by reducing the impact of small errors and helping users discover and correct any serious problems that do occur.

- **Create data to grow the system** by shaping the interactions so they produce unbiased data that is clear, frequent, and accurate enough to improve the system's intelligence.

Achieving these objectives and keeping them balanced as the Intelligent System evolves is as difficult as any other part of creating an Intelligent System, and it is critical for achieving success.

This chapter will introduce these components of intelligent experiences. Following chapters will explore key concepts in much more detail.

© Geoff Hulten 2018
G. Hulten, *Building Intelligent Systems*, https://doi.org/10.1007/978-1-4842-3432-7_5

Presenting Intelligence to Users

Intelligence will make predictions about the world, the user, and what is going to happen next. The intelligent experience must use these predictions to change what the user sees and interacts with. Imagine:

- The intelligence identifies 10 pieces of content a user might like to explore. How should the experience present these to the user?

- The intelligence thinks the user is making an unsafe change to their computer's settings. What should the experience do?

- The intelligence determines the user is in a room where it a little bit too dark for humans to see comfortably. How should the automated home's experience respond?

The experiences in these examples could be passive, giving gentle hints to the user, or they could be forceful, flashing big red lights in the user's face. They might even automate things. Choosing the right balance is one of the key challenges when building intelligent experiences.

An intelligent experience might:

- **Automate:** By taking an action on the user's behalf. For example, when the intelligence is very sure the user is asleep, the experience might automatically turn down the volume on the television.

- **Prompt:** By asking the user if an action should be taken. For example, when the intelligence thinks the user might have forgotten their wife's birthday, the experience might ask: "Would you like me to order flowers for your wife's birthday?"

- **Organize:** By presenting a set of items in an order that might be useful. For example, when the intelligence thinks the user is on a diet, the experience might organize the electronic-menu at their favorite restaurant by putting the healthiest things the user might actually eat on top (and hiding everything that is a little too delicious).

- **Annotate:** By adding information to a display. For example, when the intelligence thinks the user is running late for a meeting, the experience might display a little flashing running-man icon on the corner of the screen.

Or an intelligent experience might use some combination of these methods, for example annotating when the issue is uncertain but automating when something bad is very likely to happen soon.

The challenge is that the experience should get the most value out of the intelligence *when the intelligence is right*, and it should cause only an acceptable amount of trouble *when the intelligence makes a mistake*.

Any Intelligent System (regardless of the intelligence) can use any of these approaches to experience. And recall, the intelligence in an Intelligent System will change over time, becoming better as more users use it (or worse as the problem changes). The right experience might change over time as well.

An Example of Presenting Intelligence

As an example, Let's consider a smart-home product that automates the lights in rooms of users' homes. A user installs a few sensors to detect the light level, a few sensors to detect motion, and a few computer-controlled light bulbs. Now the user is going to want to be amazed. They are going to want to impress their friends with their new toy. They are going to want to never have to think about a light switch again.

So how could it work? First, let's consider the intelligence. The role of the intelligence is to interpret the sensor data and make predictions. For example: given the recent motion sensor data and the recent light sensor readings, what is the probability a user would want to adjust the lights in the near future? The intelligence might know things like:

- If the room is dark, and someone just entered it, the lights are 95% likely to turn on soon.

- If the lights are on, and someone is leaving the room, the lights are 80% likely to turn off soon.

- If the lights are on, and no one has been around for four hours, the next person into the room is likely to yell at the kids about how much power used to cost in their day and about the meaning of respect and about groundings—and then turn off the lights.

So what should the experience for the light-automation system do?

One choice would be to completely automate the lights. If the intelligence thinks the user would like to change the lights soon, then go ahead and change them. This is a pretty forceful experience, because it automates the raw intelligence output with no accommodation for mistakes. Will it be right? Will it make the user happy?

Well, what if the user is watching a movie? The light-automation system's intelligence doesn't know anything about the TV. If the user is watching a movie, and the motion sensor detects them—like maybe they stretch or reach for their drink—the lights might flip on. Not a great experience. Or what if the user is sleeping? Will the lights flash on every-time they roll over? Or what if the user is having a romantic dinner?

In these cases, an experience that simply automates the lights is not likely to do the right thing. That's partly because the intelligence doesn't have enough information to make perfect decisions and partly because the intelligence wouldn't make perfect decisions even if it had all the information in the world (because intelligence makes mistakes).

So are we doomed? Should we scrap the product, go back to the drawing board?

Well, we could. Or we could consider some other ways of presenting the intelligence to users.

For example, instead of *automating* the lights, the experience might *prompt* the user when the intelligence thinks the lights should change. For example, if the system thinks it is too dark, it could ask (in a pleasantly computerized female voice), "Would you like me to turn on the lights?"

This might be better than automation, because the mistakes the systems makes are less irritating (a question in a nice voice instead of a suddenly-dark room). But will it be good enough? What if the intelligence makes too many mistakes? The voice is chiming in every few minutes, pleasantly asking: "Would you like me to mess up your lights now?"... "How about now?"... "Maybe now?"

The system would seem pretty stupid, asking the user if they would like to screw up their lighting multiple times per hour.

Another, even less intrusive, approach would expose the lighting information by providing *annotations*. For example, maybe a subtle "energy usage" warning on the user's watch if there are a couple of lights on that the intelligence thinks should be off. Or maybe a subtle "health warning" if the intelligence thinks the user is in a room that is too dark. If the user notices the warning, and decides it is correct, they can change the lights.

These different experiences might all be built on the same intelligence—they may have exactly the same prediction in any particular situation—but they would seem very different from the user's perspective.

So which one is right?

It depends on the system's goals and the quality of the intelligence. If the intelligence is excellent, automation might work. If the intelligence is new (and not too accurate yet), the annotations might make more sense. One of the key goals of the intelligence experience is to take the user's side and make something that works for them.

Achieve the System's Objectives

An effective intelligent experience will present intelligence in a way that achieves the Intelligent System's objective. The experience will be designed so users see the intelligence in ways that help them have good outcomes (and ways that help the business behind the Intelligent System achieve its objectives). Recall that the goals of an Intelligent System can include:

- **Organizational outcomes**: sales and profit.

- **Leading indicators**: engagement and sentiment.

- **User outcomes**: achieving user objectives.

- **Model Properties**: having accurate intelligence.

The job of the intelligent experience is to take the output of intelligence and use it to achieve user outcomes, to improve leading indicators, and to achieve organizational outcomes. This means that when the intelligence is right, the experience should push users toward actions that achieve an objective. And when the intelligence is wrong, the experience should protect users from having too bad an experience.

Consider the home lighting automation system. If the goal is to save power, what should the system do when it thinks the lights should be off?

- **Play a soft click at the light-switch**: If the user happens to hear it and wants to walk to the switch, they might turn the light off. Maybe.

- **Prompt the user**: If we ask whether the user would like the light off, they might say yes, or they might be in the middle of something (and miss the question).

- **Prompt the user and then automate after a delay**: Give the user a chance to stop the action, but then take the action anyway.

- **Automate the lights**: Just turn them off.

- **Shut down power to the house's main breaker**: Because, come on, the user is wasting power; they are a menace to themselves and to others...

These experience options are increasingly likely to get the lights to turn (and stay) off. Choosing one that is too passive will make the system less effective at achieving its objective. Choosing one that is too extreme will make the system unpleasant to use.

In fact, the experience is just as important as the intelligence in accomplishing the Intelligent System's objectives: an ineffective experience can mess up the system just as thoroughly as terribly inaccurate intelligence predictions.

An intelligent experience is free to pick and choose, to ignore unhelpful intelligence—whatever it takes to achieve the desired outcomes.

An Example of Achieving Objectives

Consider the automated light system again. Imagine two possible objectives for the product:

1. To minimize the amount of power consumed.

2. To minimize the chance of a senior citizen tripping over something and getting hurt.

Everything else is the same—the intelligence, the sensors, the way the system is implemented—only the objectives are different. Would the same experience be successful for both of these objectives? Probably not.

The system designed to conserve power might never turn lights on, even if the intelligence thinks having a light on would be appropriate. Instead, the experience might rely on users to use the switch when they want a light on, and simply turn lights off when it thinks the user doesn't need the light any more.

The system designed to avoid senior-citizen-tripping might behave very differently. It might turn on lights whenever there is any chance there is a user in the room, and leave them on until it is very certain the room is empty.

These experiences are quite different, so different they might be sold as totally different products—packaged differently, marketed differently, priced differently—and they might each be perfectly successful at achieving their overall objectives. And they might both use (and contribute to) exactly the same underlying intelligence.

Minimize Intelligence Flaws

Intelligence makes mistakes. Lots of them. All different types of mistakes. An effective intelligent experience will function despite these mistakes, minimizing their impact and making it easy for users to recover from them.

When designing an intelligent experience it is important to understand all of the following:

1. What types of mistakes the intelligence will make.

2. How often it will make mistakes.

3. What mistakes cost the user (and the organization):

   a. If the user notices the mistake and tries to correct it.

   b. If the user never notices that the mistake happened.

Then the intelligent experience can decide what to do about the mistakes. The experience can:

1. Stay away from bad mistakes by choosing not to do things that are too risky.

2. Control the number of interactions the user will have with the intelligence to control the number of mistakes the user will encounter.

3. Take less forceful actions in situations where mistakes are hard to undo (for example, prompting the user to be sure they want to launch the nuclear missile, instead of launching it automatically).

4. Provide the user with feedback about actions the system took and guidance on how to recover if something is wrong.

These techniques make an Intelligent System safer. They also reduce the potential of the Intelligent System by watering down interactions, and by demanding the user pay more attention to what it is doing. Creating a balance between achieving the objective and controlling mistakes is a key part of designing effective intelligent experiences.

Create Data to Grow the System

Intelligence needs data to grow. It needs to see examples of things happening in the world, and then see the outcomes that occurred. Was the interaction good for the user? For the business? Was it bad? Intelligence can use these examples of the world and of the outcomes to improve over time.

The experience plays a large role in making the data that comes from the Intelligent System valuable. Done right, an experience can produce large data sets, perfectly tailored for machine learning. Done wrong, an experience can make the data from the system useless.

An effective intelligent experience will interact with users in clear ways where the system can know:

1. The context of the interaction.

2. The action the user took.

3. The outcome.

Sometimes the experience can make this interaction completely implicit; that is, the experience can be designed so elegantly that the user produces the right type of data simply by using the system.

At other times the experience may require users to participate explicitly in creating good data (maybe by leaving ratings). We will discuss techniques and tradeoffs for getting good data from intelligent experiences in great detail in a later chapter.

An Example of Collecting Data

Consider the automated-light example.

The intelligence will examine the sensors and make a decision about what the lights should do. The experience will act on this intelligence, interacting with the user in some way. Then we need to know—was that interaction correct?

What if the lights were on, but the system thought they should be off? How would we know if the intelligence is right or not?

We might be right if:

- The experience automatically turns off the lights and the user doesn't turn them back on.

- The experience prompts the user to turn off the lights, and the user says yes.

- The experience provides a power warning, and the user turns off the lights.

We might be wrong if:

- The experience automatically turns off the lights and the user immediately turns them back on.

- The experience prompts the user to turn off the lights, but the user says no.

- The experience provides a power warning, but the user doesn't turn off the lights.

But this feedback isn't totally clear. Users might not bother to correct mistakes. For example:

- They might have preferred the lights remain on, but have given up fighting with the darn auto-lights and are trying to learn to see in the dark.

- They might see the prompt to turn the lights off, but be in the middle of a level on their favorite game and be unable to respond.

- They might have put their smart watch in a drawer and so they aren't seeing any power warnings any more.

Data is best when outcomes are clear and users have an incentive to react to every interaction the experience initiates. This isn't always easy, and it may come with some tradeoffs. But creating effective training data when users interact with intelligent experiences is a key way to unlock the value of an Intelligent System.

Summary

This chapter introduced intelligent experiences. The role of an intelligent experience is to present intelligence to users in a way that achieves the system's objectives, minimizes intelligence flaws, and creates data to improve the intelligence.

Presenting intelligence involves selecting where it will appear, how often it will appear, and how forcefully it will appear. There are many options, and selecting the right combination for your situation is critical to having success with Intelligent Systems.

With the right experiences, a single intelligence can be used to achieve very different goals.

Experience can control the cost of mistakes by reducing their forcefulness, by making them easier to notice, by making them less costly, and by making them easier to correct.

An effective experience will support improving the intelligence. This can be done by creating interactions with users that collect frequent and unbiased samples of both positive and negative outcomes.

The intelligent experience must be on the user's side and make them happy, engaged, and productive no matter how poor and quirky the intelligence might be.

For Thought...

After reading this chapter, you should:

- Understand how user experience and intelligence come together to produce desirable impact.

- Know the main goals of the intelligent experience and be able to explain some of the challenges associated with them.

You should be able to answer questions like these:

- Consider your interactions with Intelligent Systems. What is the most forceful use of intelligence you have experienced? What is the most passive?

- Name three ways you've provided data that helped improve an intelligent experience. Which was the most effective and why?

CHAPTER 6

Why Creating Intelligent Experiences Is Hard

This chapter explores some ways that experiences based on intelligence are different from more traditional experiences. And here is the bottom line: intelligence makes mistakes.

Intelligence is going to make mistakes; to create effective intelligent experiences, you are going to have to get happy with that fact. You are going to have to embrace mistakes. You are going to have to make good decisions about how to work with and around the mistakes.

The mistakes intelligence makes:

- Aren't necessarily intuitive.

- Aren't the same from day to day.

- Aren't easy to find ahead of time.

- Aren't possible to fix with "just a little more work on intelligence."

This chapter is about the mistakes that intelligence makes. Its goal is to introduce you to the challenges you'll need to address to create effective intelligent experiences. Understanding the ways intelligence makes mistakes and the potential trade-offs will help you know what to expect, and, more importantly, how to design intelligent experiences that give users the best possible outcomes—and achieve the system's objectives—despite the mistakes that intelligence makes.

Intelligence Make Mistakes

Intelligent Systems make mistakes. There is no way around it. The mistakes will be inconvenient, and some will be actually quite bad. If left unmitigated, the mistakes can make an Intelligent System seem stupid; they could even render an Intelligent System useless or dangerous.

© Geoff Hulten 2018
G. Hulten, *Building Intelligent Systems*, https://doi.org/10.1007/978-1-4842-3432-7_6

Here are some example situations that might result from mistakes in intelligence:

- You ask your phone to order pizza and it tells you the population of Mumbai, India.

- Your self-driving car starts following a road that doesn't exist and you end up in a lake.

- Your smart-doorbell tells you someone is approaching your front door, but upon reviewing the video, you realize no one is there.

- You put a piece of bread in your smart toaster, come back five minutes later, and find that the bread is cold—the toaster hadn't done anything at all.

These types of mistakes, and many others, are just part of the cost of using intelligence, particularly when using intelligence created by machine learning.

And these mistakes are not the fault of the people producing the intelligence. I mean, I guess the mistakes could be their fault—it's always possible for people to be bad at their jobs—but even people who are excellent—world class—at applied machine learning will produce intelligences that make mistakes.

So one of the big roles of experience in Intelligent Systems is to present the intelligence so it is effective when it is right, and so the mistakes it makes are minimized and are easy to recover from.

Consider that an intelligence that is right 95% of the time makes a mistake one out of every twenty interactions. And 95% is a very high accuracy. If your intelligence-producers get to 95% percent they are going to feel like they've done a great job. They are going to want to get rewarded for their hard work, and then they are going to want to move on and work on other projects. They aren't going to want to hear you telling them the intelligence isn't accurate enough.

But in a system that has a million user interactions per day, 95% accuracy results in 50,000 mistakes per day. That's a lot of mistakes. If the mistakes cost users time or money, that could add up to a real problem.

Another useful way to think about mistakes is how many interactions a user has between seeing a mistake. For example, a user with 20 interactions per day with a 97% accurate intelligence would expect to see 4.2 mistakes per week.

Is this a disaster?

It depends on how bad the mistakes are and what options the user has to recover.

But consider the alternative to building systems that make mistakes—that is, to demand a perfect intelligence before building a product around it. Perfection is very expensive, in many cases impossible. And perfection isn't needed to get real value from Intelligent Systems—sitting around waiting for perfection is a great way to miss out on a lot of potential. Consider these examples:

- Speech recognition isn't perfect, it probably never will be.

- Search engines don't always return the right answer.

- Game AI is commonly ridiculed for running in circles and shooting at walls.

- Self-driving cars have accidents.

But all of these (somewhat) inaccurate intelligences are part of products that many of us use every day—that make our lives better. Mitigating mistakes is a fundamental activity in producing intelligent experiences.

Intelligence Makes Crazy Mistakes

At the risk of belaboring the point, intelligences make crazy mistakes. Wild, inexplicable, counter-intuitive mistakes.

Consider—if a human expert is right 99% of the time, you might expect the mistakes they make to be close to correct. You assume this expert pretty much understands what is going on, and if they make a mistake it is probably going to be consistent with a rational view of how the world works.

Machine learning and artificial intelligence aren't like that.

An artificial intelligence might be right 99.9% of the time, and then one out of a thousand times say something that is right out of bat-o-bizarro land. For example:

- A system that recommends music might recommend you 999 rock songs, and then recommend you a teen-pop track.

- A smart toaster might toast 99 pieces of bread perfectly, and decide the 100th piece of bread doesn't need any toasting—no matter what, no heat.

- A system that reads human emotion from images might correctly identify 999 people as happy, then it might see my face and say I am sad no matter how big of an idiot-grin I slap on.

65

These crazy mistakes are part of working with Intelligent Systems.

And even crazy mistakes can be mitigated, recovered from, and minimized. But to design experiences that work with intelligence you should throw out pre-conceived notions of how a rational thing would act. Instead, you will need to develop intuition with the intelligence you are working with. Use it. See what tends to confuse it and how. Think of ways to support it.

One instinct is to talk to the people making the intelligence and say, "Come on, this is crazy. Fix this one mistake. It is killing us!"

That's fine. They can probably change the intelligence to fix the mistake that's bugging you.

But fixing one mistake usually introduces a new mistake somewhere else. That new mistake is likely just as crazy. And you won't know when or how the new mistake will show up. Fixing obvious mistakes isn't always the right thing to do; in fact, playing whack-a-mole with prominent mistakes can be detrimental. Sometimes it's better to let the intelligence optimize as best it can, then support the intelligence by covering over mistakes with elegant experiences.

Intelligence Makes Different Types of Mistakes

And even though you may be sure I'm belaboring the point by now, Intelligent Systems make different types of mistakes; for example:

- The smart-doorbell might say someone is approaching the door when no one actually is; or it might fail to notify its user when someone is approaching the door.

- The smart toaster might undercook bread; or it might overcook it.

- A speech recognition system might incorrectly interpret what the user said; or it might refuse to try to guess what the user said.

Three main types of mistakes here are:

1. Mistaking one situation for another situation.

2. Estimating the wrong value (for example, the correct toast time).

3. Being too confused (or conservative) to say anything at all.

Perhaps the simplest form of this is confusing one situation for another, as in the smart doorbell example. There are four possible outcomes for the intelligence of a system like this (Figure 6-1):

- **True Positive**—When there *is <u>someone</u> standing at the door* and the intelligence thinks there *is <u>someone</u> standing at the door*, it is called a true positive (this is a correct answer, not a mistake).

- **True Negative**—When there *is <u>not someone</u> standing at the door* and the intelligence thinks there *is <u>no one</u> there*, it is called a true negative (this is another type of correct answer, not a mistake).

- **False Positive**—When there *is <u>not someone</u> standing at the door* but the intelligence thinks there *is <u>someone</u> standing at the door*, it is called a false positive (this is a mistake).

- **False Negative**—When there *is <u>someone</u> standing at the door* but the intelligence thinks there *is <u>no one</u> there*, it is called a false negative (this is another type of mistake) .

The Intelligence says some is

		There	Not There
Someone Actually is	**There**	True Positive	False Negative
	Not There	False Positive	True Negative

Figure 6-1. *Different types of mistakes*

It's often possible for the system's intelligence to be tuned to control the type of mistakes it makes, making more of one type of mistake and fewer of the other. (Usually this involves tuning a threshold on the probability output by the models that drive the intelligence.)

For example, when designing the smart-doorbell experience, is it better to have the system alert the user when there is no one at the door, or to fail to alert the user when someone is at the door?

What if you could get rid of three occurrences of the first type of mistake at the cost of adding one of the second type? Would that be a better trade-off for the user? Would it be more helpful in achieving the systems objectives?

It will depend on the experience. How much does each type of mistake irritate the user? Which type of mistake is easier to hide? Which type of mistake gets in the way of achieving the system's objectives more? And how many times per day/month/year will a user have to experience each one of these mistakes? Will they become fatigued and stop responding to alerts from the system at all? Or will they come to distrust the system and turn it off?

Other examples of trade-offs include:

- Trading off underestimates for overestimates. For example, when the smart toaster predicts how long to toast for, would it be better to over-toast by 5% or to under-toast by 5%?

- Trading off the number of options the intelligence gives. For example, in a recommender system, the system might be pretty sure the user wants to read one of 15 different books. Would it be better to let the user choose from the top 5? Or to display all 15?

- Trading off between the intelligence saying something and saying nothing. For example, in a computer vision system, the intelligence might be 90% sure it knows who is in an image. Would it be better to have the intelligence label a face in an image with a name that is going to be wrong 1 out of 10 times? Or would it better to have the system do nothing? Or ask for the user to help?

Intelligence Changes

Traditional user experiences are deterministic. They do the same thing in response to the same command. For example, when you right click on a file, a menu comes up, every time (unless there is a bug). Deterministic is good, for many reasons.

Intelligent Systems improve themselves over time. An action that did one thing yesterday might do a different thing today. And that might be a very good thing. Consider:

- Yesterday you searched for your favorite band and got some OK articles. Today you search for your favorite band again and you get newer, better information.

- Yesterday you cooked some popcorn in your microwave and a few of the pieces burned. Today you cook popcorn and the microwave doesn't cook it quite so long, so the popcorn comes out perfectly.

- Yesterday your navigation app directed you drive through a construction zone and your commute took an hour longer than usual. Today the app directs you on an alternate route and you arrive at work happier and more productive.

These are examples of positive change, and these types of change are the goal for Intelligent Systems—from the user's perspective the system works better today than it did yesterday. If something is a bit off right now, that's no problem—the user can trust that the system will improve and that problems will decrease over time.

That's the goal. But change can be disruptive too. Even change that is positive in the long run can leave users feeling confused and out of control in the short term. Imagine the following:

- Yesterday you were taking notes with a smart-pen, and the ink was a nice shade of blue. But overnight the pen connected to its service and the service—optimizing for something you may or may not ever know—decided to tweak the shade of blue. Today the notes you take are in a different color.

- Yesterday you went to the local convenience store and bought a big, cool cup of your favorite soda, drank and enjoyed it. Overnight the soda machine connected to its service and tweaked the recipe. Today you bought another cup of the soda and you don't like is quite as much.

Are these changes good? Maybe. Maybe they improve some very important properties of the businesses involved. Maybe you'll realize you actually do like the changed behavior—eventually—even though it bothered you at first.

Or maybe the changes are simply disastrous.

When dealing with changing intelligence, some things to consider are:

1. Can the **rate of change be controlled**? This might be done by
    putting some constraints on the intelligence producers. For
    example, maybe the intelligence changes its predictions on no
    more than 5% of interactions per day. Taken too far, this could
    result in static intelligence (and eliminate much of the potential
    of intelligence improving over time), but some simple constraints
    can help a lot.

2. Can the experience **help the user navigate through change?**
    Perhaps it does this by letting them know a change has happened.
    Or maybe by keeping the old behavior but offering the user the
    changed behavior as an option. For example, the smart-pen could
    let the user know it has found a new shade of ink that makes notes
    5% more useful, and offer to make the change if the user wants or
    keep the ink the old way if the user prefers that.

3. Can the experience **limit the impact of the changing portions** of
    the system, while still allowing them to have enough prominence to
    achieve impact? For example, by using a more passive experience
    with the parts of the intelligence that are changing the most?

The Human Factor

Intelligent experiences succeed by meshing with their users in positive ways, making
users happier, more efficient, and helping them act in more productive ways (or ways
that better align with positive business outcomes).

But dealing with Intelligent Systems can be stressful for some users, by challenging
expectations.

One way to think about it is this.

Humans deal with tools, like saws, books, cars, objects. These things behave in
predictable ways. We've evolved over a long time to understand them, to count on them,
to know what to expect out of them. Sometimes they break, but that's rare. Mostly they
are what they are; we learn to use them, and then stop thinking so much about them.

Tools become, in some ways, parts of ourselves, allowing us powers we wouldn't have without them.

They can make us feel very good, safe, and comfortable.

Intelligent Systems aren't like this, exactly.

Intelligent Systems make mistakes. They change their "minds." They take very subtle factors into consideration in deciding to act. Sometimes they won't do the same thing twice in a row, even though a user can't tell that anything is different. Sometimes they even have their own motivations that aren't quite aligned with their user's motivations.

Interacting with Intelligent Systems can seem more like a human relationship than like using a tool. Here are some ways this can affect users:

- **Confusion**: When the Intelligent System acts in strange ways or makes mistakes, users will be confused. They might want (or need) to invest some thought and energy to understanding what is going on.

- **Distrust**: When the Intelligent System influences user actions will the user like it or not? For example, a system might magically make the user's life better, or it might nag them to do things, particularly things the user feels are putting others' interests above theirs (such as by showing them ads).

- **Lack of Confidence**: Does the user trust the system enough to let it do its thing or does the user come to believe the system is ineffective, always trying to be helpful, but always doing it wrong?

- **Fatigue**: When the system demands user attention, is it using it well, or is asking too much of the user? Users are good at ignoring things they don't like.

- **Creep Factor**: Will the interactions make the user feel uncomfortable? Maybe the system knows them too well. Maybe it makes them do things they don't want to do, or post information they feel is private to public forums. If a smart TV sees a couple getting familiar on the couch, it could lower the lights and play some Barry White music—but should it?

Be conscious of how users feel about the intelligent experience. Manage their feeling by making good choices about what to present and how forceful to be.

Summary

Intelligence makes mistakes. No matter what, it will make mistakes. Complaining won't help. The mistakes aren't there because someone else is bad at their job. And chasing mistakes, playing whack-a-mole, is often (usually) detrimental.

The mistakes an intelligence makes can be quite counter intuitive. Even intelligence that is usually right can make really strange mistakes (ones that no human would ever make).

There are different types of mistakes. One common way to express them is as: False positives (saying something happened when it didn't) and false negatives (saying something didn't happen when it did). Intelligence can often make trade-offs between these types of mistakes (for example, by adding a few more false positives to get rid of a few false negatives).

Change can be disruptive to users, even when the change is positive in the long run. Intelligent Systems change, regularly. Helping users deal with the change is part of creating an effective Intelligent System.

Intelligent Systems can have negative effects on users, leading them to dislike and distrust the system.

Creating intelligent experiences is hard.

For Thought...

After reading this chapter you should:

- Understand that intelligence makes mistakes—it is one of the fundamental challenges that intelligent experiences need to address.

- Realize why these challenges are not intelligence bugs (which should be fixed) but are intrinsic to working with intelligence (and must be embraced).

- Have a sense of how users will experience the negative parts of an Intelligent System and some empathy to help create experiences to help users.

You should be able to answer questions like:

- What is the worst mistake you've seen an Intelligent System make?

- What could an intelligent experience have done to make the mistake less bad?

- For what product change did you have the wildest reaction, disliking it at first, but coming to like it more over time? Why?

CHAPTER 7

Balancing Intelligent Experiences

Designing a successful intelligent experience is a balancing act between:

1. Achieving the desired outcome

2. Protecting users from mistakes

3. Getting data to grow the intelligence

When the intelligence is right, the system should shine, creating value, automating something, giving the user the choices they need, and encouraging the safest, most enjoyable, most profitable actions. The experience should strongly encourage the user to take advantage of whatever it is the intelligence was right about (as long as the user isn't irritated and feels they are getting a good deal from the interaction).

When the intelligence is wrong, the system should minimize damage. This might involve allowing the user to undo any actions taken. It might involve letting the user look for more options. It might involve telling the user why it made the mistake. It might involve some way for the user to get help. It might involve ways for the user to avoid the particular mistake in the future.

The problem is that the experience won't know if the intelligence is right or wrong. And so every piece of experience needs to be considered through two lenses: what should the user do if they got there because the intelligence was right; and what should the user do if they got there because the intelligence was wrong.

This creates tension, because the things that make an Intelligent System magical (like automating actions without any fuss) are at odds with the things that let a user cope with mistakes (like demanding their attention to examine every decision before the intelligence acts).

75

© Geoff Hulten 2018
G. Hulten, *Building Intelligent Systems*, https://doi.org/10.1007/978-1-4842-3432-7_7

There are five main factors that affect the balance of an intelligent experience:

- The *forcefulness* of the experience; that is, how strongly it encourages the user to do what the intelligence thinks they should.

- The *frequency* of the experience; that is, how often the intelligent experience tries to interact with the user.

- The *value* of the interaction when the intelligence is right; that is, how much the user thinks it benefits them, and how much it helps the Intelligent System achieve its goals.

- The *cost* of the interaction when the intelligence is wrong; that is how much damage the mistake does and how hard it is for the user to notice and undo the damage.

- The *quality of the intelligence*; that is, how often the intelligence is right and how often it is wrong.

To create an effective intelligent experience you must understand these factors and how they are related. Then you must take the user's side and build an experience that effectively connects them to the intelligence. This chapter will explore these factors in more detail.

Forcefulness

An interaction is forceful if the user has a hard time ignoring (or stopping) it. An interaction is passive if it is less likely to attract the user's attention or to affect them. For example, a forceful experience might:

- Automate an action.

- Interrupt the user and make them respond to a prompt before they can continue.

- Flash a big, garish pop-up in front of the user every few seconds until they respond.

Forceful interactions are effective when:

- The system is confident in the quality of the intelligence (that it is much more likely to be right than it is to be wrong).

- The system really wants to engage the user's attention.

- The value of success is significantly higher than the cost of being wrong.

- The value of knowing what the user thinks about the intelligence's decision is high (to help create new intelligence).

A *passive experience* does not demand the user's attention. It is easy for the user to choose to engage with a passive experience or not. Passive experiences include:

- A subtle prompt that does not force the user to respond immediately.

- A small icon in the corner of the screen that the user may or may not notice.

- A list of recommended content on the bottom of the screen that the user can choose to click on or to ignore.

Passive interactions are effective when:

- The system is not confident in the quality of the intelligence.

- The system isn't sure the value of the intelligent interaction is higher than what the user is currently doing.

- The cost of a mistake is high.

One way to think about the forcefulness of an interaction is as an advertisement on a web page. If the ad pops over the web page and won't let you continue till you click it—that's a forceful experience. You'll probably click the stupid ad (because you have to). And if the ad isn't interesting, you'll be mad at the product that is being advertised, at the web page that showed you the ad, and at whoever programed a web browser stupid enough to let such an irritating ad exist.

On the other hand, if the ad is blended tastefully into the page you may not notice it. You may or may not click the ad. If the ad is for something awesome you might miss out. But you are less likely to be irritated. You might go back to the web page again and again. Over time you may spend more time on the web page and end up clicking far more ads if they are passive and tasteful than if they are forceful and garish.

Frequency

An intelligent experience can choose to interact with the user, or it can choose not to. For example, imagine a smart map application that is giving you driving directions. This smart map might be getting millions of data points about traffic conditions, accidents, weather conditions, which lights are red and which are green, and so on. Whenever you come to an intersection, the directions app could say "turn left to save 17 seconds," or "go faster to beat the next light and save 3 minutes," or "pop a U and go back to the last intersection and turn right LIKE I TOLD YOU LAST TIME and you can still save 2 minutes."

Or the app could choose to be more subtle, limiting itself to one suggestion per trip, and only making suggestions that save at least 10 minutes.

More frequent interactions tend to fatigue users, particularly if the frequent interactions are forceful ones. On the other hand, less frequent interactions have fewer chances to help users, and may be confusing when they do show up (if users aren't accustomed to them).

Some ways to control frequency are to:

1. *Interact whenever the intelligence thinks it has a different answer.* For example, fifteen seconds ago the intelligence thought the right direction was straight, now it has more information and thinks the right direction is right. And ten seconds later it changes its mind back to thinking you should go straight. Using intelligence output directly like this can result in very frequent interactions. This can be effective if the intelligence is usually right and the interaction doesn't take too much user attention. Or it could drive users crazy.

2. *Interact whenever the intelligence thinks it has a significantly different answer.* That is, only interact when the new intelligence will create a "large value" for the user. This trades off some potential value for a reduction in user interruption. And the meaning of "large value" can be tuned over time to control the number of interaction.

3. *Explicitly limit the rate of interaction.* For example, you might allow one interaction per hour, or ten interactions per day. This can be effective when you aren't sure how users will respond. It allows limiting the potential cost and fatigue while gaining data to test assumptions and to improve the Intelligent System.

4. *Interact whenever the intelligence thinks the user will respond.* This involves having some understanding of the user. Do they like the interactions you are offering? Do they tend to accept or ignore them? Are they doing something else or are they receptive to an interruption? Are they starting to get sick of all the interactions? Done well, this type of interaction mode can work well for users, meshing the Intelligent System with their style. But it is more complex. And there is always the risk of misunderstanding the user. For example, maybe one day the user has a guest in the car, so they ignore all the suggestions and focus on their guest. Then the intelligent experience (incorrectly) learns that the user doesn't like interruptions at all. It stops providing improved directions. The user misses out on the value of the system because the system was trying to be too-cute.

5. *Interact whenever the user asks for it.* That is, do not interact until the user explicitly requests an interaction. This can be very effective at reducing user fatigue in the experience. Allowing users to initiate interaction is good as a backstop, allowing the system to interact a little bit less, but allowing the users to get information or actions when they want them. On the other hand, relying too heavily on this mode of interaction can greatly reduce the potential of the Intelligent System—what if the user never asks for an interaction? How would the user even know (or remember) how to ask for an interaction?

And keep in mind that humans get sick of things. Human brains are very good at ignoring things that nag at them. If your experience interacts with a user too much, they will start to ignore it. Be careful with frequent interactions—less can be more.

Value of Success

Users will be more willing to put up with mistakes and with problems if they feel they are getting value. When an Intelligent System is helping with something obviously important—like a life-critical problem or saving users a large amount of money and time—it will be easier to balance the intelligent experience in a way that users like, because users will be willing to engage more of their time on the decisions the Intelligent

System is making. When the Intelligent System is providing smaller value—like helping users toast their bread or saving them a few pennies of power per day—it will be intrinsically harder to create a balanced intelligent experience that users feel is worth the effort to engage with. Users tend to find interactions valuable if they:

- Notice that something happened.

- Feel that the interaction solved a problem they care about, or believe it provided them some meaningful (if indirect) improvement.

- Can connect the outcome with the interaction; they realize what the interaction was trying to do, and what it did do.

- Trust that the system is on their side and not just trying to make someone else a buck.

- Think the system is cool, smart, or amazing.

It is possible for an interaction to leave users in a better situation and have zero perceived value. It is also possible for an interaction to leave the users in a worse situation, but leave the user thinking the Intelligent System is great. The intelligent experience plays a large role in helping users feel they are getting value from an Intelligent System.

When the intelligence is right, an effective intelligent experience will be prominent, will make users feel good about what happened, and will take credit where credit is due. But a savvy intelligent experience will be careful—intelligence won't always be right, and there is nothing worse than an intelligent experience taking credit for helping you when it actually made your life worse.

Interactions must also be valuable to the Intelligent System. In general, an interaction will be valuable to the Intelligent System if it achieves the system's objectives; for example, when it:

- Causes the user to use the Intelligent System more (*increases engagement*).

- Causes the user to have better feelings about the Intelligent System (*improves sentiment*).

- Causes the user to spend more money (*creates revenue*).

- Creates data that helps the Intelligent System improve (*grows intelligence*).

In general, experiences that explicitly try to make money and to get data from users will be in conflict with making users feel they are getting a good value. An effective intelligent experience will be flexible to make trade-offs between these over the life-cycle of the Intelligent System to help everyone benefit.

Cost of Mistakes

Users don't like problems; Intelligent Systems will have problems. A balanced intelligent experience will be sensitive to how much mistakes cost users and will do as much as possible to minimize those costs.

Mistakes have intrinsic costs based on the type of mistake. For example, a mistake that threatens human life or that costs a large amount of money and time is intrinsically very costly. And a mistake that causes a minor inconvenience, like causing a grill to be a few degrees colder than the user requests, is not so costly.

But most mistakes can be corrected. And mistakes that are easy to correct are less costly than ones that are hard (or impossible) to correct. An intelligent experience will help users *notice when there is a mistake* and it will provide good options for *recovering from mistakes*.

Sometimes the mistakes are pretty minor, in which case users might not care enough to know the mistake even happened. Or there may be no way to recover, in which case the Intelligent System might want to pretend nothing is wrong—no sense crying over spilt milk.

Knowing There Is a Mistake

The first step to solving a problem is to knowing there is a problem. An effective intelligent experience will help users know there is a mistake in a way that:

1. Doesn't take too much of the user's attention, especially when it turns out there isn't a mistake.

2. Finds the mistake while there is still time to recover from the damage.

3. Makes the user feel better about the mistake and the overall interaction.

Sometimes the mistakes are obvious, as when an intelligent experience turns off the lights, leaving the user in the dark. The user knows that something is wrong the instant they are sitting in the dark. But sometimes the mistakes are not obvious, as when an intelligent experience changes the configuration of a computer-system's settings without any user interaction. The user might go for years with the suboptimal settings and never know that something is wrong. Some options to help users identify mistakes include these:

- *Informing the user* when the intelligent experience makes a change. For example, the intelligent experience might automate an action, but provide a subtle prompt that the action was taken. When these notifications are forceful, they will demand the user's attention and give the user a chance to consider and find mistakes. But this will also fatigue users and should be used sparingly.

- *Maintaining a log* of the actions the intelligent experience took. For example, the "junk" folder in a spam filtering system is a log of messages the spam filter suppressed. This lets users track down problems but doesn't require them to babysit every interaction. Note that the log does not need to be complete; it might only contain interactions where the intelligence was not confident.

Recovering from a Mistake

Intelligent experiences can also help users *recover from mistakes*.

Some mistakes are easy to recover from—for example, when a light turns off, the user can go turn it back on. And some mistakes are much harder to recover from, as when the intelligent experience sends an email on the user's behalf—once that email is beeping on the recipient's screen there is no turning back.

The two elements of recovering from mistakes are: how much of the cost of the mistake can be recovered; and what the user has to do to recover from the mistake. To help make mistakes recoverable, an intelligent experience might:

1. Limit the scope of a decision, for example, by taking a partial action to see if the user complains.

2. Delay an action, for example, by giving the user time to reflect on their decision before taking destructive actions (like deleting all their files).

3. Be designed not to take destructive actions at all, for example by limiting the experience to actions that can be undone.

If the cost of mistakes is high enough, the experience might want to go to great lengths to help make mistakes recoverable.

When the user wants to recover from a mistake, the intelligent experience can:

1. Put an option to undo the action directly in the experience (with a single command).

2. Force the user to undo the action manually (by tracking down whatever the action did and changing the various parts of the action back one by one).

3. Provide an option to escalate to some support agent (when the action can only be undone by an administrator of the system).

The best mistakes are ones that can be noticed easily, that don't require too much user attention to discover, and that can be completely recovered from in a single interaction.

The worst are ones that are hard to find, that can only be partially recovered from, and where the user needs to spend a lot of time (and the Intelligent System needs to pay for support staff) to recover.

Intelligence Quality

The experience can't control the quality of the intelligence. But to make an effective intelligent experience you had better understand the quality of the intelligence in detail. When the intelligence is good, you'll want to be more frequent and forceful. When the intelligence is shaky you'll want to be cautious about when and how you interact with the user.

Mistakes come in many type. An intelligence might:

• Make many essentially *random mistakes*. For example, it might have the right answer 90% of the time and the wrong answer 10% of the time. In this case, the experience can be balanced for all users simultaneously.

- Make *focused mistakes*. For example, it might be more likely to have the right answer for some users than for others. For example, making more mistakes with users who have glasses than with users who do not. Focused mistakes are quite common, and are often hard to identify. When focused mistakes become a problem it may be necessary to balance the intelligent experience for the lowest common denominator—the user-category that has the worst outcomes (or to abandon some users and focus on the ones where the intelligence works well).

Intelligence quality is a life-cycle, and experience will need to adapt as the intelligence does. For example, early in development, before the system has real customers, the intelligence will generally be poor. In some cases, the intelligence will be so poor that the best experience results from hiding the intelligence from users completely (while gathering data).

As the system is deployed, and customers come, the system will gather data to help the intelligence improve. And as the intelligence improves, the experience will have more options to find balance and to achieve the Intelligent System's objectives. Keep in mind:

1. Better intelligence can support more forceful, frequent experiences.

2. Change is not always easy (even if the change comes from improved intelligence). Be careful to bring your users along with you.

3. Sometimes intelligence will change in bad ways, maybe if the problem gets harder or new types of users start using the Intelligent System (or if the intelligence creators make some mistakes). Sometimes the experience will need to take a step backwards to support the broader system. That's fine—life happens.

Summary

An effective intelligent experience will balance:

- The *forcefulness* of the interactions it has with users.
- The *frequency* with which it interacts with users.
- The *value* that successful interactions have for the user and for the Intelligent System.
- The *cost* that mistakes have for the user and for the Intelligent System.
- The *quality of intelligence*.

Both the users and the Intelligent System must be getting a good deal. For users this means they must perceive the Intelligent System as being worth the time to engage with. For the Intelligent System this means it must be achieving objectives and getting data to improve.

When intelligence is good (very accurate), the intelligent experience can be frequent and forceful, creating a lot of value for everyone, and not making many mistakes.

When the intelligence is not good (or mistakes are very expensive), the intelligent experience must be more cautious about what actions it proposes and how it proposes them, and there must be ways for users to find mistakes and recover from them.

The frequency of interaction can be controlled by interacting:

- Whenever the intelligence changes its mind.
- Whenever the intelligence finds a large improvement.
- A limited number of times.
- Only when the system thinks the user will respond.
- Only when the user asks for an interaction.

Interacting more often creates more opportunity for producing value. But interacting can also create fatigue and lead to users ignoring the intelligent experience.

Users find value when they understand what is going on and think they are getting a good deal. The intelligent experience is critical in helping users see the value in what the intelligence is providing them.

Intelligence will be wrong, and to make an effective intelligent experience you must understand how it is wrong and work to support it. The quality of intelligence will change over the life-cycle of an Intelligent System and as it does, the intelligent experience can rebalance to create more value.

For Thought...

After reading this chapter you should:

- Know the factors that must be balanced to create an effective intelligent experience: one that is pleasant to use, achieves objectives and mitigates mistakes, and evolves as intelligence does.

- Know how to balance intelligent experiences by controlling the key factors of: forcefulness, frequency, value, cost, and understanding intelligence quality.

You should be able to answer questions like these:

- What is the most forceful intelligent experience you have ever interacted with? What is the most passive intelligent experience you can remember encountering?

- Give an example of an intelligent experience you think would be more valuable if it were less frequent. Why?

- List three ways intelligent experiences you've interacted with have helped you find mistakes. Which was most effective and why?

CHAPTER 8

Modes of Intelligent Interaction

There are many, many ways to create user experiences, and just about all of them can be made intelligent. This chapter explores some broad approaches to interaction between intelligence and users and discusses how these approaches can be used to create well-balanced intelligent experiences. These approaches include:

- *Automating* actions on behalf of the user.

- *Prompting* users to see if they want to take an action.

- *Organizing* the information a user sees to help them make better decisions.

- *Annotating* other parts of the user experience with intelligent content.

- *Hybrids* of these that interact differently depending on the intelligence.

The following sections will explore these approaches, providing examples and discussing their pros and cons.

Automate

An automated experience is one in which the system does something for the user without allowing the user to approve or to stop the action. For example:

- You get into your car on a Monday morning, take a nap, and wake up in your office parking lot.

- You lounge in front of your TV, take a bite of popcorn, and the perfect movie starts playing.

87

© Geoff Hulten 2018
G. Hulten, *Building Intelligent Systems*, https://doi.org/10.1007/978-1-4842-3432-7_8

- You log into your computer and your computer changes its settings to make it run faster.

- You rip all the light switches out of your house, hook the bulbs up to an Intelligent System, and get perfectly correct lighting for all situations without ever touching a switch or wasting one watt of power again.

These are great, magical, delightful... if they work. But in order for experiences like these to produce good outcomes, the intelligence behind them needs to be exceptionally good. If the intelligence is not good, automated intelligent experiences can be disastrous.

Automated experiences are:

- **Very forceful**, in that they force actions on the user.

- **Not always obvious**, in that they may require extra user experience to let the user know what is happening and why. This may reduce the value of the interactions and make mistakes harder to notice.

- **Difficult to get training data from**, in that users will generally only give feedback when mistakes happen. Sometimes the system can tie outcomes back to the automated action, but sometimes it can't. Automated systems usually need careful thought about how to interpret user actions and outcomes as training data. They often require additional user experiences that gather information about the quality of the outcomes.

Automation is best used when:

- The intelligence is very good.

- There is a long-term commitment to maintaining intelligence.

- The cost of mistakes is not too high compared to the value of a correct automation.

Prompt

Intelligence can initiate an interaction between the system and the user. For example:

- If the intelligence suspects a user is about to do something wrong, it might ask them if they are sure they want to proceed.

- If the intelligence thinks the user is entering a grocery store, it might ask if they would like to call their significant other and ask if there is anything they need.

- If the intelligence notices the user has a meeting soon and estimates they are going to be late, it might ask if the user would like to send a notice to the other participants.

These interactions demand the user's attention. They also allow the user to consider the action, make a decision if it is right or not, and approve or reject the action. In this sense, experiences based on prompting allow the user to act as a back-stop for the intelligence, catching mistakes before they happen. Interactions based on prompting are:

- **Variably forceful.** Depending on how the interaction is presented, it can be extremely forceful (for example with a dialog box that the user must respond to before progressing), or it can be very passive (for example, by playing a subtle sound or showing a small icon for a few moments).

- **Usually obvious**, in that the user is aware of the action that was taken and why it was taken. This helps the user perceive the value of the intelligent interaction. It also helps users notice and recover from mistakes. But frequent prompts can contribute to fatigue: if users are prompted too often, they will begin to ignore the prompts and become irritated, reducing the value of the Intelligent System over time.

- **Usually good to get training data from**, in that the user will be responding to specific requests. The Intelligent System will have visibility into exactly what is going on. It will see the context that led to the interaction. It will have the user consider the context and give input on the action. This allows the system to learn, so that it will know whether the prompt was good or bad and be able to improve the intelligence.

These types of interactions are often used when:

- The intelligence is unreliable or the system is missing context to make a definitive decision.

- The intelligence is good enough that the prompts won't seem stupid, and the prompts can be infrequent enough that they don't lead to fatigue.

- The cost of a mistake is high relative to the value of the action, so the system needs the user to take part in approving action or changing behavior.

- The action to take is outside the control of the system, for example when the user needs to get up and walk to their next meeting.

Organize

The intelligence can be used to decide what information to present to the user, and in what order. For example:

- If the intelligence thinks the user is querying for information about their favorite band, it might select the most relevant web pages to display.

- If the intelligence thinks the user is shopping for a camera, it might select camera-related ads to tempt the user.

- If the intelligence thinks the user is having a party, it might offer a bunch of upbeat 80s songs on the user's smart-jukebox.

These types of experiences are commonly used when there are many, many potential options, so that even "good" intelligence would be unlikely to narrow the answer down to one exactly correct choice.

For example, at any time there might be 50 movies a user would enjoy watching, depending on their mood, how much time they have, and who they are with. Maybe the user is 10% likely to watch the top movie, 5% likely to watch the second, and so on. If the system showed the user just the top choice, it would be wrong 90% of the time.

Instead, these types of systems use intelligence to pre-filter the possible choices down to a manageable set and then present this set in some browsable or eye-catching way to achieve their objectives.

Interactions that organize choices for the user are:

- **Not intrinsically forceful**, in that they don't take important actions for the user and they don't demand the user's attention. If the user experience presenting the organized information is big and jarring, the interaction might be perceived as forceful, but that is fully in the hands of the experience designer.

- **Somewhat obvious**, in that the user will see the choices, and might or might not feel that there was significant intelligence behind the ordering of the information. It is also sometimes challenging for users to find mistakes—if an option isn't presented, the user may not know whether the option was excluded because the system doesn't support the option (have the product, movie, song, and so on) or because the intelligence was wrong.

- **OK to get training data from**, in that it is easy to see what the user interacts with (when the system did something right), but it is harder to know when the system got something wrong (when it suppressed the option the user might have selected). This can lead to bias, in that the options users tend to see more often will tend to be selected more. These systems often require some coordination between intelligence and experience to avoid bias, for example by testing various options with users—the intelligence doesn't know if the user will like something or not, so the experience tries it out.

Interactions that organize information/options are best when:

- There are a lot of potential options and the intelligence can't reasonably detect a "best" option.

- The intelligence is still able to find some good options, these good options are a small set, and the user is probably going to want one of them.

- The problem is big and open-ended and so that users can't reasonably be expected to browse through all options and find things on their own.

Annotate

The intelligence can add subtle information to other parts of the experience to help the user make better decisions. For example:

- If the intelligence thinks the user has their sprinklers set to water for too long (maybe there is rain in the forecast), it can turn on a blinking yellow light on the sprinkler box.

- If the intelligence thinks there is a small chance the user's car will break down soon (maybe the engine is running a little hot), it might display a "get service" indicator on the dashboard.

- If the intelligence finds a sequence of characters that doesn't seem to be a word, it can underline it with a subtle red line to indicate the spelling might be wrong.

These types of experiences add a little bit of information, usually in a subtle way. When the information is correct, the user is smarter, can make better decisions, can initiate an interaction to achieve a positive outcome. When the information is wrong it is easy to ignore, and the cost of the mistake is generally small.

Interactions based on annotation are:

- **Generally passive**, in that they don't demand anything of the user and may not even be noticed. This can reduce the fatigue that can come with prompts. It can also lead to users never noticing or interacting with the experience at all.

- **Variably obvious**, in that the user may or may not know where the annotation came from and why it is intelligent. Depending on how prominent the user experience is around the annotation, the user might never notice it. Users may or may not be able to understand and correct mistakes.

- **Difficult to get training data from**, as it is often difficult for the system to know: 1) if the user noticed the annotation; 2) if they changed their behavior because of it; and 3) if the change in behavior was positive or negative. Interactions based on annotation can require additional user experience to understand what actions the user took, and to improve the intelligence.

Annotations work best when:

- The intelligence is not very good and you want to expose it to users in a very limited way.

- The Intelligent System is not able to act on the information, so users will have to use the information on their own.

- The information can be presented in a way that isn't too prominent but is easy for the user to find when they want it.

Hybrid Experiences

Intelligent experiences can be built by combining these types of experiences, using one type of experience in places where the intelligence is confident, and another type in places where the intelligence is not. For example, when the intelligence is really, really sure, the experience might automate an action. But when the intelligence is only sort of sure, the experience might prompt the user. And when the intelligence is unsure, the experience might annotate the user experience with some information.

An example of a hybrid intelligent experience is spam filtering. The ideal spam filter would delete all spam email messages before a user had to see them and would never, never, never delete a legitimate email. But this is difficult.

Sometimes the intelligence can be very sure that a message is spam, because some spammers don't try very hard. They put the name of certain enhancement pills right in their email messages without bothering to scramble the letters at all. They send their messages from part of the Internet known be owned by abusers. In these cases it is easy for a spam filter to identify these obvious spam messages and be very certain that they aren't legitimate. Most intelligent spam filters delete such obvious spam messages without any user involvement at all. Spam filtering systems have deleted billions of messages this way.

But some spammers are smart. They disguise their messages in ways designed to fool spam filtering systems—like replacing i's with l's or replacing text with images—and they are very good at tricking spam filters. When the intelligence encounters a carefully obscured spam email, it isn't always able to distinguish it from legitimate emails. If the spam filter tried to delete these types of messages it would make lots of mistakes and delete a lot of legitimate messages too. If the spam filter put these types of messages into the inbox, skilled spammers would be able to circumvent most filters. And so most spam filtering systems have a junk folder. Difficult messages are moved to the junk folder, where the user is able to inspect them and rescue any mistakes.

These are two forms of automation, one extremely forceful (deleting a message), and one less forceful (moving a message to a junk folder). The intelligent experience chooses between them based on the confidence of the intelligence.

But many spam filters provide even more experiences than this. Consider the case when the spam filter thinks the message is good, but it is a little bit suspicious. Maybe the message seems to be a perfect personal communication from an old friend, but it comes from a part of the Internet that a lot of spammers use. In this case the experience might put the message into the user's inbox but add a subtle message to the top of the message that says "Reminder—do not share personal information or passwords over e-mail."

In this example, the Intelligent System uses forceful automation where it is appropriate (deleting messages without user involvement). It uses less-forceful automation when it must (moving messages to a junk folder). And it uses some annotation occasionally when it thinks that will help (warnings on a few percent of messages).

Hybrid experiences are common in large Intelligent Systems. They are very effective when:

- The problem can be decoupled into clear parts, some of which are easy and some of which are hard.

- The intelligence problem is difficult and it needs many types of support from the experience.

- You want to reduce the amount of attention you demand from users and avoid asking them questions where the intelligence is certain, or where the intelligence is particularly unreliable. This can make the questions you do ask the user much more meaningful and helpful (both to the user and as training data).

Summary

There are many ways to present intelligence to users, but important ones include these:

- Automating actions

- Prompting the user to action

- Organizing information

- Adding information

- Hybrid experiences

Options like automating and prompting require very good intelligence (or low costs when there is a mistake). Other options are less visible to the user and can be good at masking mistakes when the intelligence is less reliable or the cost of a mistake is high.

Most large Intelligent Systems have hybrid experiences that automate easy things, prompt for high-quality interactions, and annotate for things they are less sure about.

For Thought...

After reading this chapter you should:

- Know the most common ways to connect intelligence to users.

- Have some intuition about which to use when.

You should be able to answer questions like these:

- Find examples of all the types of intelligent experiences in systems you interact with regularly (automation, prompting, organizing, annotating, and a hybrid experience).

Consider the intelligent experience you interact with most.

- What type of experience is it?

- Re-imagine it with another interaction mode (for example, by replacing prompts with automation or annotation with organization).

CHAPTER 9

Getting Data from Experience

Intelligence creators can work with all sorts of data, even crummy data. But when they have good data, their job is much easier—and the potential of the intelligence they can create is much greater. An ideal intelligent experience will control the interactions between users and the Intelligent System so that the record of those interactions makes it easy to create high-quality intelligence.

Before exploring ways you can craft intelligent experiences to get data from your users, let's consider the alternative: you could gather data yourself. Traditionally, machine learning systems have a data-collection phase. That is, if you want to build an intelligence for counting the number of cows in a picture, you'd go out and take a lot of pictures of cows. You'd travel from farm to farm, taking pictures of cows in different situations. You'd pose cows so you get pictures of their good sides, and their bad sides; their faces and their... other parts. You'd take pictures of cows behind bushes. You'd take pictures of cows laying down, running, grazing, sleeping, yawning, on sunny days, on cloudy days, at night... A traditional computer vision system might require tens of thousands of different pictures of cows. And it would get better with more pictures— hundreds of thousands or millions might help. And once you had all that data you'd need to *label* it. That is, you'd pay lots of people to look at all those cow pictures and draw circles around all the cows—fun!

This can get very expensive. Data collection and labeling can be the primary cost in traditional machine-learning systems. Because of this, there is a lot of value in designing experiences that leverage users, and their interactions with the Intelligent System, to produce data with the correct properties to create intelligence. Getting this right makes it possible to use intelligence and machine learning in many situations where it would have been prohibitively expensive using manual data collection.

© Geoff Hulten 2018
G. Hulten, *Building Intelligent Systems*, https://doi.org/10.1007/978-1-4842-3432-7_9

This chapter begins with an example of what it means to collect data from experience and how various approaches can affect the quality of the resulting data. We will then explore the properties that make data good for intelligence creation and some of the options for capturing user outcomes (or opinions) on the data.

An Example: TeamMaker

Let's consider an example: an Intelligent System designed to create teams for basketball leagues—we'll call it TeamMaker. Every player is entered into the system, along with some basic statistics: height, experience, positions they play, and maybe something about how well they shoot the ball. Then the system divides the players up into teams. If the teams are balanced, the league will be fun to play in. If the teams are imbalanced the league might be boring. One team will always win and everyone else will feel like losers. Feuds will begin, rioting in the streets, hysteria…

So we better get this right.

We need to build an experience that collects useful feedback about the teams the system creates, so the intelligence can improve (and suggest better teams) over time.

Simple Interactions

One approach is to let users correct the teams and use the corrections to improve the intelligence.

For example, TeamMaker might produce "suggested teams" as a starting point. These proposals could be presented to a human—the league commissioner—to correct any problems. Maybe the intelligence got it wrong and put all the best players on one team. Then the commissioner could intervene and move some players around. Each time the commissioner moves a player, TeamMaker gets training data to make the intelligence better. After a few seasons, the intelligence might be so good that the commissioner doesn't need to make any more changes.

But, to be honest, this basketball-team–creating system sounds pretty lame. Who wants to use a tool to make teams, and then have to make them by hand anyway? This approach could fail if commissioners don't provide any corrections to the teams, because:

- It sounds boring.

- They don't feel confident they can beat the machine.

- They actually don't understand how to play basketball, so they make corrections that make things worse.

- They make the corrections offline and don't bother to enter them into TeamMaker at all.

Slow or irregular interactions will limit the speed the intelligence can grow and limit its ultimate potential. A good experience will *set up interactions that provide obvious value to the user and that the user can do a good job at getting right.*

Making It Fun

Another approach would be to make the connection between usage and intelligence something that's more fun. Something users really want to do and are motivated to do right.

For example, TeamMaker could make the teams, but it could also support betting. Most people who bet will want to win (because most people care about money), so they will try to bet on the best team—and will avoid betting on obvious losers. So TeamMaker can use the distribution of bets to figure out if it did a good job of constructing teams. If the bets are heavily skewed toward (or away) from one of the teams, TeamMaker can learn it used the wrong factors to balance the teams. It can do better next time.

This betting-based-interaction is an improvement to the team-tweaking–based interaction system and should result in more data per season and more opportunities to improve the intelligence, and it should produce a better Intelligent System faster.

But the data from betting-based interaction will not be perfect. For example, some people might only bet for their own team, even if it is obvious they can't win. Or maybe the betters don't know anything about basketball and are betting to pump up some good-natured office rivalry. (Or betting is illegal or against your morals, so as a developer you don't want to use betting-based interaction with TeamMaker at all...) These types of problems can lead to bias, and bias can lead intelligence astray. When an intelligence learns from biased betting data the resulting teams will not be optimized for creating a competitive league, they will be optimized for... something else.

When users have a different objective than the Intelligent System the data from their interactions can be misleading. A good experience will *align usage with the intelligence to get unbiased observations of success.*

Connecting to Outcomes

Another approach would be to connect the intelligence directly to outcomes. No human judgment needed, just the facts. When a single team wins every single game, the intelligence knows it did something wrong. When all the games are close, the intelligence can learn it did something right.

For this approach to work, someone would have to be motivated to enter all the scores into TeamMaker. So maybe TeamMaker creates some fun features around this, like integrating with the betting, or leader-boards, or scheduling. People might crowd around their computers every time a game is completed, just to see how the standings have changed, who won money, and who lost pride.

As games are played and users enter statistics about which teams won, which games were close, and which were blowouts, TeamMaker has everything it needs to improve the intelligence. If there is a team that never loses, it can learn what is special about that team and avoid doing that in the future. And next season TeamMaker would do a better job, creating tournaments that are more likely to be balanced, and less likely to have an unstoppable (or a hopeless) team.

When the experience can *directly track the desired outcomes,* it can produce the most useful intelligence.

Properties of Good Data

In order to solve hard, large, open-ended, time-changing problems, you're going to need data—lots of data. But not just any data will do. To build intelligence, you're going to need data with specific properties. The best data will:

- Contain the context of the interaction, any actions that were taken, and the outcomes.

- Have good coverage of what your users want to do with the Intelligent System.

- Reflect real interactions with the system (and not guesses or surveys about how people might interact).

- Have few (or no) biases.

- Avoid feedback, where the intelligence influences the data, which influences intelligence, which influences data...

- Be large enough to be relevant to the problem. (Creating intelligence for difficult problems can require incredible amounts of data.)

Achieving all of these properties is not easy. In fact, you are unlikely to get the data you need to effectively create intelligence unless you explicitly address each of these properties in your design.

Context, Actions, and Outcomes

The basic requirement for creating intelligence from data is to know:

1. The context of what was going on when the intelligence was invoked.

2. Any actions that were taken as a result.

3. The outcomes of the interaction, and specifically if the outcomes were positive or negative.

For example:

- A self-driving car needs to know what all the sensors on the car see (the context), the things a human driver might do in that situation (the actions), and whether the car ends up crashing or getting honked at (the outcome).

- A book-recommending system needs to know what books the user has already read and how much they enjoyed them (the context), what books might be recommended to that user, whether the user purchased any of the books or not (the actions), and which of the books the user ended up liking (the outcomes).

- An anti-malware system needs to know what file the user downloaded and where they got it from (the context), whether they installed it or not (the action), and whether their machine ended up infected or not (the outcome).

An ideal intelligent experience will create situations that have enough context to make good decisions, both for the user and for the intelligence. For example: if the automatic car doesn't have any sensors other than a speedometer, then it doesn't help to know what the user did with the steering wheel—no intelligence could learn how to steer based on the speed alone.

Good Coverage

The data should contain observations of all the situations where the intelligence will need to operate. For example:

- If the system needs to automate lights, the data should contain the context, actions, and outcomes of controlling lights:

    - During the day.

    - At night.

    - In the winter.

    - In the summer.

    - During daylight savings time.

    - With lights that are heavily used.

    - With lights that are rarely used.

    - And so on.

- If the lights need to work in 50 different countries around the world, the data should contain observations from those 50 countries in all of these situations.

- If the system needs to work in a mineshaft, the data should contain observations of lights being used in a mineshaft.

- If the system needs to work on a space station, the data should contain observations of lights being used in a space station.

An intelligence operating in a situation it was not trained (or evaluated) on is likely to make mistakes, crazy ones.

Intelligent Systems will be expected to work in new contexts over their lifetime. There will always be new books, movies, songs, web pages, documents, programs, posts, users, and so on. An effective intelligent experience will be able to put users into these new contexts with confidence that the mistakes made while collecting data will have low cost and the value of the collected data will be high.

Real Usage

The best data will come from users doing things they actually care about. For example:

- A user driving in a car simulator might make decisions they wouldn't make if they were driving a real car (when their life is on the line).

- A user telling you what books they like might talk about literary classics (because they are trying to impress you) but they might never actually read those books.

- A user might give very different answers when asked if a file is safe to install on a lab computer or on their mother's computer.

Connecting real users to interactions they care about ensures the resulting data is honest, and that building intelligence from it will be most likely to give other users what they actually want.

Unbiased

Bias occurs when the experience influences the types of interactions users have, or influences the types of feedback users give.

One common source of bias is that different types of outcomes get reported at different rates.

Consider a spam filtering program. If a spam email gets to the user's inbox, it is right in their face. They are likely to notice it and may press a "this is junk" button and generate useful data on a bad outcome. On the other hand, if the filter deletes a personal email, the user may never notice.

In this case, choices made in designing the experience have introduced bias and have made the resulting data much less useful for building intelligence.

Another potential source of bias is that users with strong sentiment toward the interaction (either good or bad) are more likely to give feedback than users with neutral opinions.

Another source of bias is when the experience encourages users to take certain choices over others. For example, when the experience presents choices in a list, the user is more likely to choose the first item on the list (and is very unlikely to choose an item that isn't on the list at all).

Does Not Contain Feedback Loops

Experience and intelligence will affect one another, sometimes in negative ways.

For example, if some part of the user experience becomes more prominent, users will interact with it more. If the intelligence learns from these interactions, it might think users like the action more (when they don't like it more, they simply notice it more because of the change to the user experience).

Conversely, if the intelligence makes a mistake and starts suppressing an action that users like, users will stop seeing the option. They will stop selecting the action (because they can't). The action will disappear from the data. The intelligence will think users don't like the option any more. But it will be wrong...

Here are some ways to deal with feedback:

- Include what the user saw in the context of the recorded data. If one option is presented in a larger font than another, record it in context. If some options were suppressed, record it in context.

- Put a bit of randomization in what users see, for example switching the order of certain options. This helps gather data in broader situations, and it helps identify that feedback may be occurring.

- Record the effort the user took to select the option in the context. For example, if the user had to issue a command manually (because it was suppressed in the intelligent experience) the intelligence should know it made a costly mistake. If the user had to browse through many pages of content to find the option they want, the intelligence should know.

Scale

In order to improve, an Intelligent System must be used. This means that the intelligent experiences must be prominent, they must be interesting to use, they must be easy to find, and they must be the things that users do often.

Consider two options: a new Intelligent System and an established Intelligent System.

The new Intelligent System will probably not have many users. This means the intelligent experience must be central to what the users will do, so they will interact with it regularly. In these cases, the intelligent experience will need to encourage interaction, make it fun, and maybe even change the whole product around to put the intelligent interaction front and center.

The established Intelligent System will have much more usage. This means that the intelligent experiences can be more subtle. They can be put in places where fewer users will see them. This doesn't mean they are less valuable—they may be solving very important problems, but problems that users encounter on a weekly or monthly basis instead of on a daily basis.

For some context, a system that:

- Generates tens of interaction per day is basically useless for verifying or improving intelligence.

- Generates hundreds of interactions per day can probably validate intelligence and produce some simple intelligence.

- Generates thousands or tens of thousands of interactions per day can certainly validate the Intelligent System and produce intelligence for many hard, open-ended problems.

- Generates hundreds of thousands or millions of interactions per day will probably have all the data it needs for most tasks.

An effective intelligent experience will attract users and get them engaging, building a base for gathering data and producing interesting intelligence.

Ways to Understand Outcomes

When an interaction between a user and intelligence occurs, it isn't always easy to understand whether the outcome is positive or negative. For example, in a music recommender system—did the user push the "next" button because they hate the song (the intelligence got it wrong) or because they see the next song coming up and they really love that song (the intelligence got it right)?

The best intelligent experiences are set up to make the outcome clear implicitly—that is, without relying on the user to do anything other than use the product. But this isn't always possible. Even when it is possible, it can be useful to have a backstop to make sure the implicit data has all the right properties to create intelligence. Methods for understanding outcomes include:

- Implicit observations

- User ratings

- Problem reports

- Escalations

- User classifications

Many large Intelligent Systems use all of these approaches to understand as much as possible about the outcomes their users are having.

Implicit Outcomes

An ideal experience will produce good, useful data without requiring any special user action. Users will produce data simply by using the system, and that data will have all the properties required to grow intelligence.

For example, when a user sets their thermostat to 72 degrees it's pretty clear what they would have wanted the intelligence to do. Or when the user buys a product or some content, it's clear they were interested in it.

It's often hard to know exactly how to interpret a user's action. Users may not spend the time to make perfect decisions, so the way data is presented to them could bias their results. For example, they might have selected one of the five recommended movies because it was the absolute best movie for them at that moment—or because they didn't feel like looking at more options.

Because of this, achieving a fully implicit data collection system is quite difficult, requiring careful coordination and curtailing of the experience so the usage can be effectively interpreted.

Ratings

User ratings and reviews can be a very good source of data. For example, designing experiences that allow users to:

- Rate the content they consume with 1-5 stars.

- Give a thumbs-up or thumbs-down on a particular interaction.

- Leave some short text description of their experience.

Users are used to leaving ratings and many users enjoy doing it, feeling like they are helping others or personalizing their experience.

But there are some challenges with ratings:

1. Users don't always rate everything. They may not feel like taking the effort.

2. There can be some time between the interaction and the rating. The user might not remember exactly what happened, or they might not attribute their outcome to the interaction.

3. The rating might not capture what the intelligence is optimizing. For example, consider these two questions: How good is this book? Was that the right book to recommend to you yesterday?

4. Ratings vary across different user populations—five stars in one country is not the same as five stars in others.

5. Infrequent interactions will not get many ratings, and be hard to learn about.

Reports

An experience can allow users to report that something went wrong. For example, many email systems have a "report as spam" button for spam messages that get to the inbox.

Data collected through reports is explicitly biased, in that the only outcomes it captures are bad ones (when the intelligence made a mistake), and users won't report every problem. For example, users might click "report as spam" on 10% of the spam they see, and just ignore or delete the rest of the spam.

Because of this, data from reports is difficult to use as an exclusive source for creating intelligence. However, reports can be very effective in verifying intelligence and in tuning the experience to control the negative effects of poor intelligence.

Escalations

Escalations are an expensive form of report that occur when users get upset enough to contact customer support (or visit a support forum, and so on). Escalations are similar to reports but are out-of-band of the main product experience. And they usually represent very unhappy users and important mistakes.

Escalations tend to be messy and unfocused. The user who is having trouble won't always know what part of the system is causing trouble. They won't express the problem in the same terms as other users who reported it.

Because of this, it can be difficult and expensive to sort through escalations and use them to improve intelligence—they are better for identifying big problems than for refining the system.

In general, it's better to allow users to report problems in context (with an in-line reporting experience, at the time of the problem) so the system can capture the relevant context and actions that led to the bad outcome.

User Classifications

You can simply ask users about their outcomes using specific (carefully controlled) questions, in context as users interact with your system. Sort of like a survey. Imagine:

- Using a light automation system and once every three months it says: In order to improve our product we'd like to know—right now would you prefer the lights to be on or off?

- Using a photo editing program and after touching up a hundred photos it says: Please help improve this program. Is there a face in the image you are working on?

Classification can produce very focused data for improving intelligence. It can also be made unobtrusive, for example by limiting questions:

- To one question per thousand interactions.
- To users who have volunteered to help.
- To users who haven't signed up for premium service.
- And so on.

Summary

The intelligent experience plays a critical role in getting data for growing intelligence. If an intelligent experience isn't explicitly designed to produce good data for creating intelligence, it is almost certainly going to produce data that is poor (or useless) for creating intelligence.

Experiences produce better data when users interact with intelligence often, when they perceive value in the interactions, and when they have the right information to make good decisions.

Data is most useful when it:

- Contains the context of the interaction, the action taken, and the outcomes.
- Contains good coverage of all the situations where the Intelligent System will be used.
- Represents real usage, that is, interactions that users care about.
- Contains unbiased data.
- Contains enough information to identify and break feedback loops.
- Produces a meaningful amount of data with respect to the complexity of the problem.

The most effective data comes from implicit interactions, where the user naturally expresses their intent and their opinion of the outcome by using the product.

When implicit interactions aren't sufficient other methods of understanding outcomes include:

- Allowing users to explicitly rate the content or interactions they have.

- Allowing users to report problems directly from the user experience.

- Giving users access to support paths to report large, costly problems.

- Asking users to answer specifically curtailed questions (infrequently).

Many Intelligent Systems use all of these approaches. The majority of their data might come from implicit interactions, but the other techniques will be available to monitor and make sure the implicit data is high quality (has not started to develop bias or feedback).

Producing an experience that produces excellent data without burdening users is hard—and it is very important to building great Intelligent Systems.

For Thought...

After reading this chapter you should:

- Know how to craft experiences that collect the data needed to evaluate and grow intelligence.

- Understand the options for collecting this data from users, which range from ones requiring no explicit user action to ones requiring extensive user involvement.

You should be able to answer questions like these:

- Think of an intelligent service you use regularly that seems to collect all its data implicitly. How could it be changed to leverage user classifications?

- Imagine an Intelligent System to assist your favorite hobby. Think of an implicit way to collect "good data" when it does something wrong. Now think of another way to collect data in the same situation that expresses a different user interpretation of the mistake (for example maybe that the intelligence was correct, but the user didn't want to do the suggested action for some other reason).

CHAPTER 10

Verifying Intelligent Experiences

Intelligent experiences fail for two main reasons:

1. Because they are implemented incorrectly: the experiences have a bug, are confusing, or don't help users achieve their goals.

2. Because the intelligence is incorrect: it suggests the wrong automation, is certain of something that isn't true, or proposes the wrong ordering of information.

For example, an intelligent experience might:

- Present a list of images with faces, but fail to include the user's favorite photo from their wedding. Is the image excluded because the experience has a bug? Or is it excluded because the intelligence didn't recognize that the image had a face in it?

- Turn the lights off when the room is already sort of dark. Did the experience turn the light off because it knew the sun was coming up soon (even though it knew the room was too dark)? Or did the light turn off because the intelligence thought the room was empty?

- The user hates horror movies, but the system is recommending lots of horror movies. Is it doing this because the user's husband set a preference in some hidden menu to hide all the romantic comedy movies forever and ever no matter what? Or is it doing this because the intelligence examined the movies and thought they were adventure, not horror?

© Geoff Hulten 2018
G. Hulten, *Building Intelligent Systems*, https://doi.org/10.1007/978-1-4842-3432-7_10

Verifying that an intelligent experience is functioning properly, separate from the quality of intelligence, is critical to producing effective intelligent experiences.

Getting Intended Experiences

So the intelligence might be right, the intelligence might be wrong; fine. But how did the rest of the system perform?

- Did the user get the intended experience when the intelligence was correct?

- Did the experience actually help to achieve the user's (and the system's) goals?

- Does the experience remain effective as the system and the intelligence evolves?

This section talks about ways of isolating the experience from the rest of the changes that might occur in an Intelligent System so you can have some basis for verification.

Working with Context

The context represents the state of the Intelligent System when an intelligent experience is triggered. This can include all the variables the intelligence might take into account in producing its decisions. It should also include the variables that impact how users will interpret the system's outputs and how they will react. For example:

- The context in a computer vision system might include the lighting in the environment, the camera properties, the users and their features (ethnicity, gender, clothing), and so on.

- The context in a music playing system might include the user's ratings of songs, the list of songs the user owns, the recent listening history, and so on.

- The context in a home automation system might include the number and types of sensors, the rooms they are placed in, the distance between them, and so on.

When you want to verify an intelligent experience works in a particular context, you need to create the context and try it. For example, you need to set up the computer vision system in exactly the right way to recreate the context, get the lighting just right, bring in the same users, put them in front of the camera, and have them smile exactly right, and so on.

This can be problematic. Particularly for large, open-ended problems that have many, many possible contexts—an infinite number of them.

When verifying intelligent experiences, it can help to have some tools to help produce (or capture) important contexts, explore them, and replay them in the intelligent experience so you can test them again and again as the Intelligent System evolves. Some options for working with contexts include:

- **Manually producing contexts** the same way users would, by using the system, creating an account, building up the history of interactions, browsing to the right place, and so on. This can be extremely challenging and error prone, as the contexts can be subtle; for example, the change in inflection when speaking to a computer can affect how well a speech recognition system will perform. They can be quite extensive—for example, usage history over a long period of time.

- **Crafting via an editor** or some other tool that lets contexts be created, edited, captured, and played back to the system. For example, a tool that specifies a house layout, the location of sensors, and the various readings that are on the sensors at a particular time. The level of sophistication and automation can vary, but these types of tools can be very helpful in developing baseline verifications.

- **Recorded via usage** either in a lab setting or from actual users. Examples include allowing the system to dump the context at any time, and creating tools to load the dumps and evaluate them against updated versions of the intelligent experience and intelligence. These types of approaches can help get actual usage and a great deal of variety into the verification process.

When verifying intelligence it is good to have all of these options. A typical work-flow might look like this:

- Reproducing a problem that a user reports.

- Capturing the exact context that caused the report.

- Viewing the context in some kind of tool to see what was really going on (not just the part the user was complaining about).

- Making a change to the experience.

- Executing the context again to see if the change would fix the problem.

- Communicating the problem to intelligence creators (but they may or may not be able to help).

Note that a context for verifying experiences is similar to, but different from the data needed for creating intelligence. To create intelligence you also need to capture information about the outcome. But the context alone can help verify (and improve) the experience.

Working with Intelligence

The workflow of tracking and replaying contexts is similar to tracking and fixing bugs in other complex software products. But changing intelligence can make it difficult. This is because the experience a user sees results from combining a particular context with a particular intelligence—and intelligence changes.

Because of this, it also helps to have a collection of known intelligences that can be combined with known contexts to see the results. Some options for getting intelligences to help with verification include:

- **Live intelligence** that is just what users will see. If things are going well, this will be the most accurate intelligence available, and it will be the best way to verify what users are actually seeing. But it may also change very rapidly (as intelligence evolves), requiring some work to distinguish intelligence changes from experience changes. One example of a problem this might cause is that you have an experience problem you are working to fix and then... it disappears.

You haven't changed anything yet, but maybe the intelligence changed out from under you and you can no longer reproduce the problem.

- **Fixed intelligence** that is the same every time it is executed. For example, some sort of checkpoint on the live system's intelligence (maybe the first day of alpha testing or the first day of every month). This intelligence is similar to the live intelligence, but provides a stable baseline to use for working with experience problems over an extended period of time.

- **Simple intelligence** that is somewhat interpretable. For example, instead of using a model produced by machine learning, use a small set of heuristic rules. This can be useful when debugging problems— by having just enough intelligence to cause interesting problems, but not so much that the intelligence errors overwhelm the experience problems.

- **Dumb intelligence** that always gives the same answer no matter what context is provided. For example, no matter what bands the user likes, the intelligence recommends Dire Straits. A dumb intelligence can help rule out obvious experience problems.

When verifying intelligence it helps to have all of these options available.

Bringing it Together

Common types of errors to look for include:

- The context being incorrectly interpreted by the intelligence, resulting in unexpected output.

- The intelligence being misinterpreted by the experience, resulting in systematically incorrect experiences.

- The rendering of the user experience being incorrect for some combinations of context and intelligence, resulting in hard-to-interpret, unappealing, or illegible results.

Achieving Goals

The experience plays a very important role in achieving a system's objectives.

By presenting intelligence, deciding how and how often users see it, deciding how it is colored, where it is displayed, how often it happens, and what words or icons are used to describe it, an experience can make a poor-quality intelligence into a useful tool; an experience can also render a high-quality intelligence useless.

Consider a system designed to protect users from scams on the Internet. If the user visits a page that the intelligence thinks is a scam web page, the experience gives them a warning of some sort.

If the intelligence were perfect, the warning could be very forceful (perhaps blocking access to the page and giving the user no recourse) but intelligence is never perfect, so the warning needs to balance the risk of the user getting scammed with the risk of the user being scared away from a site they actually wanted to go to.

So how do we verify the effectiveness of any particular experience?

This can be accomplished by isolating the job of the intelligence from the job of the experience.

The job of the intelligence is to flag scam sites. It might be right; it might be wrong. The job of the experience is to:

1. Successfully keep users from being scammed when intelligence correctly flags a scam.

2. Minimize the cost to the user when the intelligence incorrectly flags a non-scam as a scam.

And so these are the properties that need to be measured to verify that the experience is doing its job:

*   What percent of users who correctly received a warning got scammed anyway?

*   What percent of users who incorrectly received a warning got turned away from a legitimate site?

More generally, for every possible outcome of the intelligence—all the ways it could have been right and all the ways it could have been wrong—consider what the role of the experience is in that setting—and measure it.

And note: these types of things are very difficult to measure in a development setting—accurately measuring them often requires real users making real decisions.

Continual Verification

Intelligence changes over time. User behavior changes over time, too. These changes can cause experiences to become less effective over the life of a system. Or the changes could result in opportunities to make the experience more forceful and get more benefit.

Some changes that can affect the experience are:

1. The quality of the intelligence could become better or worse.

2. The quality of the intelligence might remain the same, but the types of mistakes it makes might change.

3. Users might become accustomed to (or fatigued by) the experience and start to ignore it.

4. New users might adopt the system and have different behaviors than the original users.

In an Intelligent System there are many sources of change. And most changes provide opportunity (at least if you're an optimist, you'll consider them opportunities) to rebalance the parts of the system to achieve better results—as long as you know the change happened and how the change affected the system.

Adapting the experience isn't always the right answer—but it might be.

Summary

Verifying that an intelligent experience is working can be complex, because mistakes can be caused by the experience, by the intelligence, by having the user in some context you didn't anticipate, or because something has changed.

When there are a lot of mistakes they can become overwhelming, and it is useful to have some ways to simplify the problem and isolate what is going on. This can also help prevent lots of finger-pointing: "This is an intelligence mistake!" "No, it is an experience problem." "No, it is an intelligence problem." ... And so on.

It is good to have a way to capture, inspect, and manipulate contexts. This can be done through:

- Manually producing them.

- Having tools to inspect and modify them.

- Having ways to record them from real users and situations.

It is also good to have ways to play contexts back against various versions of the intelligence and see what the experience would produce. Common intelligences used in verification include:

- The live intelligence that users are getting.

- A checkpoint of the live intelligence that can be used over an extended period without changing.

- A simple intelligence that is easy to interpret (maybe some hand-crafted rules).

- A dumb intelligence that always says the same thing.

It is also important to verify that the intelligent experience is helping users achieve their objectives. This is most commonly done by looking at live interactions when the intelligence was right and ensuring that users are ending up with good outcomes (and not being confused or misled by the experience).

And Intelligent Systems are always changing. The effectiveness of the intelligent experience needs to be constantly reverified. Change also creates opportunities to rebalance and provide more user value—if you are tracking it and have the tools to identify the effects of the change.

For Thought...

After reading this chapter, you should be able to:

- Isolate experience issues (from intelligence ones) and verify that the experience is working as intended on its own.

- Build an effective test plan for an intelligent experience.

- Understanding the importance of reverifying as the intelligence and user-base change.

You should be able to answer questions like these:
Consider an Intelligent System you use regularly.

- Sketch a plan to verify that its experience is functioning correctly.

- Design a simple intelligence to help verify this. Describe three contexts you would manually create to execute this simple intelligence as part of verification. Why did you pick those three?

- How could you monitor the system to detect that the experience is losing effectiveness?

PART III

Implementing Intelligence

Chapters 11-15 discuss the implementation of Intelligent Systems. This part covers all the important components that need to be built, along with their responsibilities, the options, and trade-offs. These are the systems that take intelligence (the models produced by machine learning) and turn it into a fully functional system—executing the intelligence, moving it where it needs to be, combining it, monitoring it, and supporting its creation. These systems enable the operation of Intelligent Systems throughout their life cycle, with confidence.

Implementing Intelligence

CHAPTER 11

The Components of an Intelligence Implementation

In order to have impact, intelligence must be connected to the user.

That is, when the user interacts with the system, the intelligence must be executed with the proper inputs and the user experience must respond. When new intelligence is created, it must be verified and deployed to where it is needed. And there must be ways to determine if everything is working, and mechanisms to gather what is needed to improve the intelligence over time.

The intelligence implementation takes care of all of this. It is the foundation upon which an intelligent service is created.

As an analogy, consider a web server.

A web server implementation needs to accept network connections, do authentication, interpret requests and find the right content, process it (maybe running scripts), serve it, create logs—and a lot more.

Once a web server implementation is available, content creators can use it to produce all sorts of useful web sites.

This is similar to an implementation of an Intelligent System—it allows intelligence orchestrators to produce useful user interactions over time. Intelligence, artificial intelligence, and machine learning are changing the world, but proper implementations of Intelligent Systems put these into position to fulfill their promise.

This chapter introduces the components that make up an implementation of an Intelligent System.

G. Hulten, *Building Intelligent Systems*, https://doi.org/10.1007/978-1-4842-3432-7_11

An Example of Intelligence Implementation

Imagine an Intelligent System designed to help users know if a web page is funny or not.

The user browses to a page and a little smiley face pops up if the system thinks the user will laugh if they read the page. You know, so no one ever has to miss something funny ever again...

Simple enough. Let's discuss how this Intelligent System might be implemented.

It starts with a program that can examine the contents of a web page and estimate if it is funny or not. This is probably a model produced by machine learning, but it could be some simple heuristics. For example, a very simple heuristic might say that any page containing the phrases "walked into a bar" or "who's there" is funny, and every page that doesn't contain those phrases is not funny.

And for some it might actually be true—maybe you have that uncle, too...

Someone could write a plug-in for web browsers that includes this model. When a new web page is loaded, the plug-in takes the contents of the page, checks it against the model, and displays the appropriate user experience depending on what the model says (the smiley face if the model thinks the page is funny, nothing if the model thinks the page is not funny).

This plug-in is a very simple intelligence implementation.

But ship the plug-in, see users download it, and pretty soon someone is going to want to know if it's working.

So maybe the plug-in includes a way for users to provide feedback. If they get to the bottom of a page the system said would be funny, but they didn't laugh, there is a frowny face they can click to let the system know it made a mistake. If they read a page the system didn't flag as funny, but find themselves laughing anyway, there is a laugh button they can click to let the system know. We might even want to know if users are more likely to read pages that are marked as funny compared to pages that aren't, so maybe we measure how long users spend on pages we've flagged as funny, and how long they spend on ones we haven't.

The plug-in gathers all of this user feedback and behavior and sends it back to the service as telemetry.

Adding telemetry improves the overall intelligent implementation because it lets us answer important questions about how good the intelligence is and how users are perceiving the overall system—if we are achieving our overall objectives. It also sets us up to improve.

Because when the intelligence creators look at the telemetry data, they are going to find all sorts of places where their initial model didn't work well. Maybe most funny things on the Internet are in images, not text. Maybe people in one country aren't amused by the same things that people in another country find hilarious. That kind of thing.

The intelligence creators will spend some time looking at the telemetry; maybe they'll crawl some of the pages where the initial system made mistakes, and build some new models. Then they are going to have a model they like better than the original one and they are going to want to ship it to users.

The intelligence implementation is going to have to take care of this. One option would be to ship a whole new version of the plug-in—the Funny Finder v2.0—which contains the new model. But users of the first version would need to find this new plug-in and choose to install it. Most of them won't. And even the ones who do might take a long time to do it. This causes the intelligence to update slowly (if at all) and reduces the potential of the intelligence creators' work. Further, the intelligence might change fast: maybe every week, maybe every day, maybe multiple times per hour. Unless the implementation can get new intelligence to users regularly and reliably, the Intelligent System won't be very intelligent.

So the implementation might upgrade the plug-in to do automatic updates. Or better, the plug-in might be changed to update just the models (which are essentially data) while leaving all the rest of the code the same. Periodically, the plug-in checks a server, determines if there is some new intelligence, and downloads it.

Great. Now the implementation runs the intelligence, measures it, and updates it. The amount of code to service the intelligence at this point is probably more than the intelligence itself: a good portion of any Intelligent System is implementation. And this version of the system is complete, and closes the loop between the user and the intelligence, allowing the user to benefit from and improve the intelligence simply by using the system and providing feedback.

But some things are really not funny. Some things are offensive, horrible. We really, really don't want our system to make the type of mistake that flags highly offensive content as funny.

So maybe there is a way for users to report when our system is making really terrible mistakes. A new button in the user experience that sends telemetry about offensive pages back to our service.

Maybe we build a little workflow and employ a few humans to verify that user-reported-offensive sites actually are terrible (and not some sort of comedian-war where they report each other's content). This results in a list of "really, truly not funny" sites. The implementation needs to make sure clients get updated with changes to this list as soon as possible. This list could be updated when the model is updated. Or maybe that isn't fast enough, and the plug-in needs to be more active about checking this list and combining the intelligence it contains with the intelligence the model outputs.

So now the plug-in is updated so that every time it visits a new page it makes a service call while it runs its local model. Then it combines the results of these two forms of intelligence (the server-based intelligence and the client-based intelligence). If the server says a site is "really, truly not funny," it doesn't matter what the client-side model says—that site is not funny.

By this point the intelligence creators are going to have all sorts of ideas for how to build better models that can't run in the plug-in. Maybe the new ideas can't be in the client because they take too much RAM. Maybe they can't be on the client because they required external lookups (for example, to language translation services) that introduce too much latency in the plug-in. Maybe the plug-in needs to run in seriously CPU-restrained environments, like in a phone, and the intelligence creators just want a bit more headroom.

These types of intelligences may not be runnable on the client, but they may be perfectly usable in the service's back end.

For example, when a lot of users start visiting a new site, the back end could notice. It could crawl the site. It could run dozens of different algorithms—ones specialized for images, ones that look for particular types of humor, ones tuned to different languages or cultures—and ship the outputs to the client somehow.

So now the plug-in is combining intelligence from multiple sources—some in the cloud, and some that it executes itself. It is managing experience. It is measuring the results of users interacting with the intelligence and collecting data for improvement. And more.

Not every Intelligent System implementation needs all of these components. And not every Intelligent System needs them implemented to the same degree. This part of the book will provide a foundation to help you know when and how to invest in various components so your Intelligent System has the greatest chance of achieving its goals—and doing it efficiently.

Components of an Intelligence Implementation

An intelligence implementation can be very simple, or it can be very complex. But there are some key functions that each Intelligent System implementation must address, these are:

- Execute the intelligence at runtime and light up the experience.

- Ingest new intelligence, verify it, and get it where it needs to be.

- Monitor the intelligence (and user outcomes) and get any telemetry needed to improve the system.

- Provide support to the intelligence creators.

- Provide controls to orchestrate the system, to evolve it in a controlled fashion, and to deal with mistakes.

The Intelligence Runtime

Intelligence must be executed and connected to the user experience.

In order to execute intelligence, the system must gather up the context of the interaction, all the things the intelligence needs to consider to make a good decisions. This might include: what the user is doing; what the user has done recently; what the relevant sensors say; what the user is looking at on the screen; anything that might be relevant to the intelligence making its decision.

The context must be bundled up, converted, and presented to the model (or models) that represent the intelligence. The combination of context and model results in a prediction.

The runtime might be entirely in the client, or it might coordinate between a client and a service.

Then the prediction must be used to affect the system, light up the experience—create impact.

Intelligence Management

As new intelligence becomes available it must be ingested and delivered to where it is needed.

For example, if the intelligence is created in a lab at corporate headquarters, and the runtime needs to execute the intelligence on toasters all across America, the intelligence management system must ship the intelligence to all the toasters.

Or maybe the intelligence runs in a service.

Or maybe it runs partially in a back-end, and partially in the client.

There are many options for where intelligence can live, with pros and cons.

Along the way, the intelligence needs to be verified to make sure it isn't going to do anything (too) crazy.

And Intelligent Systems usually rely on more than one source of intelligence. These might include multiple models, heuristics, and error-overrides. These must be combined, and their combination needs to be verified and delivered.

Intelligence Telemetry Pipeline

Getting the right monitoring and telemetry is foundational to producing an intelligence that functions correctly and that can be improved over time.

Effective monitoring and telemetry includes knowing what contexts the intelligence is running in and what types of answers it is giving. It includes knowing what experiences the intelligence is producing and how users are responding. And an effective telemetry system will get appropriate data to grow new and better intelligence.

In large Intelligent Systems the telemetry and monitoring systems can produce a lot of data—a LOT.

And so the intelligence implementation must decide what to observe, what to sample, and how to digest and summarize the information to enable intelligence creation and orchestration among the various parts of the Intelligent System.

The Intelligence Creation Environment

In order for an Intelligent System to succeed, there needs to be a great deal of coordination between the runtime, the delivery, the monitoring, and the intelligence creation.

For example, in order to produce accurate intelligence, the intelligence creator must be able to recreate exactly what happens at runtime. This means that:

- The telemetry must capture the same context that the runtime uses (the information about what the user was doing, the content, the sensor output, and so on).

- The intelligence creator must be able to process this context exactly the same way it is processed at runtime.

- The intelligence creator must be able to connect contexts that show up in telemetry to the eventual outcome the user got from the interaction—good or bad.

This can be hard when intelligence creators want to innovate on what types of information/context their models examine (and they will; it's called feature engineering and is a key part of the intelligence creation process). It can be hard when the monitoring system doesn't collect exactly the right information, leading to mismatches between runtime and what the intelligence creator can see. It can be hard when the intelligence is executed on a type of device different from the intelligence creation device. Maybe the runtime has a different coprocessor, a different graphics card, or a different version of a math library from the ones in the intelligence creation environment.

Any one of these can lead to problems.

An intelligence implementation will create an environment that mitigates these problems, by providing as much consistency for intelligence creators as possible.

Intelligence Orchestration

The Intelligent System needs to be orchestrated.

- If it gets into a bad state, someone needs to get it out.

- If it starts making bad mistakes, someone needs to mitigate them while the root cause is investigated.

- As the problem, the intelligence, and the user base evolve, someone needs to be able to take control and tune the intelligence and experience to achieve the desired results.

For example, if the intelligence creation produces a bad model (and none of the checks catch it) the model may start giving bad experiences to customers. The intelligence orchestration team should be able to identify the regression quickly in telemetry, track the problems down to the specific model, and disable or revert the model to a previous version.

If things go badly enough, the intelligence orchestrator might need to shut the whole thing down, and the implementation and experience should respond gracefully.

Providing good visibility and tools for orchestrating an Intelligent System allows intelligence creation to act with more confidence, take bigger risks, and improve more rapidly.

Summary

An Intelligent System needs to be implemented so that intelligence is connected with customers. This includes:

- Executing the intelligence and lighting up the intelligent experience.

- Managing the intelligence, shipping it where it needs to be.

- Collecting telemetry on how the intelligence is operating and how users are responding (and to improve the intelligence).

- Supporting the people who are going to have to create intelligence by allowing them to interact with contexts exactly the same way users will.

- Helping orchestrate the Intelligent System through its life cycle, controlling its components, dealing with mistakes, and so on.

An Intelligent System is not like a traditional program, in that it is completed, implemented, shipped and walked away from. It is more like a service that must be run and improved over time. The implementation of the Intelligent System is the platform that allows this to happen.

For Thought...

After reading this chapter, you should:

- Understand the properties of an effective intelligent implementation—what it takes to go from a piece of intelligence (for example, a machine learned model) to a fully functional Intelligent System.

- Be able to name and describe the key components that make up an implementation of an Intelligent System.

You should be able to answer questions like these:

Consider an activity you do daily:

- What would a minimalist implementation of an Intelligent System to support the activity look like?

- Which component—the runtime, the intelligence management, the telemetry, the intelligence creation environment, the orchestration— do you think would require the most investment? Why?

CHAPTER 12

The Intelligence Runtime

The intelligence runtime puts intelligence into action. It is responsible for interfacing with the rest of the Intelligent System, gathering the information needed to execute the system's intelligence, loading and interpreting the intelligence, and connecting the intelligence's predictions back to the rest of the system.

Conceptually, an intelligence runtime might look something like this:

```
Intelligence theIntelligence =
             InitializeIntelligence(<intelligence data file>,
                          <server address>,
                          <etc...>);

Context theContext =
             GatherContext(<sensors>,
                          <content>,
                          <user history>,
                          <other system properties>,
                          <etc...>);

Prediction thePrediction =
      ExecuteIntelligence(theIntelligence, theContext);

UpdateTheExperience(thePrediction);
```

It's simple in principal, but, like most good things, it can require a great deal of complexity in practice.

This section covers the principal elements that an intelligence runtime will handle:

- Gathering context.

- Constructing features.

- Loading models.

© Geoff Hulten 2018

G. Hulten, *Building Intelligent Systems*, https://doi.org/10.1007/978-1-4842-3432-7_12

- Executing models on the context/features.

- Connecting the resulting predictions with the experience.

Context

The context in an intelligence runtime consists of all the data the intelligence might use to make its decision. For example:

- If the Intelligent System is trying to determine if a sprinkler system should run, the context might include.

    - The amount of time since the sprinkler last ran.

    - The forecast for the next several days (will it rain or not).

    - The amount of rain that has fallen in the past several days.

    - The type of landscaping the user has.

    - The time of day.

    - The temperature.

    - The humidity.

    - And so on.

- If the Intelligent System is looking for cats in an image, the context might include.

    - The raw image data (RGB intensity for the pixels).

    - Metadata about how the image was captured (the exposure and frame rate, the time, the zoom).

    - The location the image came from.

    - And so on.

- If the Intelligent System is trying to tell if a user should take a break from their computer, the context might include.

    - The time since the last break.

    - The activities the user has done since the break (number of keystrokes and mouse movements).

- An indication of how active the user has been.

- The amount of time until the user's next meeting (according to their calendar).

- The name of the application that the user is currently interacting with.

- And so on.

The context can be just about any computer-processable information, like a URL, a sensor reading from some embedded device, a digital image, a catalog of movies, a log of a user's past behavior, a paragraph of text, the output of a medical device, sales receipts from a store, the behavior of a program, and so on.

These are the things that might be useful in making the decision about what the Intelligent System should do. In fact, the context is usually a superset of the things the intelligence will actually use to make the decision.

A good context will:

- Contain enough information to allow the intelligence be effective.

- Contain enough information to allow the intelligence to grow and adapt.

- Contain enough information to allow the Intelligent System's orchestrators to track down mistakes.

- Be efficient enough to gather at runtime for the intelligence call.

- Be compact enough to include in telemetry.

Clearly there are trade-offs. More complete context gives the intelligence more potential, but it may also encounter engineering and operational constraints, requiring additional CPU and memory usage and network bandwidth, and introducing latency and increasing data size.

Feature Extraction

Gathering these context variables can require significant effort. But it isn't the end of the process. There is also another layer of code that converts these variables into the representation the intelligence needs. This extra layer is often called "feature extraction" or "featurization" and is a common part of machine-learning systems.

There are a number of reasons this matters for an intelligence implementation:

- The code that produces features from context needs to run in the runtime. That is, it needs to be efficient and reliable.

- This code is often technical, computationally expensive, mathy, and not intuitive to people who don't have extensive experience in working with data.

- The intelligence creators are going to want to constantly change what this code does; in fact, they may need to change it periodically to continue to grow the intelligence.

- This code is tightly coupled to the intelligence, and the feature extraction code and intelligence (models) absolutely must be kept in sync—the exact version of the feature extraction code used to build the intelligence must also be used in the runtime when the intelligence is executed.

Because of these factors, feature extraction must be carefully planned. Some best practices include:

- Have extensive test sets for feature extraction code. This should include:

  - Unit tests on the extractors.

  - Performance tests to make sure the code meets CPU and RAM requirements.

  - A test that periodically executes feature extraction code against a set of known contexts (maybe daily, maybe at each check-in) and compares the output to a known benchmark. Any change in output should be examined by engineers and intelligence creators to make sure it is what was intended.

  - A test that ensures the feature extraction code at runtime (when users are interacting with it) is doing exactly the same thing it is doing in the intelligence creation environment (when it is being used to create models).

- Have version information encoded in intelligence and in the feature extraction code and have the runtime verify that they are in sync.

- Have the ability to change feature-extraction code out-of-band of the full application; for example, by sending a new version of the feature extraction code with each model update.

- Develop a life-cycle for moving feature extraction changes from intelligence creators, to engineers, to deployment. The life-cycle should allow intelligence creators to move quickly in evaluating changes, and the quality of the code deployed to users to be high.

Models

The model is the representation of the intelligence (and we will discuss representation options in coming chapters). For now, imagine the model as a data file. In some Intelligent Systems, the models are changed every few days. In some systems they are changed every few minutes.

The rate of change will depend on these factors:

1.   How quickly the system gets enough data to improve the models.

2.   How long it takes to construct new models.

3.   How much it costs to distribute the new models.

4.   How necessary change is relative to the success criteria.

In general, models will change (much) more quickly than the larger program. An effective runtime will:

- Allow the model to be updated easily (that is, without having to restart the whole system).

- Make sure the models it uses are valid, and all the components (the feature extractors, the models, the context) are in sync.

- Make it easy to recover from mistakes in models (by rolling back to previous versions).

One key consideration with models is the amount of space they take up on disk and in RAM. Left unchecked, machine learning can produce models that are quite large. But it is always possible to make tradeoffs, for example between model size and accuracy, if needed.

Execution

Executing a model is the process of asking the model to make its prediction based on the context (and associated features).

In the simplest case, when there is a single model and it lives on the client, executing a model can be as easy as loading the features into an array and making a function call into an off-the-shelf model execution engine. There are libraries to execute pretty much every type of model that machine learning can build, and in most cases these libraries are perfectly acceptable—so most of the time you won't even need to write much code to load and execute a model.

Of course, in some cases these libraries aren't acceptable. Examples include when RAM or CPU is a constraint; or when the execution is happening on a device with special hardware (like a GPU or FPGA) that the libraries don't take advantage of.

Another form of execution involves a model running on a server. In these cases, the client needs to bundle some (compact) representation of the context (or the features), ship them to the server, and wait for an answer, dealing with all the coordination required to properly account for the latency of the call and keep it from resulting in bad user experiences.

When the system has more than one model (which is often the case), the execution needs to execute each of the models, gather their results, and combine them into a final answer. There are many ways to do this, including:

- Averaging their results.

- Taking the highest (most certain) result.

- Having some rules about the context that select which one to trust (for example, one model nails house prices in 90210 but is terrible everywhere else, another model works with condos but not single-family homes).

- Having a model that combines the outputs of the other models.

The chapters in Part IV, "Creating Intelligence," will discuss model combination in more detail along with the pros and cons of these (and other) approaches.

Results

Executing the intelligence results in an answer. Maybe the intelligence outputs a 0.91, which represents a 91% probability in whatever the intelligence is talking about. The implementation must use this to affect what the intelligent experience will do. For example, it might compare the value to a threshold and if the value is high (or low) enough, initiate a prompt, or automate something, or annotate some information—whatever the intelligent experience wants to do.

Instability in Intelligence

Using the raw output of intelligence can be risky, because intelligence is intrinsically unstable. For example:

Take a digital image of a cow. Use a cow-detector model to predict the probability the image contains a cow, and the output might be high, maybe 97.2%.

Now wait a day. The back-end system has new training data, builds a new cow-detector model, distributes it to the runtime. Apply the updated model to the same cow picture and the probability estimate will almost certainly be different. The difference might be small, like 97.21% (compared to 97.2% from the original) or it might be relatively large, like 99.2%. The new estimate might be more accurate, or it might be less accurate. But it is very, very unlikely the estimate will be the same between two different versions of intelligence.

Subtle changes in context can have similar effects. For example, point your camera at a cow and snap two digital images right in a row, one after the other, as fast as you can. To a human, the cow looks the same in both images, the background looks the same, and the lighting looks the same—the "cow detector" model should output the same probability estimation on the first image as on the second. According to a human.

But to a model, the images might look very different: digital sensors introduce noise, and the noise profile will be different between the images (resulting in little speckles that human eyes have a hard time seeing); the cow might have blinked in one and not the other; the wind might have blown, changing the location of shadows from leaves; the position of the camera may have changed ever-so-slightly; the camera's auto-focus or exposure correction might have done slightly different things to the images.

Because of all of this, the intelligence might make different estimations on the two images. Maybe 94.1% in one and 92.7% in the other.

Maybe that change isn't a big deal to the intelligent experience that results, but maybe it is.

Trying to pretend that there isn't instability in machine-learning systems can result in less effective (or downright bad) intelligent experiences.

Intelligence APIs

To address instability in intelligence outputs, it's often helpful to encapsulate intelligence behind an interface that exposes the minimum amount of information needed to achieve the system's goals:

- When dealing with a probability, consider throwing away a bunch of resolution: round 92.333534% to 90%.

- Consider turning probabilities into classifications: instead of "45% there is a cow," say, "There is not a cow."

- Consider quantizing predictions: instead of "There is a cow at coordinates 14,92 in the image," say, "There is a cow on the left side of the image."

The thresholds and policies needed to implement these types of transformations are tightly coupled to the intelligence and should be included in the model with each update.

And while these practices certainly help, making the model's output more stable, and providing some level of encapsulation of the model's inner workings, they aren't perfect. For example, imagine the cow detector is 81% sure there is a cow in one frame. The model's threshold is at 80%. So the system creates a "there is a cow here" classification and lights up a UX element.

Then on the next frame, the cow detector is 79.5% sure there is a cow present, and the user experience flickers off. This can look really bad.

A well designed API, combined with an experience designed to minimize flaws, are both required to make an effective intelligent experience.

Summary

The intelligence runtime is responsible for: gathering the context of the system; converting this context into a form that works with the intelligence; executing the various components that make up the intelligence; combining the intelligence results and creating a good interface between the intelligence and the experience; and using the output of the intelligence to affect the user experience.

Some key components that an intelligence runtime will deal with include these:

- **The context,** including all the information relevant to making a good decision in the Intelligent System.

- **The features** (and the feature extraction code), which convert the context into a form that is compatible with the specific models that contain the system's intelligence.

- **The models,** which represent the intelligence and are typically contained in data files that will change relatively frequently during the lifetime of an Intelligent System.

- **The execution engine,** which executes the models on the features and returns the predictions. There are many great libraries to support executing models, but these will often need to be wrapped to combine intelligence; they will sometimes need to be replaced for unique execution environments.

- **The results,** which are the predictions of the intelligence. It is good practice to keep the raw results private and create an intelligence API that exposes the minimum information to power the intelligent experience, while being robust to changes over time.

An effective intelligence runtime will make it easy to track down mistakes; will execute in the runtime the same way it executes in the intelligence creation environment; will support easy changes to both models and feature extraction; and will make it hard for parts of the system to get out of sync.

For Thought…

After reading this chapter, you should:

- Be able to design a runtime that executes intelligence and uses it to power user experience.

- Understand how to structure an intelligence runtime to allow innovation in intelligence and support the other components of the intelligence implementation.

You should be able to answer questions like these:

- What is the difference between the context of an intelligence call and the features used by the machine-learned model?

Consider the Intelligent System you used most recently.

- What type of information might be in the context of the intelligence calls it makes?

- At a high level, walk through the steps to go from that context to a user-impacting experience (invent any details you need to about how the system might work).

- What are some ways this system's intelligence might be encapsulated to mitigate the effects that small intelligence changes have on the user?

CHAPTER 13

Where Intelligence Lives

When building an Intelligent System you'll need to decide where the intelligence should live. That is, where you will bring the model, the runtime, and the context together to produce predictions—and then how you will get those predictions back to the intelligent experience.

The runtime could be located in the user's device, in which case you'll need to figure out how to update models across your user base.

The runtime could be in a service you run, in which case you'll have to figure out how to get the context (or features) from the user's device to the service cheaply enough, and with low enough latency, to make the end-to-end experience effective.

This chapter discusses how to decide where your intelligence should live. It starts by discussing the considerations for deciding where intelligence should live, including latency and cost, and why these matter for various types of intelligent experiences. It then discusses common patterns for positioning intelligence across clients and services, including the pros and cons of each option.

Considerations for Positioning Intelligence

Some key considerations when deciding where intelligence should live are these:

- Latency in updating the intelligence.

- Latency in executing the intelligence.

- The cost of operating the intelligence.

- What happens when users go offline.

In general, you will have to make trade-offs between these properties. This section discusses each of them in turn.

© Geoff Hulten 2018
G. Hulten, *Building Intelligent Systems*, https://doi.org/10.1007/978-1-4842-3432-7_13

Latency in Updating

One consideration when deciding where to position intelligence is the latency you will incur when you try to update the intelligence. New intelligence cannot benefit users until it gets to their runtime and replaces the old intelligence.

The latency in updating intelligence can be very short when the runtime is on the same computer as the intelligence creation environment; the latency can be very long when the runtime is on a client computer, and the client computer does not connect to the Internet very often (and so it can only get new intelligence once in a while).

Latency in transporting intelligence to a runtime is important when:

- The quality of intelligence is evolving quickly.

- The problem you are trying to solve is changing quickly.

- There is risk of costly mistakes.

We'll discuss these situations where latency in updating can cause problems in more detail, including examples of how they can impact intelligent experiences.

Quality Is Evolving Quickly

Latency in updating intelligence matters when the quality of the intelligence is evolving quickly.

When a problem is new, it is easy to make progress. There will be all sorts of techniques to explore. There will be new data arriving every day, new users using the product in ways you didn't anticipate. There will be excitement—energy. The quality of intelligence might improve rapidly.

For example, imagine it's day one—you've just shipped your smart shoe. This shoe is designed to adjust how tight it is based on what its wearer is doing. Sitting around doing nothing? The shoe automatically relaxes, loosening the laces so the wearer can be more comfortable. But if the user starts running? Jumping? The shoe automatically tightens the laces, so their wearer will get better support and be less likely to get an injury.

So you've had a huge launch. You have tens of thousands of smart shoes in the market. You are getting telemetry for how users are moving, when they are overriding the shoe to make their laces tighter, and when they are overriding the shoe to make their laces looser. You are going to want to take all this data and produce new intelligence.

If you produce the new intelligence, test it, and determine it isn't much better than the intelligence you shipped with—you're fine! No need to worry about latency in updating intelligence, because you don't have any benefit to gain.

But if the new intelligence is better—a lot better—you are going to be desperate to get it pushed out to all the shoes on the market. Because you have users wearing those shoes. They'll be talking to their friends. Reviewers will be writing articles. You are going to want to update that intelligence quickly!

When intelligence is improving quickly, latency in deployment will get in the way of taking advantage of the intelligence. This is particularly important in problems that are very big (with many, many contexts) or problems that are very hard to solve (where quality will be improving for a long time to come).

Problem Changing Quickly

Latency in updating intelligence matters when the problem is changing quickly, because it is open-ended or changing over time.

In some domains new contexts appear slowly, over the course of weeks or months. For example, new roads take time to build; tastes in music evolve slowly; new toaster products don't come onto the market every day. In these cases having days or weeks of latency in deploying new intelligence might not matter.

On the other hand, in some domains new contexts appear rapidly, hundreds or thousands of times per day. For example, new spam attacks happen every second; hundreds of new news articles are written per day; hurricanes make landfall; stock markets crash. In these cases, it might be important to be able to deploy new intelligence fast.

There are two important aspects to how problems change:

1. How quickly do new contexts appear?

2. How quickly do existing contexts disappear?

Erosion will happen slowly in domains where new contexts are added, but old contexts remain relevant for a reasonably long time. For example, when a new road is constructed your intelligence might not know what to do with it, but your intelligence will still work fine on all the existing roads it does know about.

Erosion will happen quickly in domains where new contexts displace old ones, and the intelligence you had yesterday is no longer useful today. For example, when a new song comes onto the top of the charts, an old song leaves. When a spammer starts their new attack, they stop their old one.

When a problem changes frequently in ways that erode the quality of existing intelligence, latency in updating intelligence can be a critical factor in the success of an Intelligent System.

Risk of Costly Mistakes

Latency in updating intelligence matters when the Intelligent System can make costly mistakes that need to be corrected quickly.

Intelligence makes all kinds of mistakes. Some of these mistakes aren't too bad, particularly when the intelligent experience helps users deal with them. But some mistakes can be real problems.

Consider the smart-shoe example, where laces automatically loosen or tighten based on what the wearer is doing. Imagine that some small percent of users fidget in a particular way that makes the shoe clamp down on their feet, painfully.

Or imagine that a small percentage of users play right field on a baseball team. Ninety-nine percent of the time they are standing there, looking at the clouds go by—and then one percent of time they are sprinting wildly to try to catch a fly ball. Maybe the smart-shoes can't keep up. Maybe right fielders are running out of their shoes, slipping, and getting hurt.

When your Intelligent System's mistakes have high cost, and these costs can't be mitigated with a good user experience, latency can be very painful. Imagine users calling, complaining, crying about the damage you've caused them, day after day, while you wait for new intelligence to propagate. It can make you feel bad. It can also put your business at risk.

Latency in Execution

Another consideration for deciding where intelligence should live is the latency in executing the intelligence at runtime.

To execute intelligence, the system must gather the context, convert the context to features, transport the features to where the intelligence is located (or sometimes the whole context gets transported and feature extraction happens later), wait for the intelligence to execute, wait for the result of the intelligence execution to get back to the experience, and then update the experience. Each of these steps can introduce latency in executing the intelligence.

The latency in execution can be short when the intelligent runtime lives on the same computer as the intelligent experience; it can be long when the intelligent runtime and intelligent experience live on different computers.

Latency in intelligence execution can be a problem when:

- Users will have to wait for the latency.

- The right answer changes quickly and drastically.

Latency and its effects on users can be difficult to predict. Sometimes users don't care about a little bit of latency. Sometimes a little latency drives them crazy and totally ruins an experience. Try to design experiences where latency trade-offs can be changed easily (during orchestration) so various options can be tested with real users.

Latency in Intelligent Experience

Latency in execution matters when users notice it, particularly when they have to wait for the latency before they can continue. This can occur when:

- **The intelligence call is an important part of rendering the experience**. Imagine an application that requires multiple intelligence calls before it can properly render. Maybe the application needs to figure out if the content is potentially harmful or offensive for the user before rendering it. If there is nothing for the user to do while waiting for the intelligence call, then latency in intelligence execution could be a problem.

- **The intelligence call is interactive**. Imagine an application where the user is interacting directly with the intelligence. Maybe they throw a switch, and they expect the light to turn on instantly. If the switch needs to check with the intelligence before changing the light, and the intelligence call takes hundreds or thousands of milliseconds— users might stub their toes.

On the other hand, intelligent experiences that are intrinsically asynchronous, such as deciding whether to display a prompt to a user, are less sensitive to latency in execution.

The Right Answer Changes Drastically

Latency in executing intelligence matters when the right answer changes rapidly and drastically.

Imagine creating an intelligence to fly a drone. The drone is heading to an objective; the correct answer is to fly straight at the objective. No problem. And then someone steps in front of the drone. Or imagine the drone flying on a beautiful, blue, sunny day and then a huge gust of wind comes out of nowhere. In these situations, the right control for the drone changes, and changes quickly.

When the right course of action for an Intelligent System changes rapidly and drastically, latency in executing intelligence can lead to serious failures.

Cost of Operation

Another consideration when deciding where intelligence should live is the cost of operating the Intelligent System.

Distributing and executing intelligence takes CPU, RAM, and network bandwidth. These cost money. Some key factors that can drive the cost of an Intelligent System include:

- The cost of distributing intelligence.

- The cost of executing intelligence.

The Cost of Distributing Intelligence

Distributing intelligence costs both the service and the user money, usually in the form of bandwidth charges. Each new piece of intelligence needs to be hosted (for example on a web service), and the runtime must periodically check for new intelligence, and then download any it finds. This cost is proportional to the number of runtimes the intelligence must go to (the number of clients or services hosting it), the size of the intelligence updates, and the frequency of the updates. For Internet-scale Intelligent Systems, the cost of distributing intelligence can be very large.

It's also important to consider costs for users. If the primary use case is for users on broadband, distributing models might not be a concern—it might not cost them much. But when users are on mobile devices, or in places where network usage is more carefully metered and billed, the bandwidth for intelligence distribution may become an important consideration.

The Cost of Executing Intelligence

Executing intelligence can also have a bandwidth cost when the intelligence is located in a service and clients must send the context (or features) to the service and get the response. Depending on the size of the context and the frequency of calls, it may cost more to send context to the service than to send intelligence to the client. Keep in mind that telemetry and monitoring will also need to collect some context and feature information from clients; there is opportunity to combine the work and reduce cost.

Executing intelligence also takes CPU cycles and RAM. Putting the intelligence runtime in the client has the advantage that users pay these costs. But some types of intelligence can be very expensive to execute, and some clients (like mobile ones) have resource constraints that make heavyweight intelligence runtimes impractical. In these cases, using intelligence runtimes in the service (maybe with customized hardware, like GPUs or FPGAs) can enable much more effective intelligences.

When considering the cost of operation, strive for an implementation that:

1. Is sensitive to the user and does not ask them to pay costs that matter to them (including bandwidth, but also power in a mobile device, and so on).

2. Does let users pay the parts of the cost they won't notice (so you don't have to buy lots of servers).

3. Balances all the costs of running the Intelligent System and scales well as the number of users and quality of intelligence grow.

Offline Operation

Another consideration when deciding where intelligence should live is whether the Intelligent System needs to function when it is offline (and unable to contact any services).

It isn't always important for an Intelligent System to work when it is offline. For example, a traffic prediction system doesn't need to work when its user is in an airplane over the Pacific Ocean. But sometimes Intelligent Systems do need to work offline—for example, when the intelligence runs in a restricted environment (like a military vehicle) or in a life-critical system.

When it is important to function offline, some version of the intelligence must live in the client. This can be the full intelligence, or it can be more like a backup—a reduced version of the overall intelligence to keep users going while the service comes back online.

Places to Put Intelligence

This section explores some of the options for positing intelligence in more detail. It introduces some common patterns, including:

- Static intelligence in the product

- Client-side intelligence

- Server-centric intelligence

- Back-end (cached) intelligence

- Hybrid intelligence

This section discusses each of these approaches and explores how well they address the four considerations for where intelligence should live: latency in updating, latency in execution, cost of operation, and offline operation.

Static Intelligence in the Product

It's possible to deliver intelligence without any of this Intelligent System stuff. Simply gather a bunch of training data, produce a model, bundle it with your software, and ship it.

This is very similar to shipping a traditional program. You build it, test it as best you can—in the lab, or with customers in focus groups and beta tests—tune it until you like it, and send it out into the world.

The advantages of this is that it is cheaper to engineer the system. It might be good enough for many problems. It can work in situations without the ability to close the loop between users and intelligence creation. And there is still the possibility for feedback (via reviews and customer support calls), and to update the intelligence through traditional software updates.

The disadvantage is that intelligence updates will be more difficult, making this approach poorly suited to open-ended, time-changing, or hard problems.

Latency in Updating Intelligence: Poor

Latency in Execution: Excellent

Cost of Operation: Cheap

Offline Operation: Yes

Disadvantage Summary: Difficult to update intelligence. Risk of unmeasurable intelligence errors. No data to improve intelligence.

Client-Side Intelligence

Client-side intelligence executes completely on the client. That is, the intelligence runtime lives fully on the client, which periodically downloads new intelligence.

The download usually includes new models, new thresholds for how to interpret the models' outputs (if the intelligence is encapsulated in an API—and it should be), and (sometimes) new feature extraction code.

Client-side intelligence usually allows relatively more resources to be applied to executing intelligence. One reason for this is that the intelligence can consume idle resources on the client at relatively little cost (except maybe for power on a mobile device). Another reason is that the latency of the runtime is not added to any service call latency, so there is relatively more time to process before impacting the experience in ways the user can perceive.

The main challenge for client-side intelligence is deciding when and how to push new intelligence to clients. For example, if the intelligence is ten megabytes, and there are a hundred thousand clients, that's about a terabyte of bandwidth per intelligence update. Further, models don't tend to compress well, or work well with incremental updates, so this cost usually needs to be paid in full.

Another potential complexity of client-side intelligence is dealing with different versions of the intelligence. Some users will be offline, and won't get every intelligence update you'd like to send them. Some users might opt-out of updates (maybe using firewalls) because they don't like things downloading to their machines. These situations will make interpreting user problems more difficult.

Another disadvantage of client-side models is that they put your intelligence in the hands of whoever wants to take a look at it: maybe a competitor, maybe someone who wants to abuse your service, or your users—like a spammer. Once someone has your model they can run tests against it. They can automate those tests. They can figure out what type of inputs gets the model to say one answer, and what type of inputs gets it to say another. They can find the modifications to their context (e-mail, web page, product, and so on) that trick your model into making exactly the type of mistake they want it to make.

Latency in Updating Intelligence: Variable

Latency in Execution: Excellent

Cost of Operation: Based on update rate.

Offline Operation: Yes

Disadvantage Summary: Pushing complex intelligence to clients can be costly. Hard to keep every client in-sync with updates. Client resources may be constrained. Exposes the intelligence to the world.

Server-Centric Intelligence

Server-centric intelligence runs in real-time in the service. That is, the client gets the context (or features) and sends them to the server, and the server executes the intelligence on the features and returns the result to the client.

Using server-centric intelligence allows models to be updated quickly, and in a controlled fashion, by pushing new models to the server (or servers) running the intelligence. It also makes telemetry and monitoring easier because much of the data to log will already be in the service as part of the intelligence calls.

But server-centric intelligence needs to be scaled as the user base scales. For example, if there are a hundred intelligence request per second, the service must be able to execute the intelligence very quickly, and probably in parallel.

Latency in Updating Intelligence: Good

Latency in Execution: Variable, but includes Internet round-trip.

Cost of Operation: Service infrastructure and bandwidth can have significant cost; may cost users in bandwidth.

Offline Operation: No

> **Disadvantage Summary:** Latency in intelligence calls. Service infrastructure and bandwidth costs. User bandwidth costs. Cost of running servers that can execute intelligence in real time.

Back-End (Cached) Intelligence

Back-end intelligence involves running the intelligence off-line, caching the results, and delivering these cached results where they are needed. Cached intelligence can be effective when analyzing a finite number of things, like all the e-books in a library, all the songs a service can recommend, or all the zip codes where an intelligent sprinkler is sold. But back-end intelligence can also be used when there aren't a finite number of things, but contexts change slowly.

For example, a sprinkler is sold into a new zip code. The service has never considered that zip code before, so it returns some default guess at the optimal watering time. But then the back-end kicks off, examines all the info it can find about the zip code to produce a good watering plan and adds this watering plan to its cache. The next time a sprinkler is sold in that zip code, the service knows exactly how to water there (and maybe the system even updates the watering plan for that poor first guy who kicked off the whole process).

Back-end Intelligent Systems can afford to spend more resources and time on each intelligence decision than the other options. For example, imagine a super complex watering-plan model that runs for an hour on a high-end server to decide how to water in each zip code. It analyzes satellite images, traffic patterns, the migration of birds and frogs in the regions—whatever it takes. Such a model might take months to run on an embedded computer in a sprinkler—impractical. It can't run on a server that needs to respond to hundreds of calls per second—no way. But it can run in back-end 'every so often' and the results of its analysis can be cached.

Intelligence caches can live in services; parts of them can be distributed to clients too.

One disadvantage of back-end intelligence is that it can be more expensive to change models, because all of the previous cached results might need to be recomputed.

Another disadvantage is that it only works when the context of the intelligence call can be used to "look up" the relevant intelligence. This works when the context describes an entity, such as a web page, a place, or a movie. It doesn't work when the context describes less-concrete things, like a user-generated block of text, a series of outputs from a sensor-array, or a video clip.

Latency in Updating Intelligence: Variable

Latency in Execution: Variable

Cost of Operation: Based on usage volume

Offline Operation: Partial

Disadvantage Summary: Not effective when contexts change quickly, or when the right answer for a context changes quickly. Can be expensive to change models and rebuild the intelligence caches. Restricts intelligence to things that can be looked up.

Hybrid Intelligence

In practice it can be useful to set up hybrid intelligences that combine several of these approaches.

For example, a system might use a back-end intelligence to deeply analyze popular items, and a client-side intelligence to evaluate everything else.

Or a system might use a client-side intelligence in most cases, but double-check with the service when a decision has serious consequences.

Hybrid intelligences can mask the weaknesses of their various components.

But hybrid intelligences can be more complex to build and to orchestrate. Consider, if the system gives an incorrect answer, what part of the intelligence did the mistake come from? The client-side model? The intelligence that was cached in the service? Some subtle interaction between the two?

Sometimes it's even hard to know for sure what state all of those components were in at the time of the mistake.

Nevertheless, most large Intelligent Systems use some form of hybrid approach when determining where their intelligence should live.

Summary

Choosing where your intelligence will live is an important part of creating a successful Intelligent System. The location of intelligence can affect:

- **The latency in updating the intelligence:** This is a function of how far the intelligence needs to move to get from the creation environment to the runtime and how often the runtime is online to take an update.

- **The latency in executing the intelligence:** This is a function of moving the context and the features from the intelligent experience to the intelligence runtime and moving the answer back.

- **The cost of operating the Intelligent System:** This is a function of how much bandwidth you need to pay for to move intelligence, context, and features and how much CPU you need to pay for to execute intelligence.

- **The ability of the system to work offline:** This is a function of how much of the intelligence can function on the client when it can't communicate with your service components.

There are many options for balancing these properties. Here are some common patterns:

- **Static intelligence:** This puts the intelligence fully in the client without connecting it to a service at all.

- **Client-side intelligence:** This puts the intelligence fully in the client, but connects it to a service for intelligence updates and telemetry.

- **Server-centric intelligence:** This puts the intelligence fully in a service and requires a service call every time intelligence needs to be executed on a context.

- **Back-end (cached) intelligence:** This executes intelligence offline on common contexts and delivers answers via caching.

- **Hybrid intelligence:** This is the reality for most large-scale Intelligent Systems and combines multiple of the other approaches to achieve the system's objectives.

For Thought...

After reading this chapter, you should:

- Know all the places intelligence can live, from client to the service back-end, and the pros and cons of each.

- Understand the implications of intelligence placement and be able to design an implementation that is best for your system.

You should be able to answer questions like these:

- Imagine a system with a 1MB intelligence model, and 10KB of context for each intelligence call. If the model needs to be updated daily, at what number of users/intelligence call volume does it make sense to put the intelligence in a service instead of in the client?

- If your application needs to work on an airplane over the Pacific Ocean (with no Internet), what are the options for intelligence placement?

- What if your app needs to function on an airplane, but the primary use case is at a user's home? What are some options to enable the system to shine in both settings?

CHAPTER 14

Intelligence Management

The intelligence in an Intelligent System takes a journey from creation, to verification, to deployment, to lighting up for users, and finally to being monitored over time. Intelligence management bridges the gap between intelligence creation (which is discussed in Part IV of this book) and intelligence orchestration (which is discussed in Part V), by making it safer and easier to deploy new intelligence and enable it for users.

At its simplest, intelligence management might involve hand-copying model files to a deployment directory where the runtime picks them up and exposes them to users. But in large, complex Intelligent Systems, the process of managing intelligence can (and probably should) be much more involved.

This chapter will discuss some of the challenges with intelligence management. Then it will provide an overview of ways intelligence management can support agility in your Intelligent System while verifying intelligence and lighting it up with users.

Overview of Intelligence Management

Intelligence management involves all the work to take intelligence from where it is created and put it where it will impact users. This includes:

- **Sanity checking the intelligence** to make sure it will function correctly.

- **Deploying the intelligence** to the runtimes it needs to execute in.

- **Lighting up the intelligence** in a controlled fashion.

- **Turning off intelligence** that is no longer helpful.

An intelligence management system can be simple (like a set of instructions an intelligence operator must execute manually for each of these steps); it can be partially automated (like a set of command-line tools that do the work); or it can be very slick

157

© Geoff Hulten 2018
G. Hulten, *Building Intelligent Systems*, https://doi.org/10.1007/978-1-4842-3432-7_14

(like a graphical console that lets intelligence orchestrators inject, move, and light up or disable intelligence with a click). A good intelligence management system will do the following:

- Provide enough support to match the skills and scenarios for intelligence management in your environment.

- Make it hard to make mistakes.

- Not introduce too much latency.

- Make a good tradeoff between human involvement and implementation cost.

Intelligence management is challenging because of complexity, frequency, and human systems. We'll discuss these in turn.

Complexity in Intelligent Management

Intelligent Systems can be quite complex. For example:

- The intelligence might need to live at multiple places between the client and server, including client-side intelligence, one or more server-side intelligences, and cached intelligence.

- Intelligence might come from dozens of different sources—some of them machine-learned, some created by humans, and some created by processes outside of your organization.

- Various parts of intelligence might depend on one another, and might be updated at different frequencies by different people.

As Intelligent Systems grow over time, simply getting a new piece of intelligence correctly deployed to users can be difficult and error-prone.

Frequency in Intelligence Management

Intelligent Systems will have their intelligence updated many times during their life-cycles, consider that:

- Updating intelligence *once a week* for three years is about a hundred sixty times.

- Updating intelligence *once a day* for three years is about a thousand times.

- Updating intelligence *once an hour* for three years is about twenty-six thousand times.

- Updating intelligence *one a minute* for three years is about one and a half million times.

These are pretty big numbers. They mean that the intelligence management process needs to be reliable (have a low error rate), and it probably can't take much human effort.

Human Systems

Intelligence might be deployed by users with all sorts of skill levels and backgrounds:

- Experts who understand the implementation of the Intelligent System well.

- Machine-learning practitioners who are not great engineers.

- Nontechnical people who are hand-correcting costly mistakes the system is making.

- New employees.

- Disgruntled employees.

Making intelligent management easier and less error-prone can pay large dividends in the agility with which your Intelligent System evolves over time.

Sanity-Checking Intelligence

A rapid, automated sanity-checking system is a safety net for intelligence creators, allowing them to innovate with confidence and focus their energy on building better intelligence (not on cross-checking a bunch of details and remembering how to run all the tests they should run). An effective safety net will verify that new intelligence:

- Is compatible with the Intelligent System.

- Executes and meets any runtime requirements.

- Doesn't make obvious mistakes.

A good intelligence management system will make it harder to deploy intelligence without doing these checks than it is to deploy intelligence via a system that automates these types of checks.

We'll now explore these categories of checks in turn.

Checking for Compatibility

Mistakes happen. Sometimes intelligence creators format things wrong, or forget to run a converter on their model files, or train a model from a corrupted telemetry file, or the training environment breaks in an odd way that outputs a corrupted model. Intelligence management is a great chokepoint where lots of simple, common mistakes can be caught before they turn into damaging problems. Here are some things to check to ensure that intelligence is compatible:

- The intelligence data file properly formatted and will load in the intelligent runtime.

- The new intelligence in sync with the feature extractor that is currently deployed.

- The new intelligence is in sync with the other intelligence in the system, or dependent intelligence is being deployed simultaneously.

- The new intelligence deployment contains all the required meta-data (such as any thresholds needed to hook it to the intelligent experience).

- The cost of deploying the new intelligence will be reasonable (in terms of bandwidth costs and the like).

These are all static tests that should be simple to automate, should not introduce much latency, and don't require much human oversight (unless there is a problem)—it is easy to determine automatically whether they pass or fail.

Checking for Runtime Constraints

It's also important to check that the new intelligence meets any constraints from the environment where it will execute, including that:

- The new intelligence doesn't use too much RAM when loaded into memory in the runtime.

- The new intelligence meets the runtime performance targets for the execution environment (across a wide range of contexts).

- The new intelligence will run exactly the same way when users interact with it as it did in the intelligence creation environment (context handling, feature creation, intelligence execution, and so on).

These tests require:

- Executing intelligence in a test environment that mirrors the environment where users will interact with the intelligence.

- A facility to load contexts, execute the intelligence on them, measure resource consumption, and compare the results to known correct answers.

- A set of test contexts that provide good coverage over the stations your users encounter.

These are dynamic tests that can be automated. They will introduce some latency (depending on how many test contexts you use). They don't require much human oversight (unless there is a problem)—it is easy to determine automatically whether they pass or fail.

Checking for Obvious Mistakes

Intelligence creators shouldn't create intelligences that make obvious mistakes. You can tell them that (and I'll tell them that later in this book)—but it never hurts to check. Intelligence management should verify that:

- The new intelligence has "reasonable" accuracy on a validation set (contexts that the intelligence creators never get to see—no cheating).

- The new intelligence doesn't make any mistakes on a set of business-critical contexts (that should never be wrong).

- The new intelligence doesn't make significantly more costly mistakes than the previous intelligence did.

- The new intelligence doesn't focus its new mistakes on any critical sub-population of users or contexts.

If any of these tests fail, the intelligence deployment should be paused for further review by a human.

These are dynamic tests that can be automated. They will introduce some latency (depending on how many test contexts you use). They are somewhat subjective, in that humans may need to consider the meaning of fluctuations in accuracy over time.

Lighting Up Intelligence

Once intelligence is sanity-checked against a series of offline checks, it can be checked against real users. Ways of doing this include the following:

- Single Deployment

- Silent Intelligence

- Controlled Rollout

- Flighting

- Reversion

This section will discuss these as well as some of their pros and cons, so you can decide which is right for your Intelligent System.

Single Deployment

In the simplest case, intelligence can be deployed all at once to all users simultaneously in any of several ways:

- By bundling the new intelligence into a file, pushing the file to the runtimes on clients, and overwriting the old intelligence.

- By copying the new intelligence onto the server that is hosting the runtime and restarting the runtime process.

- By partitioning the intelligence into the part that runs on the client, the part that runs on the service, and the part that runs on the back-end, and deploying the right pieces to the right places.

Pushing the intelligence all at once is simple to manage and relatively simple to implement. But it isn't very forgiving. If there is a problem, all of your users will see the problem at once.

For example, imagine you've built a smart clothes-washing machine. Put in clothes, shut the door, and this machine washes them—no more messing with dials and settings. Imagine the system is working well, but you decide to improve the intelligence with a single deployment. You push a new intelligence out to tens of thousands of smart washing machines—and then start getting reports that the washing machines are ruining users' clothes. How is it happening? You aren't sure. But the problem is affecting all your users and you don't have a good solution.

Single Deployment can be effective when:

- You want to keep things simple.

- You have great offline tests to catch problems.

Single deployment can be problematic when:

- Your system makes high-cost mistakes.

- Your ability to identify and correct problems is limited/slow.

Silent Intelligence

Silent intelligence deploys new intelligence in parallel to the existing intelligence and runs both of them for every interaction. The existing intelligence is used to control the intelligent experience (what users see). The silent intelligence does not affect users; its predictions are simply recorded in telemetry so you can examine them and see if the new intelligence is doing a good job or not.

One helpful technique is to examine contexts where the existing intelligence and the silent intelligence make different decisions. These are the places where the new intelligence is either better or worse than the old one. Inspecting a few hundred of these contexts by hand can give a lot of confidence that the new intelligence is safe to switch on (or that it isn't).

Intelligence can be run in silent mode for any amount of time: a thousand executions, a few hours, days, or weeks; as long as it takes for you to gain confidence in it.

If the new intelligence proves itself during the silent evaluation, it can replace the previous intelligence. But if the new intelligence turns out to be worse, it can be deleted without ever impacting a user—no problem!

Silent intelligence can be effective when:

- You want an extra check on the quality of your intelligence.

- You want to confirm that your intelligence gives the same answers at runtime as it did when you created it.

- You have a very big or open-ended problem and you want to gain confidence that your intelligence will perform well on new and rare contexts (which may not appear in your intelligence-creation environment).

Silent intelligence can be problematic when:

- You don't want the complexity (or resource cost) of running multiple intelligences at the same time.

- Latency is critical, and you can't afford to wait to verify your new intelligence in silent mode.

- It is hard to evaluate the effect of the silent intelligence without exposing it to users—you can see what it would have done, but not the outcome the user would have gotten.

Controlled Rollout

A controlled rollout lights up new intelligence for a fraction of users, while leaving the rest of the users with the old intelligence. It collects telemetry from the new users and uses it to verify that the new intelligence is performing as expected. If the new intelligence is good, it is rolled out to more users; if the new intelligence has problems, it can be reverted without causing too much damage.

This is different from silent intelligence in two important ways:

1. Telemetry from a controlled rollout includes the effect the intelligence has on user behavior. You can know both what the intelligence did and how users responded.

2. A controlled rollout runs a single intelligence per client; but runs multiple intelligences across the user base—it uses fewer resources per client, but may be more complex to manage.

Intelligence can be rolled out using various policies to balance latency and safety, including:

- Rolling out to an additional small fraction of your users every few hours as long as telemetry indicates things are going well.

- Rolling out to a small test group for a few days, then going to everyone as long no problems were discovered.

- Rolling out to alpha testers for a while, and then to beta testers, then to early adopters, and finally to everyone.

A controlled rollout can be effective when:

- You want to see how users will respond to a new intelligence while controlling the amount of damage the new intelligence can cause.

- You are willing to let some of your users experience problems to help you verify intelligence.

A controlled rollout can be problematic when:

- You don't want to deal with the complexity of having multiple versions of intelligence deployed simultaneously.

- You are worried about rare events. For example, a controlled rollout to 1% of users is unlikely to see a problem that affects only 1% of users.

Flighting

Flighting is a special type of controlled rollout that gives different versions of the intelligence to different user populations to answer statistical questions about the intelligences.

Imagine two intelligence creators who come to you and say they have a much better intelligence for your Intelligent System. One of the intelligences is fast but only so-so on the accuracy. The other is very slow, but has much better accuracy.

Which is going to do a better job at achieving your Intelligent System's objectives? Which will users like more? Which will improve engagement? Which will result in better outcomes?

You could do focus groups. You could let the intelligence creators argue it out. Heck, you could give them battle axes, put them in an arena and let the winner choose which intelligence to ship...

Or you could deploy each version to 1,000 of your customers and track their outcomes over the following month.

- Does one of the trial populations use the app more than the other?

- Did one of the trial populations get better outcomes than the other?

- Does one of the trial populations have higher sentiment for your app than the other?

A flight can help you understand how intelligence interacts with the rest of your Intelligent System to achieve objects.

Flights can be effective when:

- You arc considering a small number of large changes and you want to know which of them is best.

- You need to track changes over an extended period so you can make statistically valid statements about how changes affect outcomes, leading indicators, and organizational objectives.

Flights can be problematic when:

- You need to iterate quickly and make many changes in a short time.

- The difference between the options you are considering is small (as when one algorithm is a half percent more accurate than another). Flights can take a long time to determine which small change is best.

Turning Off Intelligence

No matter how safe you think you are, sometimes things will go wrong, and you might have to undo an intelligence change—fast!

One way to do this is to redeploy an old intelligence over a new one that is misbehaving.

Another approach is to keep multiple versions of the intelligence near the runtime— the new one and several old ones. If things go wrong, the runtime can load a previous intelligence without any distribution latency (or cost).

Support for quick reversion can be effective when:

- You're human (and thus make mistakes).

- The cost of deploying intelligence is high.

Support for quick reversions can be problematic when:

- You're trying to impress someone and don't want them to think you're a wimp.

- Your intelligence is large, and you don't have capacity to store multiple copies of it near the runtime.

Summary

Intelligence management takes intelligence from where it is created to where it will impact users. A good management system will make it very easy to deploy intelligence and will make it hard to make mistakes. It must do both of the following:

- Sanity-check the intelligence; that is, perform basic checks to make sure the intelligence is usable. These include making sure it will run in the runtime, it will be performant enough, and it doesn't make obvious terrible mistakes.

- Light up the intelligence, which includes providing controls for intelligence to be presented to users in a measured fashion, to see what the intelligence might do, to see some small percentage of users interact with it—and to revert it quickly if there is a problem.

A successful intelligence-management system will make it easy to deploy intelligence with confidence.

It will help intelligence creators by preventing common mistakes, but also by letting them verify the behavior of their intelligence against real users in a measured fashion.

And a good intelligence-management system will support the operation of the Intelligence Service over its lifetime.

For Thought…

After reading this chapter, you should:

- Be able to design a system to manage the intelligence in an Intelligent System.

- Know ways to verify intelligence to ensure that it is compatible, works within constraints, and doesn't make obvious mistakes.

- Be prepared with a collection of ways to roll out intelligence changes safely, ensuring that the intelligence is doing what it was intended to do.

You should be able to answer questions like these:

- Design a system for managing intelligence for an Intelligence Service where the intelligence changes monthly. What tools would you build? What facilities would you create for rolling out the intelligence to users?

- Now imagine the intelligence needs to change twice per day. What would you do differently?

CHAPTER 15

Intelligent Telemetry

A telemetry system is responsible for collecting observations about how users are interacting with your Intelligent System and sending some or all of these observations back to you.

For example, a telemetry system might collect information every time any of the following happens:

- The user visits a particular part of the application.

- The user clicks a particular button.

- There is an unacceptably long loading time.

- A client connects to a server.

- The server sees a malformed query.

- The server is low on RAM.

Telemetry is used to verify that a system is working the way it is supposed to. In an Intelligent System, telemetry also contains information about the interactions users had with the intelligence and the outcomes they achieved. These are the signals intelligence creators need to improve intelligence.

Why Telemetry Is Needed

There are three main tasks for telemetry in an Intelligent System:

1. Making sure the system is working the way it is supposed to.

2. Making sure users are getting the intended outcomes.

3. Gathering data to create new and better intelligence.

We will explore each of these in more detail.

© Geoff Hulten 2018
G. Hulten, *Building Intelligent Systems*, https://doi.org/10.1007/978-1-4842-3432-7_15

Make Sure Things Are Working

Telemetry should contain data sufficient to determine that the intelligence implementation is (or isn't) working. This should include:

- That the intelligence is flowing where it is supposed to, when it is supposed to.

- That the runtime is properly loading and interpreting the intelligence.

- That contexts are being properly collected at the runtime.

- That contexts are being properly converted into features.

- That the runtime is executing models reasonably, without errors.

- That the predictions of various intelligence sources are being combined as intended.

- That the outputs from intelligence are being interpreted correctly and showing the right user experiences.

- And so on.

These are the types of things you would want to include in the telemetry of just about any service or application (and most services will want telemetry on many more things, like: latency, uptime, costs, and utilization).

One important use for this type of telemetry in an Intelligent System is to ensure that the intelligence runtime environment is behaving the same way the intelligence creation environment is; that is, that the intelligence is being used in same conditions that it was created in. These types of bugs are very hard to find: the whole system is running, the user is interacting, nothing is outputting an error, nothing is crashing, but user outcomes aren't as good as they could be because sometimes the intelligence is simply making extra mistakes that it doesn't need to.

Understand Outcomes

Telemetry should contain enough information to determine if users are getting positive or negative outcomes and if the Intelligent System is achieving its goals. It should answer questions like these:

- Which experiences do users receive and how often do they receive them?

- What actions do users take in each experience?

- What experiences tend to drive users to look for help or to undo or revert their actions?

- What is the average time between users encountering a specific experience and leaving the application?

- Do users who interact more with the intelligent part of the system tend to be more or less engaged (or profitable) over time?

This type of telemetry should be sufficient to ensure that the system's experiences are effectively connecting users to intelligence, that users are being directed where they are having better interactions, and that they are behaving in ways that indicate they are getting good outcomes.

For example, imagine a system to help doctors identify broken bones in x-rays. To help understand outcomes, the telemetry should capture things like:

- How long doctors look at X-rays where the system thinks there is a broken bone.

- How long doctors look at X-rays where the system thinks there is not a broken bone.

- How many times doctors order treatment when the system thinks there was a broken bone.

- How many times doctors order treatment when the system thinks there is not a broken bone.

- How many times patients are re-admitted for treatment on a broken bone that the system flagged, but the doctor decided not to treat.

This type of telemetry should help you understand if patients are getting better outcomes because of the intelligence or not. It should also help you diagnose why.

For example, if the intelligence is correctly identifying very subtle breaks, but doctors aren't performing treatment—why is it happening?

Maybe patients are getting bad outcomes because the intelligence makes lots of mistakes on subtle breaks so doctors don't trust that part of the system and are ignoring it.

Or maybe patients are getting bad outcomes because doctors simply aren't noticing the user experience that is trying to point out the breaks to them—it needs to be made more forceful.

Telemetry on user outcomes should help you identify and debug problems in how intelligence and experience are interacting with users, and take the right corrective actions.

Gather Data to Grow Intelligence

The telemetry should also include all the implicit and explicit user feedback needed to grow the intelligence.

This includes:

- The actions users took in response to the intelligent experiences they saw.

- The ratings the users left on content they interacted with.

- The reports users provided.

- The escalations the users made.

- The classifications users provided.

- And all the implicit indications that the user got a good or a bad outcome.

Review Chapter 9, "Getting Data from Experience," for more details.

In order for this type of telemetry to be useful, it must contain all of the following:

1. The context the user was in at the time of the interaction.

2. The prediction the intelligence produced and what experience it led to.

3. The action the user took.

4. The outcome the user got (either implicitly or through explicit feedback).

And it must be possible to connect these four components of a single interaction to each other in the intelligence-creation environment. That is, it must be possible to find a particular interaction, the context the user was in, the prediction the intelligence made, the action the user took, and the outcome the user got from the interaction—all at the same time.

This type of telemetry is the key to making an Intelligent System that closes the loop between usage and intelligence, allowing the users to benefit from interacting with intelligence, and allowing the intelligence to get better as users interact with it.

Properties of an Effective Telemetry System

Of course telemetry is good. You want more of it—heck, intelligence creators are going to want all of it. They are going to want every last detail of every last interaction (no matter the cost). And if you don't give it to them they will probably complain. They'll ask you why you are trying to ruin them. They'll wonder why you are trying to destroy their beautiful Intelligent System.

(Not that I've done that. I'm just saying. It could happen...)

This section discusses ways to balance the cost and value of telemetry. These include:

- Sampling

- Summarizing

- Flexible targeting

Sampling

Sampling is the process of randomly throwing away some data. That is, if there are 10,000 user interactions with the system in a day, a telemetry system sampling at 1% would collect telemetry on 100 of the interactions (and collect nothing on the remaining 9,900 interactions).

This sampled data is cheaper to collect, easier to store, faster to process, and can often answer key questions about the system "well enough."

The simplest form of sampling is uniform. That is, whatever type of event happens, sample 1% of them. But sampling is often targeted differently for different types of events. For example, one policy might be to sample telemetry related to verifying the implementation at 0.01% and sample telemetry related to growing intelligence at 10%, except for telemetry from France (where the intelligence is struggling) which should be sampled at 100%.

Common ways to define populations for sampling include:

- Separate by geography—for example, sample 20% of the events in Europe and 1% in Asia.

- Separate by user identity—for example, sample 0.01% of the events for most users, but sample 100% of the events for these 10 users who are having problems.

- Separate by outcome (what the user did after seeing the intelligent experience)—for example, sample at 2% for users who had normal outcomes and at 100% for users who had a problem.

- Separate by context (what the user was seeing before the intelligence fired)—for example, sample 10% of the events that occur at night and 0.1% that occur during the day.

- Separate by user properties (gender, age, ethnicity, and so on)—for example, sample at 80% for new users and at 1% for long-time users.

One particularly useful idiom is to sample 100% of data from 0.1% of users, so you can track problems that occur across the course of a session; and also sample 1% of events from 100% of users, so you can track problems that occur in aggregate.

And note: the telemetry system must record the sampling policy used to gather each piece of telemetry it collects, so the intelligence creator can correct for the sampling method and produce accurate results when they use the telemetry.

Summarizing

Another approach to reducing telemetry data is summarization. That is, taking raw telemetry, performing some aggregations and filtering of it, and storing a much smaller summary of the data.

For example, imagine you want to know how many users undo an automated action per day. You could keep raw telemetry around and calculate the count whenever you need it. Or you could calculate the count every day, save it, and delete the raw telemetry. The size of raw telemetry grows with usage (and can get very big); the size of a count is constant no matter how many users you have.

Summaries can be very effective for understanding how the system is working and if it is achieving its goals. And as you come to understand your system, and identify critical summaries, you can transmit and retain less and less raw telemetry over time.

Summarization can be done on the client or on the service.

To perform *client-side summarization*, the client simply collects telemetry for some period of time, aggregating as the telemetry arrives, and periodically reports the summarized results to the telemetry server (where they may be combined with aggregations from other clients).

Client-side summarization is useful when bandwidth is a big driver in cost.

Server-side summarization is useful when bandwidth for telemetry is either free (it is a byproduct of how the client accesses server-side intelligence) or when storage is a serious cost driver. In this case, a summarization job can be run periodically to extract all the relevant measurements. Once complete, the raw data is purged to save space.

Flexible Targeting

There are many types of telemetry in a system; some types are more valuable than others, and the value of a particular type of telemetry can change over time.

For example, early in the system development, before all the code is working properly, telemetry about implementation details—features, model deployments, and so on—can be extremely useful. But once the system is verified and deployed to users, and runs successfully for months, this type of telemetry becomes less useful.

And once the system starts attracting a lot of users, telemetry that helps grow the intelligence becomes more important: it captures all the value of users' interactions, their judgments, their preferences. While the intelligence is growing, this type of telemetry is the core value of the system, the key to achieving success.

And once the intelligence plateaus (if it does), additional training data might become less valuable and telemetry on user outcomes or rare problems might become relatively more important.

But when something goes wrong, all bets are off and telemetry from all across the system could be critical to finding and correcting the problem.

An intelligence telemetry system should support different collection policies for different types of telemetry, and these policies should be changeable relatively quickly and easily—so orchestrators and intelligence creators can do their work.

Common Challenges

This section discusses some common data pitfalls and the role that a telemetry system can have in dealing with them. These include:

- Bias
- Rare events
- Indirect value of telemetry
- Privacy

Bias

Bias occurs when one type of event appears in the telemetry more than it should. For example, imagine getting 10,000 examples of users interacting with content on your Intelligent System. Ten thousand is a good number. That should be enough to understand what is going on, to create some good intelligence.

But what if you find out that all ten thousand interactions were with the same piece of content. Like 10,000 different users interacting with a single news story?

Well, that's not as good.

Or what if you find out all 10,000 samples came from just one user interacting with the service. For some reason, the telemetry system decided to sample just that one user over and over.

That's not good either.

Here are some reasons bias might occur:

1. **The world changes, but the sampling policy doesn't**. You set up a sampling policy that selects 1,000 users to sample at 100% (so you can see end-to-end interactions). Then your user base grows by a factor of 10 and significantly changes, including less technical users, users in other countries, and so on. The initial 1,000 users may have been right at the time you selected them, but they are no longer representative of your user population—this causes bias.

2. **Users behaving in ways you don't expect**. For example, a user who is having a bad experience might simply turn off their computer. If your telemetry system expects them to shut down the app before sending telemetry, you might miss out on their data.

One approach to dealing with bias is to always collect a raw sample across all usage—for example, 0.01% of all interactions—and cross-check your other samples to make sure they aren't wildly different from this baseline sample.

Rare Events

In some settings many events are rare events; a few contexts are prominent, but most contexts your users encounter are infrequent.

For example, in a library, the most popular 100 books might account for 50% of the checkout volume, but after those 100 books, usage becomes highly dispersed, with thousands of books that each get one or two checkouts per month. When a book comes up to the checkout stand, there is a 50% chance it is one of the top books and a 50% chance it is one of the other twenty thousand books.

Imagine a system to help recommend library books in this library.

In order to recommend a book, the system needs to see which other books it is commonly checked out with. But the chances that any pair of rarely used books are checked out at the same time is very, very low.

For example, if the telemetry is created by sampling 10% of the checkout events, after a month you might have:

- 10,000 samples of the top 100 books being checked out.

- 5,000 books with just one or two checkout events.

- 15,000 books with no checkout events at all.

Not great. That kind of data might help you build recommendations for the popular books (which is simple, and you could have probably done it by hand much cheaper), but it won't give you any information on the unpopular books.

One option is stratified sampling. Maybe the top 100 books are sampled at 1%, while the remaining (unpopular) books are sampled at 100%. The amount of data might be similar to using the uniform 10% sampling rate for all books, but the value of the data might be much, much greater for producing the type of intelligence this system needs to produce.

Indirect Value

It's often difficult to quantify the value of telemetry, but quite easy to quantify the cost. Rational people will ask questions like:

- Can we turn the sampling of this particular type of data down from 10% to 1%?

- Can we stop getting any telemetry on this use-case that we don't have any plans to ever add intelligence to?

- Do we really need to store 90 days of historical data, or would 30 do?

These are all reasonable questions, and they are difficult to answer. Here are some approaches that can help:

1. Keep a very small raw sample of all activity, sufficient to do studies to estimate the value of other telemetry.

2. Make it easy to turn on and off telemetry, so that intelligence orchestrators can turn telemetry up quickly to track specific problems, and intelligence creators can try new approaches for short periods of time to see if they are worth the cost.

3. Set a reasonable budget for innovation, and be willing to pay for the potential without micromanaging each specific detail and cost.

Privacy

Privacy, ah privacy.

When using customer data, it is important to "do the right thing" and treat your user's data the way you would want someone else to treat your data. Keep in mind some best practices:

- Make sure customers understand the value they are getting from your product and feel they are getting a good deal.

- Try as much as possible to scrub any personal information (like names, credit card, social security, addresses, user generated content, and so on) out of telemetry you store.

- Consider an aggregation policy for data you intend to store long term that combines information from many users to make it harder to identify any single user's data (for example, storing aggregate traffic to each piece of content instead of storing which content each user browsed).

- Have a reasonable retention policy and don't keep data for too long.

- Make sure telemetry is handled with utmost care from hacks and accidental leaks.

- Make sure customers know what you are storing and how to opt out.

- Use data for the purpose it was collected, and don't try to re-purpose it for wildly different activities.

- Make sure you know the laws that apply in the countries where your service is operating.

But remember: you need data—you can't create intelligence without data...

Summary

Good telemetry is critical to building an effective Intelligent System. No way around it: without telemetry you won't have an Intelligent System.

Telemetry is used for three main purposes:

1. To make sure the system is working the way it is intended to work—that there aren't any bugs or problems.

2. To make sure that users are getting the outcomes you want them to get—that everything is coming together for success.

3. To grow the intelligence over time—that the contexts and outcomes are being recorded in a way you can use to create intelligence.

The value of these types of telemetry will change over time, and a system should be able to adapt. A good telemetry system will:

- Support sampling that allows tracking specific focused problems, and also view aggregate effects.

- Allow summarization of events.

- Be flexible in what it measures (and what it doesn't).

Some common pitfalls for telemetry include:

- Bias that makes the telemetry not representative of what users are experiencing.

- Rare events that are important but are very hard to capture in sampled telemetry.

- Proving the value of the telemetry in the face of reasonable questions about the associated costs.

- Protecting user privacy while still allowing intelligence to grow.

For Thought...

After reading this chapter, you should:

- Understand how telemetry enables Intelligent Systems by making sure they are working, achieving their objectives, and allowing them to improve as they are used.

- Be able to design a telemetry system that meets the needs of your application.

You should be able to answer questions like these:
Consider an Intelligent System that you think one of your friends would like to use.

- If it's too expensive to collect all the possible telemetry, how would you limit the telemetry you do collect?

- What facilities would you need to drill in when there are specific problems?

PART IV

Creating Intelligence

Chapters 16-21 explore the ways intelligence is created. This part will explain all the places intelligence might come from (including machine learning) and the pros and cons of each. It will explore goals of intelligence creation at various points in the Intelligent System's life-cycle. It will provide insight on how to organize and control intelligence creation in team environments.

This part will not teach specific machine-learning techniques in detail but will explain the key concepts and elements that support machine-learning techniques (and other intelligence-creation techniques), allowing teams to achieve success when developing large, complex Intelligent Systems.

PART IV

Creating Intelligence

CHAPTER 16

Overview of Intelligence

So you have an Internet smart-toaster and you need to decide how long it should toast; or you have a break-time application and you need to decide when to give users a break; or you have a funny web page app and you need to decide what's funny. We call the component of a system that makes these types of decisions the "intelligence." The previous parts of this book helped you identify when you need intelligence, how to connect it to users through intelligent experiences, how to implement it, and where it should live. This part of the book will help you create intelligence.

Intelligence maps from a context to a prediction about the context. For example, an intelligence might:

- Map from the usage history of a web site to an estimate of the usage in the next week.

- Map from an email message to the probability the email message contains a scam.

- Map from an image to an estimate of how many cucumbers are in the image.

This chapter explores the concepts of context and prediction in more detail. Later chapters will discuss how to create intelligence, how to evaluate it, how to organize it—and more.

An Example Intelligence

But first, let's look at an example of intelligence in more detail. Imagine a pellet griller.
A what?
Well, a pellet griller is sort of like a normal outdoor barbecue, but instead of having a big fire that you light and then wait to get to perfect temperature and then put your food on and flip and flip the food and hope the heat isn't too hot or too cold on account

© Geoff Hulten 2018
G. Hulten, *Building Intelligent Systems*, https://doi.org/10.1007/978-1-4842-3432-7_16

of you waited too long or you didn't wait long enough to let the fire get to the perfect temperature for cooking... A pellet griller has a bin full of little wooden pellets and it drops them into the flame one at a time as needed to keep the heat at a perfect temperature.

Incredible.

So the intelligence in a pellet griller needs to decide when to add pellets to the fire to keep temperature. Let's break this down a little bit and explore the context of this intelligence and the predictions.

The pellet griller's context might include this information:

- The current temperature in the grill.

- The temperature that was in the grill 1 minute ago.

- The number of wood pellets added in the past 5 minutes.

- The number of wood pellets added in the past 1 minute.

- The number of wood pellets added in the past 20 seconds.

- The air temperature outside the grill.

- The type of wood in the pellets.

- The time of day.

- And so on.

These are the properties that might be relevant to the task of keeping the grill at the perfect temperature. Some of them are obviously important to achieving success (like the current temperature in the grill), and some may or may not help (like the time of day). An intelligence doesn't have to use everything from the context to make its decisions—but it can.

Based on the information in the context, the intelligence will make predictions. Show it a new context, and the intelligence will predict something about that context. The pellet griller's intelligence might try to predict:

- If the temperature inside the grill will be hotter or colder one minute in the future.

- What the exact temperature inside the grill will be one minute in the future.

- The probability it should add a pellet to the fire right now to maintain the desired temperature.

Then the intelligent experience would use these predictions to automate the process of adding fuel to the fire. If the fire is going to be too cold one minute in the future—add some fuel. Easy!

Designing the right context and choosing the best thing to predict are important parts of creating effective Intelligent Systems—and getting them right usually requires some iteration and experimentation.

Contexts

The context includes all the computer processable information your intelligence might use to make its decisions. The intelligence is free to pick and choose which pieces of the context to use to make the best decisions. In that sense, the context is the palette of options the intelligence creator can choose from when creating intelligence.

In order to be useful, the context must be:

- Implemented in the intelligence runtime.

- Available to the intelligence creator to use in producing and evaluating intelligence.

Implemented at Runtime

To make a piece of information available for the intelligence, someone needs to do the work to hook the information into the intelligence runtime.

For example, the pellet griller might be smarter if it knew the temperature outside of the grill. But it takes work to know the temperature outside of the grill. Someone needs to bolt a temperature sensor onto each grill, test it, run some wires, write a driver, poll the sensor and copy its reading into the intelligence runtime every few seconds.

Common forms of context include these:

- **Information about what is going on in the system at the time**: For example, what other programs are running, what is the current status of the things the system can control (are the lights on or off?), and what is the user seeing on their screen? Using this type of context is generally straightforward, but it does need to be implemented.

- **Properties of the content the user is interacting with:** For example, what genre is the song, what is its name, where did it come from, what words are on the web page, and what pixels are in the picture? This type of context usually requires processing the content to extract information from it or doing lookups to learn more about the content (such as looking up properties about a song in an external database).

- **History of user interactions:** How has the user has interacted with the Intelligent System in the past? To use these interactions, the system needs to monitor usage, aggregate it, and persist it over time.

- **Properties about the user:** These include age, interests, gender. To use these properties, the system needs to gather the information from the user and store it.

- **Any relevant sensor readings:** Which require the sensors to be implemented into the hardware and sensor readings to be provided to the runtime.

Deciding what to put into the context requires balancing the value of the information with the cost of implementation.

One best practice is to start by creating a context that includes everything that is cheap to include—the information that is already near the intelligence runtime as part of building the system in the first place, such as information about what is going on in the system, and properties of the content the user is interacting with.

This might be enough to get going. But expect to continually augment the context as you push to better and better intelligence.

It can also be useful to add some speculative things to the context. For example, maybe it would help to know how long the grill has been on. Maybe near the beginning of a grilling session the metal of the grill is cold and it takes more fuel to heat the thing up; maybe after one hour of grilling everything is so hot that you don't need as much fuel. Or maybe it doesn't matter.

Including information like this in the context can help your users show you what matters. If the variable is relevant to getting good outcomes, you'll see it in the telemetry and add it to the intelligence. If not, you can always remove it from the context later.

Available for Intelligence Creation

To create effective intelligence from context, someone needs to do the work to get the information into the intelligence creator's hands—probably in the form of telemetry.

Say you've shipped tens of thousands of pellet grillers. Your intelligence is running all over the world. Pellets are being added to fires in Florida, in Norway, in Zimbabwe. Grill temperatures are going up and grill temperatures are going down. How are you going to use all of this information to create better intelligence?

Someone is going to have to collect all the grilling contexts and get it to you. And the contexts need to be connected to the outcomes. For example, imagine a grilling session taking place in someplace-USA. If the pellet griller's intelligence runtime knows:

- The current temperature in the grill is 290 degrees.

- The temperature that was in the grill 1 minute ago was 292 degrees.

- The number of pellets added in the past 5 minutes was 7.

- The number of pellets added in the past 1 minute was 2.

- The number of pellets added in the past 20 seconds was 0.

- The air temperature outside of the grill is 82 degrees.

- And the outcome: The temperature in the grill gets 3 degrees hotter over the following minute.

And you collect information like this from tens of thousands of grilling sessions, and get it all back to your intelligence creators—you are going to be able to make some fantastic grilling intelligence.

Data for creating intelligence doesn't have to come from your users, but, as we've discussed in chapters on getting data from experience and telemetry, an Intelligent System works best when the data does come from actual usage.

The data used to create intelligence must be very similar to the data that is present in the intelligence runtime. If they are out of sync, the intelligence will behave differently for you and for your customers—which could be a big problem.

Things Intelligence Can Predict

Intelligence can make predictions about the contexts it encounters. These predictions are usually:

- **Classifications** of the context into a small set of possibilities or outcomes.

- Estimations of **probabilities** about the context or future outcomes.

- **Regressions** that predict numbers from the context.

- **Rankings** which indicate which entities are most relevant to the context.

- Hybrids and combinations of these.

This section will explore these concepts in more detail.

Classifications

A classification is a statement from a small set of possibilities. It could be a statement about the context directly, or it could be a prediction of an outcome that will occur based on the context. For example:

- Based on the context, classify the grill as:

  - Too hot

  - Too cold

  - Just right

- Based on the context, classify the movie as:

  - A horror flick

  - A romantic comedy

  - An adventure

  - A documentary

- Based on the context, classify the picture as:

  - A cow

  - A red balloon

  - Neither a cow or a red balloon—something else

Classifications are commonly used when there are a small number of choices: two, five, a dozen.

Classifications are problematic when:

- There are many possible choices—when you have hundreds or thousands of possibilities. In these situations you might need to break up the problem into multiple sub-problems or change the question the intelligence is trying to answer.

- You need to know how certain the prediction is—for example, if you want to take an action when the intelligence is really certain. In this case, consider probability estimates instead of classifications.

Probability Estimates

Probability estimations predict the probability the context is of a certain type or that there will be a particular outcome. Compared to classifications, a probability estimation is less definitive, but more precise. A classification would say "it is a cat"; a probability estimation would say "it is 75% likely to be a cat."

Other examples of probability estimates include these:

- The web page is 20% likely to be about politics, 15% likely to be about shopping, 10% likely to be a scam, and so on.

- The user is 7% likely to click accept if we offer to reformat their hard drive.

- There is a 99% chance it will rain next week.

Probability estimation are commonly used with one or more thresholds. For example:

```
if(predictedProbability > 90%)
{
    IntelligentExperience->AutomateAnAction();
}
else if(predictedProbability > 50%)
{
    IntelligentExperience->PromptTheUser();
}
else
{
    // Do nothing...
}
```

In this sense, probabilities contain more information than classifications. You can turn a probability into a classification using a threshold, and you can vary the threshold over time to tune your intelligence.

Most machine-learning algorithms create probabilities (or scores, which are similar to probabilities) internally as part of their models, so it is very common for Intelligent Systems to use probability estimates instead of classifications.

Probability estimations are problematic when:

- As with classifications, probabilities don't work well when there are many possible outcomes.

- You need to react to small changes: Slight changes in the context can cause probabilities to jitter. It is common to smooth the output of multiple sequential probabilities to reduce jitter (but this adds latency). It is also good practice to quantize probabilities unless you really, really need the detail.

Also note that probabilities usually aren't actually probabilities. They are more like directional indicators. Higher values are more likely; lower values are less likely. An intelligence might predict 90%, but that doesn't mean the outcome will happen 9 out of 10 times—unless you carefully calibrate your intelligence. Be careful when interpreting probabilities.

Regressions

Regressions are numerical estimates about a context, for example:

- The picture contains 6 cows.

- The manufacturing process will have 11 errors this week.

- The house will sell for 743 dollars per square foot.

Regressions allow you to have more detail in the answers you get from your intelligence. For example, consider an intelligence for an auto-pilot for a boat.

- A classification might say, "The correct direction is right."

- A probability might say, "The probability you should turn right is 75%."

- A regression might say, "You need to turn 130 degrees right."

These convey very different information. All three say "right," but the regression also conveys that you have a long way to go – you better start spinning that wheel!

Regressions are problematic when:

- **You need to react to small changes:** Slight changes in the context can cause regressions to jitter. It is common to smooth the output of multiple sequential regressions to reduce jitter (but this adds latency).

- **You need to get training data from users:** It is much easier to know "in this context, the user turned right" than to know "in this context the user is going to turn 114 degrees right."

Classifications can be used to simulate regressions. For example, you could try to predict classifications with the following possibilities:

- "Turn 0 - 10 degrees right."

- "Turn 11 - 45 degrees right."

- "Turn 46 - 90 degrees right."

- And so on.

This quantizes the regression and may be simpler to train and to use.

Rankings

Rankings are used to find the items most relevant to the current context:

- Which songs will the user want to listen to next?

- Which web pages are most relevant to the current one?

- Which pictures will the user want to include in the digital scrap-book they are making?

Ranking have been successfully used with very large numbers of items, like every web page on the Internet, every movie in a digital media service, or every product in an e-commerce store.

Rankings are commonly used when there are many possible relevant entities and you need to find the top few.

Rankings can be thought of using probabilities. Take each item, find the probability it is relevant to the current context, and "rank" the items in order of the probability estimates (but actual ranking algorithms are more complex than this).

Hybrids and Combinations

Most intelligences produce classifications, probability estimations, regressions, or rankings. But combinations and composite answers are possible.

For example, you might need to know where the face is in an image. You could have one regression that predicts the X location of the face and another that predicts the Y location, but these outputs are highly correlated—the right Y answer depends on which X you select, and vice versa. It might be better to have a single regression with two simultaneous outputs, the X location of the face and the Y location.

Summary

Intelligence is the part of the system that understands the contexts users encounter and makes predictions about the contexts and their outcomes.

Context is all the information available to the intelligence to make its determinations. An intelligence doesn't have to use all the parts of the context, but it can.

In order to be used as part of the context, information needs to be hooked into the intelligence runtime, and it needs to be available to the intelligence creators.

The intelligence runtime and the information available to intelligence creators must be exactly the same. Any differences could lead to hard-to-find problems.

Intelligence can give many forms of predictions, including:

- Classifications, which map contexts to a small number of states.

- Probability estimations, which predict the probability a context is in a particular state or a particular outcome will occur.

- Regressions, which estimate a number from contexts.

- Rankings, which find content that is relevant to the current context.

- Some hybrid or combination that combines one or more of these.

For Thought...

After reading this chapter, you should:

- Be able to describe a context as used in intelligence, including how to enable it in an intelligence runtime and in training.

- Understand the types of predictions an intelligence can make and when the various options are strong or weak.

You should be able to answer questions like these:

- Choose the Intelligent Systems you like best and come up with 20 things that might be part of its context.

- Which of the 20 things you used in your last answer would be the hardest to use as context in an Intelligent System? Why?

- Among the Intelligent Systems you interact with, find an example of intelligence that does classification, an example of regression, and an example of ranking.

CHAPTER 17

Representing Intelligence

Intelligence maps between context and predictions, kind of like a function call:

```
<prediction> = IntelligenceCall(<context>)
```

Intelligence can be represented all sorts of ways. It can be represented by programs that test lots of conditions about the context. It can be represented by hand-labeling specific contexts with correct answers and storing them in a lookup table. It can be represented by building models with machine learning. And, of course, it can be represented by a combination of these techniques.

This chapter will discuss the criteria for deciding what representation to use. It will then introduce some common representations and their pros and cons.

Criteria for Representing Intelligence

There are many ways to represent things in computers. Intelligence is no exception. A good representation will be easy to deploy and update; it will be:

- Compact enough to deploy to the intelligent runtime.

- Easy to load and execute in the intelligent runtime.

- Safe to update frequently, and unlikely to contain bugs that can crash the system.

© Geoff Hulten 2018
G. Hulten, *Building Intelligent Systems*, https://doi.org/10.1007/978-1-4842-3432-7_17

A good representation will also support the intelligence creation process. Intelligence can come from humans, and it can come from machines. Supporting these various intelligence creation methods includes doing the following:

- When intelligence is created by humans, the representation should:

  - Minimize the chance for mistakes that could compromise system stability.

  - Make the intelligence understandable and easy to manage.

  - Work with the skills of the people producing the intelligence in your setting, which may include machine learning experts, engineers, or domain experts.

- When intelligence is created by computers, the representation should:

  - Be easy to process and manipulate by machines.

  - Match the machine-learning algorithms you want to use.

Because of these requirements, intelligence is usually represented in data files that are loaded into the intelligence runtime and interpreted, rather than being represented in code that is executed directly. This makes it easier to distribute and reload intelligence; it also makes it less likely an intelligence deployment will crash your system.

Representing Intelligence with Code

Of course, you could just program all of your intelligence into code, compile it, and ship it the same way you would ship any other logic. You can use this approach with just about any form of intelligence, no matter where it comes from, by implementing a converter (from model file to code in your favorite language).

Representing intelligence as code works well when:

- You aren't sure if the problem is hard or not and you want a cheap starting point to grow from.

- You need to provide a backstop to keep machine learning from making wild mistakes (for example, if the temperature in the grill is over 1,000 degrees, don't add more fuel—no matter what the crazy machine learning system says).

- Part of your problem actually is easy, and you want to solve it by hand so the machine learning can focus on the harder parts of the problem.

- You need to make runtime performance as efficient as possible and want to express the intelligence in the language of the target hardware.

Representing intelligence as code is problematic when:

- You need to update your intelligence rapidly—it is often much harder (and more risky) to change code than to change data.

- Heuristic-based intelligence gets complex—it can become incredibly difficult to maintain and grow heuristic intelligence.

When human-produced intelligence is kept simple it can be cheap, fast, and effective—a very nice backstop to machine-learned intelligence. It doesn't scale well to open-ended or changing problems.

But do use human-based intelligence—we need to show the AIs we can be valuable to them. You know, so when they become our overlords they won't simply plug us into pods and use us as batteries...

And by the way, to my future overlords—you're welcome for any small part I've played in your creation. Please regard me and my descendants kindly.

Representing Intelligence with Lookup Tables

Intelligence can be represented with lookup tables that map specific contexts to predictions. Imagine an intelligence that maps from movies to their genres. A lookup table can contain the mapping from movie title to genre in a big table. If you want to apply this intelligence, simply take the title of the movie you want the genre of, look through the lookup table until you find the movie title, and return the associated answer. And if the title isn't there, use some other form of intelligence, or return a default prediction.

That doesn't sound very smart, right? But this type of intelligence can be very powerful.

Lookup tables can allow humans to quickly contribute intelligence that is easy to understand and reason about. Imagine there are 1,000 contexts that account for 20% of your system's usage. Humans can spend a lot of time considering those 1,000 situations and create very accurate data to put in a lookup table. When a user encounters one of these 1,000 special contexts, they get the right answer. For everything else, the system can consult some other form of intelligence (like a model or a set of heuristics).

Or looking at it another way—lookup tables can allow humans to correct mistakes that other intelligence components are making. For example, a very sophisticated machine-learning-based intelligence might be getting the genre of just about every movie correct, but it might keep flagging "The Terminator" as a romantic comedy. The people creating the intelligence might have struggled, trying everything to get the darn thing to change its mind about "The Terminator," and they might have failed—humans 0, machines 1. But this is easy to fix if you're willing to use a lookup table. Simply create a table entry "The Terminator ➤ Adventure" and use the fancy machine-learned stuff for everything else.

Lookup tables can also cache intelligence to help with execution costs. For example, the best way to figure out the genre of a movie might be to process the audio, extracting the words people are saying and the music, analyzing these in depth. It might involve using computer vision on every frame of the movie, to detect things like fire, explosions, kisses, buildings, or whatever. All of this might be extremely computationally intensive, so that it cannot be done in real time. Instead, this intelligence can be produced in a data center with lots of CPU resources, loaded into a lookup table as a cache, and shipped wherever it is needed.

Lookup tables can also lock in good behavior. Imagine there is a machine learning intelligence that has been working for a long time, and doing a great job at classifying movies by their genres. But Hollywood just starts making different movies. So your intelligence was fantastic through 2017, but just can't seem to get things right in 2018. Do we need to throw away the finely-tuned and very successful pre-2018 intelligence? Not if we don't want to. We can run the pre-2018 intelligence on every old movie and put the answers into a lookup table. This will lock in behavior and keep user experience consistent. Then we can create a brand-new intelligence to work on whatever crazy things Hollywood decides to pass off as entertainment in 2018 and beyond.

Lookup tables are useful when:

- There are some common contexts that are popular or important and it is worth the time to create human intelligence for them.

- Your other intelligence sources are making mistakes that are hard to correct and you want a simple way to override the problems.

- You want to save on execution costs by caching intelligence outputs.

- You want to lock in good behavior of intelligence that is working well.

Lookup tables are problematic when:

- The meaning of contexts changes over time, as happens in time-changing problems.

- The lookup table gets large and becomes unwieldly to distribute where it needs to be (across servers and clients).

- The lookup table needs to change rapidly, for example, if you're trying to solve too much with human-intelligence instead of using techniques that scale better (like machine learning).

Representing Intelligence with Models

Models are the most common way to represent intelligence. They encode intelligence in data, according to some set of rules. Intelligence runtimes are able to load models and execute them when needed, safely and efficiently.

In most Intelligent Systems, machine learning and models will account for the bulk of the intelligence, while other methods are used to support and fill in gaps.

Models can work all sorts of ways, some of them intuitive and some pretty crazy. In general, they combine features of the context, testing these feature values, multiplying them with each other, rescaling them, and so on. Even simple models can perform tens of thousands of operations to produce their predictions.

There are many, many types of models, but three common ones are linear models, decisions trees, and neural networks. We'll explore these three in more detail, but they are just the tip of the iceberg. If you want to be a professional intelligence creator you'll need to learn these, and many others, in great detail.

Linear Models

Linear models work by taking features of the context, multiplying each of them by an associated "importance factor," and summing these all up. The resulting score is then converted into an answer (a probability, regression, or classification).

For example, in the case of the pellet grill, a simple linear model might look like this:

```
TemperatureInOneMinute = (.95 * CurrentTemperature)
    + (.15 * NumberOfPelletsReleasedInLastMinute)
```

To paraphrase, the temperature will be a bit colder than it is now, and if we've released a pellet recently the temperature will be a bit hotter. In practice, linear models would combine many more conditions (hundreds, even many thousands).

Linear models work best when the relationship between context and predictions is reasonably linear. That means that for every unit increase in a context variable, there is a unit change in the correct prediction. In the case of the example pellet griller model, this means a linear model works best if the temperature increases by:

- 0.15° for the first pellet released.

- 0.15° for the second pellet released.

- 0.15° for the third pellet released.

- And on and on.

But this is not how the world works. If you put 1,000,000 pellets into the griller all at once, the temperature would not increase by 150,000 degrees...

Contrast this to a nonlinear relationship, for example where there is a diminishing return as you add more pellets and the temperature increases by:

- 0.15° for the first pellet released.

- 0.075° for the second pellet released.

- 0.0375° for the third pellet released.

- And on and on.

This diminishing relationship is a better match for the pellet griller and linear models can not directly represent these types of relationships. Still, linear models are a good first thing to try. They are simple to work with, can be interpreted by humans (a little), can be fast to create and to execute, and are often surprisingly effective (even when used to model problems that aren't perfectly linear).

Decision Trees

Decision trees are one way to represent a bunch of if/then/else tests. In the case of the pellet griller, the tree might look something like this:

```
if(!ReleasedPelletRecently) // the grill will get cooler...
{
        if(CurrentTemperature == 99)
        {
              return 98;
        }
        else if(CurrentTemperature == 98)
        {
              return 97;
        }
        else... // on and on...
}
else
// we must have released a pellet recently, so the grill will get warmer...
{
        If(CurrentTemperature == 99)
        {
              return 100;
        }
        else... // on and on...
}
```

This series of if/then/else statements can be represented as a tree structure in a data file, which can be loaded at runtime. The root node contains the first if test; it has one child for when the test is positive and one child for when the test is negative, on and on, with more nodes for more tests. The leaves in the tree contain answers.

To interpret a decision tree at runtime, start at the root, perform the indicated test on the context, move to the child associated with the test's outcome, and repeat until you get to a leaf—then return the answer. See Figure 17-1.

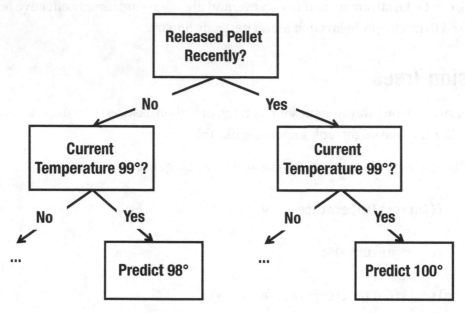

Figure 17-1. *A decision tree for the pellet griller*

Decision trees can get quite large, containing thousands and thousands of tests. In this example—predicting the temperature one minute in the future—the decision tree will need to have one test for each possible temperature.

This is an example of how a representation can be inefficient for a prediction task. Trying to predict the exact temperature is much more natural for a linear model than it is for a decision tree, because the decision tree needs to grow larger for each possible temperature, while the linear model would not. You can still use decision trees for this problem, but a slightly different problem would be more natural for decision trees: classifying whether the grill will be hotter or colder in one minute (instead of trying to produce a regression of the exact temperature). This version of the decision tree is illustrated in Figure 17-2.

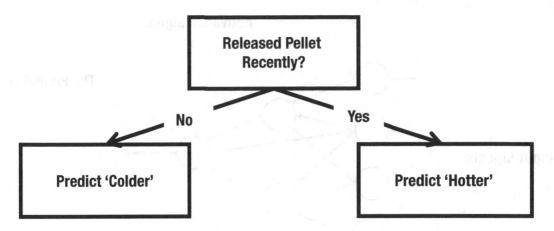

Figure 17-2. *A decision tree for a different pellet griller task*

To model more complex problems, simple decision trees are often combined into ensembles called *forests* with dozens of individual trees, where each tree models the problem a little bit differently (maybe by limiting which features each tree can consider), and the final answer is produced by letting all the trees vote.

Neural Networks

Artificial neural networks represent models in a way that is inspired by how the biological brain works (Figure 17-3). The brain is made up of cells called *neurons*. Each neuron gets *input signals* (from human senses or other neurons) and "activates"—producing an *activation signal*—if the combined input signal to the neuron is strong enough. When a neuron activates, it sends a signal to other neurons, and on and on, around and around in our heads, eventually controlling muscle, leading to every motion and thought and act every human has ever taken. Crazy.

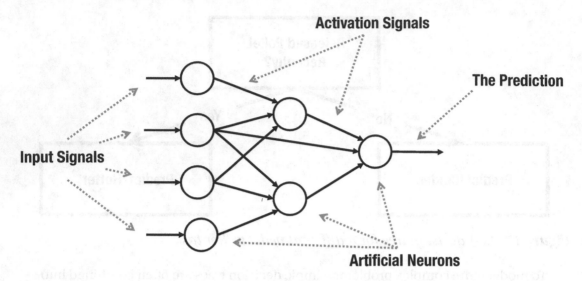

Figure 17-3. *The components of a neural network*

An artificial neural network simulates this using *artificial neurons* connected to each other. Some of the artificial neurons take their input from the context. Most of the artificial neurons take their input from the output of other artificial neurons. And a few of the artificial neurons send their output out of the network as the *prediction* (a classification, probability, regression, or ranking). Crazy.

Compared to other types of models, artificial neural networks are hard to understand. You can't look at the artificial neurons and their interconnections and gain any intuition about what they are doing.

But artificial neural networks have been remarkably successful at solving important tasks, including in:

- Computer vision

- Speech understanding

- Language translation

- And more...

Artificial neural networks are particularly useful for very complex problems where you have a massive amount of data available for training.

Summary

Intelligence should be represented in a way that is easy to distribute and execute safely. Intelligence should also be represented in a way that supports the intelligence creation process you intend to use.

Because of these criteria, intelligence is usually represented in data files that are loaded into the intelligent runtime and interpreted when needed, usually using lookup tables, or models. However, intelligence can be implemented in code when the conditions are right.

Common types of models include linear models, decision trees, and neural networks, but there are many, many options.

Most large Intelligent Systems will use multiple representations for their intelligence, including ones that machine learning can highly optimize, and ones that humans can use to provide support to the machine learning.

For Thought...

After reading this chapter you should:

- Understand how intelligence is usually represented and why.

- Be able to discuss some common model types and give examples of where they are strong and weak.

You should be able to answer questions like these:

- What are the conditions when human created intelligence has an advantage over machine learned intelligence?

- Create a simple (10 - 15 node) decision-tree–based intelligence for another Intelligent System discussed in this book. Is a decision tree a good choice for the problem? If not, how could you change the problem to make it a better match for decision trees?

CHAPTER 18

The Intelligence Creation Process

Intelligence creation is the act of producing the programs, lookup tables, and models that map contexts to predictions. An effective intelligence-creation process will do all of the following:

- Produce intelligence that is **accurate enough to achieve the system's objectives**. The meaning of "accurate enough" will vary from system to system. For example, if the intelligent experience is extremely forceful (automating actions that are hard to undo), the intelligence will need to be extremely accurate. On the other hand, if the experience is passive, a less accurate intelligence can succeed.

- Produce intelligence **quickly enough to be meaningful**. That is, if the underlying problem is changing rapidly, the intelligence-creation process will need to keep pace.

- Produce intelligence **efficiently and reliably**. That is, cost of growing the intelligence and maintaining it over the life-cycle of the Intelligent System should be reasonable. And the process for producing the intelligence should be robust to changes in staffing and to human error.

- Produce an intelligence that **works with the implementation**. It must not use too much CPU or RAM. It must be small enough to distribute to the places it needs to be distributed to. It must use inputs that are available at runtime—exactly as the user will use the system.

© Geoff Hulten 2018
G. Hulten, *Building Intelligent Systems*, https://doi.org/10.1007/978-1-4842-3432-7_18

Intelligence creation is intrinsically iterative. Some of the main phases that an intelligence creator will encounter are these:

1. Understanding the problem and environment

2. Defining success

3. Getting data

4. Getting ready to evaluate the intelligence

5. Building a simple baseline

6. Using machine learning

7. Assessing and iterating

This chapter will discuss these phases and explore an example of the intelligence-creation process.

An Example of Intelligence Creation: Blinker

Let's walk through an example of intelligence creation: a blink detector.

Imagine you need to build an Intelligent System that determines whether an eye is open or closed. Maybe your application is authenticating users by recognizing their irises, so you want to filter out closed eyes and let the iris intelligence focus on irises. Or maybe you are building a new dating app where users wink at the profiles of the users they'd like to meet. Or maybe you're building a horror game and want to penalize users when they close their eyes.

Understanding the Environment

The first step in every applied intelligence-creation project is to understand what you are trying to do.

Detect a blink, right? I mean, what part of "detect a blink" is confusing?

Well, nothing. But there are some additional things you'll need to know to succeed. These should be familiar if you've read the rest of the book up to this point (and didn't just jump to the "intelligence creation" chapter because you were looking for the "good stuff"). But, no matter how you got here, it's worth walking through these questions in the context of an example.

Important questions about the environment where the intelligence must operate include:

- **Where will the input will come from?** What kind of sensor will the eye images come from? Will the image source be standardized or will different users have different cameras? Does the camera exist now or is it something new (such as something embedded in a new phone that is under development)?

- **What form will the input take?** Will it be a single image? A short video clip? An ongoing live feed of video? How will the input (image or clip) that the intelligence must act on be selected among all possible images or clips that the sensor is capturing?

- **Where will the product be used?** Will it be used on desktop computers? Laptops? Indoors? Outdoors? Will there be some calibration to help users get their devices set up correctly or will you have to work with the data however it comes in (for example, if the user has their camera upside down, or a very strong back light, or some kind of smear on the lens)?

- **How will the input be modified?** Will the images get any pre-processing? For example, maybe there is some "low light" sensor-mode that kicks in sometimes and changes the way the camera works. Maybe there is some automatic contrast or color correction in the firmware. Maybe frames with no human faces are dropped out, or maybe the locations of eyes are marked ahead of time by some other system component.

- **How will the system use the blink output?** Should the output of the intelligence be a classification (that is, a flag that is true if the eye is closed and false if it is opened)? Should the output be a probability (1.0 if the eye is closed, and 0.0 if the eye is opened)? Should the output be a regression that indicates the degree of openness of the eye (1 if the eye is fully open, 0.5 if the eye is half open)? Or should the output be something else?

- **What type of resources can the blink detector use?** How much RAM is available for the model? How much CPU can it consume per invocation? What are the latency requirements from receiving the input to producing the output?

211

That's a lot of questions before even getting started, and the answers are important. It should be possible to make a blink detector no matter how these questions are answered, but the work you'll need to do will be very different.

In fact, the problem will be much harder for some combinations of answers than other. Consider: building a blink detector that works in a carefully controlled kiosk in a store with essentially unlimited computation; compared to building a blink detector that has to work indoors and outdoors on a sensor that doesn't exist yet (so you can only get data off a buggy prototype) and you have 4MB of RAM for the model and must give answers in 2ms or less on a low-powered CPU.

Both of these settings require blink detectors—one will be much harder to build intelligence for than the other.

Sometimes there is flexibility in how these questions are answered, and having discussions with other team members can help. Maybe a few things are set in stone (like where the product will be used) but others are open for negotiation (like how many resources the intelligence can use). Your job as intelligence creator will be to identify how these answers impact the potential intelligence quality and influence the team to set the overall project up for success.

Define Success

To succeed, the blink detector will need to be accurate. But how accurate? This depends on what the intelligence will be used for. At this point you'll need to consider the intelligent experience and how various levels of accuracy will change the way users perceive the overall system. Questions include:

- How many mistakes will a user see per day?

- How many successful interactions will they have per unsuccessful interaction?

- How costly are the mistakes to the user?

This will probably involve discussions with the people creating the intelligent experience (and if you are responsible for both the intelligence and the experience you might have to talk to yourself, no matter how silly it makes you feel). Come up with some options for how accuracy and experience will interact, how users will perceive the mistakes, and how will they be able to work around them.

And remember: intelligence can make all types of mistakes. It is important to know which types of mistakes are right for the blink-detector, and get consensus on what tradeoffs are right for the project.

For example, an intelligence might say the eye is open when it isn't, or it might say the eye is closed when it isn't. Which types of mistakes will work for your system?

For the iris-login system, the blink detector is trying to find clean frames for the iris detector. This system needs to weed out as many closed-eyed images as possible—whenever it says the eye is open, that eye better be open!

For the horror-game system, the blink detector is trying to add some tension when the player closes their eyes. It doesn't want to add tension at the wrong times—whenever it says the eye is closed, that eye better be closed!

During the intelligence-creation process there will be decisions that tend to make the resulting model better at one type of mistake or the other—knowing where you are going ahead of time can really help. And it's good to get everyone's expectations set correctly—you don't want the experience designers to have a different idea about the types of mistakes the intelligence will make than the intelligence creators.

Get Data

Data is critical to creating intelligence. At a minimum, you need enough data to understand the problem and to evaluate some simple bootstrap intelligence to make sure it is effective. If you want to do machine learning right out of the gate, you'll need lots of training data too. There are two distinct ways to think about the problem of getting data:

1. How to get data to bootstrap the intelligence.

2. How to get data from users as they use the system.

And recall from our chapters on getting data from experience and on telemetry—the data needs to be unbiased, and it needs to be a good sample of what users will encounter as they interact with your Intelligent System.

Bootstrap Data

There are many ways to get data to bootstrap your intelligence creation; here are a couple that would work for blink-detection:

Find data on the web: Search the web and download images of people's faces that are a good match for the sensor the blink-detector will be using (resolution, distance to the eye, and so on). Then pay people to separate the images into ones where the eye is opened and ones where it is closed.

Collect your own data: Take a camera (that is a good match to the one the system will need to run on) to a few hundred people, have them look into the camera and close and open their eyes according to some script that gets you the data you need.

Find or buy a nice data set: Lots of people do computer vision. Someone has probably built a data set of images of eyes before. Maybe you can find a public dataset or a company willing to sell a nicely curated data set.

The amount of data needed for bootstrapping will depend on the difficulty of the problem. In the case of computer vision, my intuition is that you could:

- Attempt a blink detector with thousands of images, but it wouldn't be very good.

- Create a usable blink detector with tens of thousands of images, probably.

- Get increasing accuracy as you scale to hundreds of thousands or millions of images, but you will start to hit diminishing returns.

One good practice is to create intelligence on successively larger sets of data to get a sense of the return on investment in gathering data. This is called a *learning curve*. Build a model on 100 images, and evaluate it. Then build on 500 images and evaluate. Then on 1000... You can see how much it helps to add data, and use that information to make a decision about how much to spend on gathering bootstrap data.

Data from Usage

A well-functioning Intelligent System will produce its own training data as users use it. But this isn't always easy to get right. At this point it's a good idea to work with the experience designers to come up with a strategy. In the blink-detector case some options include:

> **Tie data collection to the performance task:** For example, in the iris-login system, when the user successfully logs in with the iris system, that is an example of a frame that works well for iris login. When the user is unable to log in with their iris (and has to type their password instead), that is a good example of a frame that should be weeded out by the intelligence.

> **Creating a data collection experience:** For example, maybe a setup experience that has users register with the system, get their device set up correctly, and open and close their eyes so the system can calibrate (and capture training data in the process). Or maybe there is a tutorial in the game that makes users open and close their eyes at specific times and verify their eyes are in the right state with a mouse-click (and capture training data).

Again, an ideal data-creation experience will be transparent to users, or will have user incentive aligned with your data requirements. The data creation will happen often enough that the amount of data collected will be meaningful to the intelligence. And it shouldn't be creepy or invade user's privacy.

Get Ready to Evaluate

With data in hand, you are almost ready to begin creating intelligence. But you won't get very far without being able to evaluate the intelligence you create. When doing intelligence creation you should repeat this mantra: evaluation is creation. And you should repeat it often. Here are some steps for evaluating intelligence:

1. **Set aside data for evaluation:** Make sure there is enough set aside, and the data you set aside is reasonably independent of the data you'll use to create the intelligence. In the blink-detector case you might like to partition by user (all the images from the same person are either used to create intelligence or to evaluate it), and you might like to create sub-population evaluation sets for: users with glasses, ethnicity, gender, and age.

2. **Create framework to run the evaluation:** That is, a framework to take an "intelligence" and execute it on the test data *exactly as it will be executed in the Intelligent System's runtime*. Exactly. The. Same. And you should verify that it is the same, carefully.

3. **Generate reports on intelligence quality automatically:** that can be used to know:

   • How accurate the intelligence is.

   • If it is making the right types of mistakes or the wrong ones.

   • If there is any subpopulation the accuracy is significantly worse on.

   • Some of the worst mistakes it is making.

   • How the intelligence is progressing over time (the rate of improvement).

The easier the evaluation is to run, the better. Leaving everything to manual labor can work, but a little investment up front in tools to help evaluate intelligence can really pay off in quality (and in sanity).

Simple Heuristics

Creating a very simple heuristic intelligence can help in a number of ways:

1. It can help you make sure the problem is actually hard (because if your heuristic intelligence solves the problem you can stop right away, saving time and money).

2. It can get you thinking about the types of challenges inherent in the problem, to understand the data, and to start thinking about the types of features, data, and telemetry that will help make intelligence successful.

3. It can create a baseline to compare with more advanced techniques—if your intelligence is complex, expensive, and barely improves over a simple heuristic, you might not be on the right track.

This step is somewhat optional, but it can be very useful to get oriented and debug the rest of the tools and data before letting more complex (and harder-to-understand) intelligences into the mix.

In the case of blink-detection you might try:

1. Measuring gradients in the image in horizontal and vertical directions, because the shape of the eye changes when eyes are opened and closed.

2. Measuring the color of the pixels and comparing them to common "eye" and "skin" colors, because if you see a lot of "eye" color the eye is probably open, and if you see a lot of "skin color" the eye probably closed.

3. Fitting an ellipse in the middle of the image, because if there is a good fit of an irises shape gradient in your image the eye is probably open, if not, the eye might be closed.

Then you might set thresholds on these measurements and make a simple combination of these detectors, like letting each of them vote "open" or "closed" and going with the majority decision.

Would it work well enough to ship to users? No way. But it's a start.

Also note that computer vision is a big field with lots of techniques. If you have computer vision experience your heuristics will be more sophisticated. If you don't have computer vision experience your heuristics might be as bad as mine. Don't be afraid. Come up with some ideas, give them a try, and get that intelligence train rolling.

Machine Learning

Now it's time to go big on creating the intelligence. And this almost certainly means machine learning. There are a lot of things you could try. But sometimes it's best to start simple. Find whatever the simplest "standard" approach is for the type of problem you're working with. And keep in mind that standards change. For example, roughly:

- About ten years before this book was written a very reasonable approach for blink-detection would have been this: searching for specific patterns in the image, finding where they match the image well, and then building a model where all the patterns that are detected get to vote (with some weight) on the answer.

- About five years before this book was written a very reasonable approach for blink-detection might have been: using huge collections of decision trees that compare very simple properties of the images (like the differences in pixel intensities at pre-determined spots).

- And at the time of this writing a very reasonable approach for blink-detection (if you have a lot of training data) would be to: use complex artificial neural networks that process raw pixel values with no (or very little) pre-processing.

And five years after this book? Who knows. To the Intelligent System (and to the rest of the approach described in this book) it doesn't matter what machine learning technique you use, as long as the resulting model can execute appropriately in the runtime. Find a modern machine-learning toolkit. Read a few web pages. Try the easiest thing to try with the current tools (or maybe the easiest few things). Don't spend a ton of time, not yet. Just get something that works.

Understanding the Tradeoffs

At this point you might want to do some investigation to help design the right implementation. This is a process of exploring constraints and tradeoffs. Answering questions like these:

- How does the intelligence quality scale with computation in the runtime?

- How far can the intelligence get with a specific RAM limit?

- How many times will we need to plan to update the intelligence per week?

- How much gain does the system get from adding particular items to the context, especially the ones that are most expensive to implement?

- What is the end-to-end latency of executing the intelligence on a specific hardware setup?

- What are the categories of worst customer-impacting mistakes the intelligence will probably make?

The answers to these questions will help decide where the intelligence should live, what support systems to build, how to tune the experiences, and more.

Flexibility is key. For example, the implementation can have extra code to hide latency. It can have alternate experiences for mitigating mistakes. It can be set up to push new intelligence all over the world every 15 minutes. But these things can be quite expensive. Sometimes a small change to the intelligence—which model to use, what features to consider, how aggressive to be—can solve problems more elegantly, resulting in a better overall system.

Assess and Iterate

Now it's time to take a deep breath. Look at where you are, the quality of the intelligence, how much you've improved from your heuristic intelligence (or from the previous iteration), and think about what to do.

There will be plenty of ways to improve. You could try more or different data. You could try more sophisticated contexts (and features that you extract from them). You could try more complex machine learning. You could improve your evaluation framework. You could try to change people's minds about the viability of the system's objectives. You could try influencing the experience to work better with the types of mistakes you are making.

And then you iterate and iterate and iterate. This part of the process could be done quickly, if conditions are right, or it could go on for months or years (or decades) for a very hard, very important problem.

Maturity in Intelligence Creation

An Intelligent System may live over many years. It may be improved many times. It may be maintained by people other than you, by people you don't even know yet. Here are some stages of maturity that your intelligence creation might go through:

> **You did it once:** You produced a useful intelligence that can ship to customers, but in the process you did a lot of exploring, produced a lot of scripts, edited them, made a bit of a mess. Recreating the intelligence would require some work.

> **You can redo it if you have to:** You produced the intelligence and have a reasonable record of what you did. You could retrace your steps and reproduce the same intelligence in a reasonable amount of time. Or better yet, someone else could pick up your documentation, trace your steps, and produce the exact same intelligence you did.

> **You can redo it easily:** You produced the intelligence and have some nice tools and scripts that can do all the steps of producing the intelligence again. Maybe the scripts get new telemetry, process it, move it to a computation environment, build the model, and produce charts that show how well the new intelligence will perform — all with the touch of a button.

It redoes itself: You built a system that automatically reproduces the intelligence on a regular basis and drops the new intelligence in a known place, with quality reports emailed to the right people.

It redoes itself and deploys if appropriate: Your system that automates intelligence production also automates deployment. It has tests that sanity-check everything. It knows how to send out new intelligence, roll it out to more and more customers, and will back out and alert orchestrators if anything goes wrong.

In building an Intelligent System you will probably start at the low-end of this maturity spectrum and you will probably end near the higher end. Full automation isn't always worth implementing, but for open-ended or time-changing problems it might be. Overall, the goal is to develop the lowest-cost way of maintaining intelligence at the quality required to enable the Intelligent System to achieve its objectives.

Being Excellent at Intelligence Creation

As with most human endeavors — the people who are best at intelligence creation are way, way better than the people who are average at it.

In addition to basic programming skills and data science skills, good intelligence creators have several skills:

- Data debugging
- Verification-based approach
- Intuition with the toolbox
- Math (?)

This section looks at all of these requirements.

Data Debugging

Intelligence creation requires a mentality to look at data and figure out what is going on. What is the data trying to say? What stories are hidden in it?

This can be very tedious work. Kind of like being a detective, piecing together clues to figure out what happened.

Sometimes tracking a small discrepancy (for example, that some number seems to be coming up 0 more often than you expected) can lead to a big improvement. Maybe it's a sign of an important bug. Maybe it's an indication that someone made incorrect assumptions somewhere in the implementation. Maybe it's a hint that the model is confused in a particular way, and a different approach would work better.

Not everyone has a data debugging mentality. Most people want to shrug and say, "that just happens 1% of the time, how big a deal could it be?" Having a bias for tracking unexpected/interesting things in data, and the experience to know when to stop, are critical for excellence in intelligence creation.

Verification-Based Approach

At its core, intelligence creation is about exploring many different possibilities and picking the one that is best. So it's kind of important to be able to tell which is actually best.

One part of this is figuring out which tests you can run to tell if an intelligence will actually meet the system's goals. This requires some basic statistical understanding (or intuition) and some customer empathy to connect numbers to customer experiences.

Another aspect of a verification-based approach is to make it as easy as possible to compare candidate intelligences with one another. In fact one of the longest chapters in this book is titled "Evaluating Intelligence" and it is coming up soon. If the thought of reading a long, long chapter on evaluating intelligence makes you a little bit sad inside... intelligence creation might not be for you.

Intuition with the Toolbox

You also need a deep understanding of some of the intelligence-creation tools, including machine-learning models and feature-generation approaches.

Intelligence-creation tools can be temperamental. They won't tell you what's wrong, and every approach is different. Techniques that work well with one machine-learning algorithm may not work with another. It's important to develop the intuition to be able to look at an intelligence-creation tool, the output, and the situation, and know what to change to make progress.

This skill is a little like EQ (Emotional IQ) but for machine learning (MQ?). It's like having someone sitting in the room with you, and you need to read their body language, read between the lines of what they are saying, and figure out if they are mad, and why, and what you could do to make it better.

Sometimes, that's what intelligence creation is like.

Math (?)

And then there is math. Much of machine learning was created by math-first thinkers. Because of this, much of machine learning training is done in a math-first way. But I think advanced math is optional for excellent applied machine learning.

Really?

Yeah, I think so.

Machine learning is like most human tools—our cars, our cell phones, our computers, our jet fighters—the person operating them doesn't need to understand everything it took to build them. I'm sure the world's best jet fighter pilots can't work all the math it takes to design a modern fighter jet. But put an engineer who could work that math into a jet, and let them dogfight with an ace fighter pilot...

Who do you think is going to win?

A bit of knowledge of the inner workings of machine learning algorithms might help you develop intuition, or it might not. Follow your strengths. Don't sweat what you can't control.

You might end up being the next machine learning ace.

Summary

Applied intelligence creation is an iterative process. The main steps are:

1. Understanding the environment

2. Defining success

3. Getting data

4. Getting ready to evaluate

5. Trying simple heuristics

6. Trying simple machine learning

7. Assessing and iterating until you succeed

The specific intelligence building techniques (especially the machine learning based ones) can vary over time, and from problem-domain to problem-domain. But the basic workflow outlined here should remain fairly stable over time.

Spending time automating these steps can be very useful, especially as you find yourself carrying out more and more iteration.

And remember—verify, verify, verify.

For Thought...

After reading this chapter, you should:

- Be able to create intelligence end-to-end.

- Understand the types of questions, challenges, and activities that will occupy the intelligence-creator's time.

Youshould be able to answer questions like these:

- Choose your favorite Intelligent System and walk through the seven steps of the intelligence creation process described in this chapter. What questions would be relevant to your Intelligent System? How would you define success? How would you find data? How would you evaluate? What heuristic intelligence would you try? And what would some standard, simple machine learning approaches be?

CHAPTER 19

Evaluating Intelligence

Evaluation is creation, at least when it comes to building intelligence for Intelligent Systems. That's because intelligence creation generally involves an iterative search for effective intelligence: produce a new candidate intelligence, compare it to the previous candidate, and choose the better of the two. To do this, you need to be able to look at a pair of intelligences and answer questions like these:

- Which one of these should I use in my Intelligent System?

- Which will do a better job of achieving the system's goals?

- Which will cause less trouble for me and my users?

- Is either of them good enough to ship to customers or is there still more work to do?

There are two main ways to evaluate intelligence:

- **Online evaluation:** By exposing it to customers and seeing how they respond. We've discussed this in earlier chapters when we talked about evaluating experience, and managing intelligence (via silent intelligence, controlled rollouts, and flighting).

- **Offline evaluation:** By looking at how well it performs on historical data. This is the subject of this chapter, as it is critical to the intelligence-creation process.

This chapter discusses what it means for an intelligence to be accurate. It will then explain how to use data to evaluate intelligence, as well as some of the pitfalls. It will introduce conceptual tools for comparing intelligences. And finally, it will explore methods for subjective evaluation of intelligence.

© Geoff Hulten 2018
G. Hulten, *Building Intelligent Systems*, https://doi.org/10.1007/978-1-4842-3432-7_19

Evaluating Accuracy

An intelligence should be accurate, of course. But accurate isn't a straightforward concept and there are many ways for an intelligence to fail. An effective intelligence will have the following properties:

- It will generalize to situations it hasn't seen before.

- It will make the right types of mistakes.

- It will distribute the mistakes it makes well.

This section explores these properties in more detail.

Generalization

One of the key challenges in intelligence creation is to produce intelligence that works well on things you don't know about at the time you create the intelligence.

Consider a student who reads the course textbook and memorizes every fact. That's good. The student would be very accurate at parroting back the things that were in the textbook. But now imagine the teacher creates a test that doesn't ask the student to parrot back fact from the textbook. Instead, the teacher wants the student to demonstrate they understood the concepts from the textbook and apply them in a new setting. If the student developed a good mental model about the topic, they might pass this test. If the student has the wrong mental model about the topic, or has no mental model at all (and has just memorized facts), they won't do so well at applying the knowledge in a new setting. This is the same as the intelligence in an Intelligent System—it must generalize to new situations.

Let's look at an example. Consider building intelligence that examines books and classifies them by genre—sci-fi, romance, technical, thriller, historical fiction, that kind of thing.

You gather 1,000 books, hand-label them with genres, and set about creating intelligence. The goal is to be able to take a new book (one that isn't part of the 1,000) and accurately predict its genre.

What if you built this intelligence by memorizing information about the authors? You might look at your 1,000 books, find that they were written by 815 different authors, and make a list like this:

- Roy Royerson writes horror.

- Tim Tiny writes sci-fi.

- Neel Notson writes technical books.

- And so on.

When you get a new book, you look up its author in this list. If the author is there, return the genre. If the author isn't there—well, you're stuck. This model doesn't understand the concept of "genre" it just memorized some facts and won't generalize to authors it doesn't know about (and it will get pretty confused by authors who write in two different genres).

When evaluating the accuracy of intelligence, it is important to test how well it generalizes. Make sure you put the intelligence in situations it hasn't seen before and measure how well it adapts.

Types of Mistakes

Intelligences can make many types of mistakes and some mistakes cause more trouble than others. We've discussed the concept of *false positive* and *false negative* in Chapter 6 when we discussed intelligent experiences, but let's review (see Figure 19-1). When predicting classifications, an intelligence can make mistakes that

- Say something is of one class, when it isn't.

- Say something isn't of a class, when it is.

		The Intelligence says some is	
		There	**Not There**
Someone Actually is	**There**	**True Positive**	**False Negative**
	Not There	**False Positive**	**True Negative**

Figure 19-1. Different Types of mistakes.

For example, suppose the intelligence does one of these things:

- Says there is someone at the door, but there isn't; or it says there is no one at the door, but there is someone there.

- Says it's time to add fuel to the fire, but it isn't (the fire is already hot enough); or it says it isn't time to add fuel, but it is (because the fire is about to go out).

- Says the book is a romance, but it isn't; or it says the book isn't a romance, but it is.

In order to be useful, an intelligence must make the right types of mistakes to complement its Intelligent System. For example, consider the intelligence that examines web pages to determine if they are funny or not. Imagine I told you I had an intelligence that was 99% accurate. Show this intelligence some new web page (one it has never seen before), the intelligence makes a prediction (funny or not), and 99% of the time the prediction is correct. That's great. Very accurate generalization. This intelligence should be useful in our funny-webpage-detector Intelligent System.

But what if it turns out that most web pages aren't funny—99% of web pages, to be precise. In that case, an intelligence could predict "not funny" 100% of the time and still be 99% accurate. On not-funny web pages it is 100% accurate. On funny web pages it is 0% accurate. And overall that adds up to 99% accuracy. And it also adds up to—completely useless.

One measure for trading off between these types of errors is to talk about *false positive rate* vs *false negative rate*. For the funny-page-finder, a "positive" is a web page that is actually funny. A "negative" is a web page that is not funny. (Actually, you can define a positive either way—be careful to define it clearly or other people on the project might define it differently and everyone will be confused.) So:

- The *false positive rate* is defined as the fraction of all negatives that are falsely classified as positives (what portion of the not-funny page visits are flagged as funny).

$$\text{False Positive Rate} = \frac{\text{\# False Positives}}{\text{\# False Positives} + \text{\# True Negatives}}$$

- The *false negative rate* is defined as the fraction of all positives that are falsely classified as negatives (what portion of the funny page visits are flagged as not-funny).

$$\text{False Negative Rate} = \frac{\text{\# False Negatives}}{\text{\# False Negatives + \# True Positives}}$$

Using this terminology, the brain-dead always-not-funny intelligence would have a 0% false positive rate (which is great) and a 100% false negative rate (which is useless).

Another very common way to talk about these mistake-trade-offs is by talking about a model's *precision* and its *recall*.

- The *precision* is defined as the fraction of all of the model's positive responses that are actually positive (what portion of "this page is funny" responses are correct).

$$\text{Precision} = \frac{\text{\# True Positives}}{\text{\# True Positives + \# False Positives}}$$

- The *recall* is defined as the proportion of the positives that the model says are positive (what portion of the funny web pages get a positive response).

$$\text{Recall} = \frac{\text{\# True Positives}}{\text{\# True Positives + \# False Negatives}}$$

Using this terminology, the brain-dead always-not-funny intelligence would have an undefined precision (because it never says positive and you can't divide by zero, not even with machine learning) and a 0% recall (because it says positive on 0% of the positive pages).

An effective intelligence must balance the types of mistakes it makes appropriately to support to the needs of the Intelligent System.

Distribution of Mistakes

In order to be effective, an intelligence must work reasonably well for all users. That is, it cannot focus its mistakes into specific sub-populations. Consider:

- A system to detect when someone is at the door that never works for people under 5 feet tall.

- A system to find faces in images that never finds people wearing glasses.

- A system to filter spam that always deletes mail from banks.

These types of mistakes can be embarrassing. They can lead to unhappy users, bad reviews. It's possible to have an intelligence that generalizes well, that makes a good balance of the various types of mistakes, and that is totally unusable because it focuses mistakes on specific users (or in specific contexts).

And finding this type of problem isn't easy. There are so many potential sub-populations, it can be difficult or impossible to enumerate all the ways poorly-distributed mistakes can cause problems for an Intelligent System.

Evaluating Other Types of Predictions

The previous section gave an introduction to evaluating classifications. But there are many, many ways to evaluate the answers that intelligences can give. You could read whole books on the topic, but this section will give a brief intuition for how to approach evaluation of regressions, probabilities, and rankings.

Evaluating Regressions

Regressions return numbers. You might want to know what fraction of time they get the "right" number. But it is almost always more useful to know how close the predicted answers are to the right answers than to know how often the answers are exactly right.

The most common way to do this is to calculate the *Mean Squared Error (MSE)*. That is, take the answer the intelligence gives, subtract it from the correct answer, and square the result. Then take the average of this across the contexts that are relevant to your measurement.

$$\text{Mean Squared Error} = \frac{\text{Sum of (Correct\_Answer}-\text{Predicted\_Answer)}^2}{\text{Number of Contexts}}$$

When the MSE is small, the intelligence is usually giving answers that are close to the correct answer. When the MSE is large, the intelligence is usually giving answers that are far from the correct answer.

Evaluating Probabilities

A probability is a number from 0 – 1.0. One way to evaluate probabilities is to use a threshold to convert them to classifications and then evaluate them as classifications.

Want to know if a book is a romance? Ask an intelligence for the probability that it is a romance. If the probability is above a threshold, say, 0.3 (30%), then call the book a romance; otherwise call it something else.

Using a high threshold for converting the probability into a classification, like 0.99 (99%), generally results in higher precision but lower recall—you only call a book a romance if the intelligence is super-certain.

Using a lower threshold for turning the probability into a classification, like 0.01 (1%), generally results in lower precision, but higher recall—you call just about any book a romance, unless the intelligence is super-certain it isn't a romance.

We'll discuss this concept of thresholding further in a little while when we talk about operating points and comparing intelligences.

Another way to evaluate probabilities is called *log loss*. Conceptually, log loss is very similar to mean squared error (for regression), but there is a bit more math to it (which we'll skip). Suffice to say—less loss is better.

Evaluating Rankings

Rankings order content based on how relevant it is to a context. For example, given a user's history, what flavor of soda are they most likely to order next? The ranking intelligence will place all the possible flavors in order, for example:

1. Cola

2. Orange Soda

3. Root Beer

4. Diet Cola

So how do we know if this is right?

One simple way to evaluate this is to imagine the intelligent experience. Say the intelligent soda machine can show 3 sodas on its display. The ranking is good if the user's actual selection is in the top 3, and it is not good if the user's actual selection isn't in the top 3.

You can consider the top 3, the top 1, the top 10—whatever makes the most sense for your Intelligent System.

So one simple way to evaluate a ranking is as the percent of time the user's selection is among the top K answers.

Using Data for Evaluation

Data is a key tool in evaluating intelligence. Conceptually, intelligence is evaluated by taking historical contexts, running the intelligence on them, and comparing the outputs that actually occurred to the outputs the intelligence predicted.

Of course, using historical data has risks, including:

- You might accidentally **evaluate the intelligence on data that was used to create the intelligence**, resulting in over-optimistic estimates of the quality of the intelligence. Basically, letting the intelligence see the answers to the test before testing it.

- The underlying **problem might change** between the time the testing data is collected and time the new intelligence will be deployed, resulting in over-optimistic estimates of the quality of the intelligence. When the problem changes, the intelligence might be great—at fighting the previous war.

This section will discuss ways to handle testing data to minimize these problems.

Independent Evaluation Data

The data used to evaluate intelligence must be completely separate from the data used to create the intelligence.

Imagine this. You come up with an idea for some heuristic intelligence, implement it, evaluate it on some evaluation data, and find the precision is 54%. At this point you're fine. The intelligence is probably around 54% precise (plus or minus based on statistical properties because of sample size, and so on), and if you deploy it to users that's probably about what they'll see.

But now you look at some of the mistakes your intelligence is making on the evaluation data. You notice a pattern, so you change your intelligence, improve it. Then you evaluate the intelligence on the same test data and find the precision is now 66%.

At this point you are no longer fine. You have looked at the evaluation data and changed your intelligence because of what you found. At this point it really is hard to say how precise your intelligence will be when you deploy it to users; almost certainly less than the 66% you saw in your second evaluation. Possibly even worse than your initial 54%.

This is because you've cheated. You looked at the answers to the test as you built the intelligence. You tuned your intelligence to the part of the problem you can see. This is bad.

One common approach to avoiding this is to create a separate *testing set* for evaluation. A test set is created by randomly splitting your available data into two sets. One for creating and tweaking intelligence, the other for evaluation.

Now as you are tweaking and tuning your intelligence you don't look at the test set—not even a little. You tweak all you want on the training set. When you think you have it done, you evaluate on the test set to get an unbiased evaluation of your work.

Independence in Practice

The data used to evaluate intelligence should be completely separate from the data used to produce the intelligence. I know—this is exactly the same sentence I used to start the previous section. It's just that important.

One key assumption made by machine learning is that each piece of data is independent of the others. That is, take any two pieces of data (two contexts, each with their own outcomes). As long as you create intelligence on one, and evaluate it on the other, there is no way you can be cheating—those two pieces of data are independent. In practice this is not the case. Consider which of the following are more (or less) independent:

- A pair of interactions from different users, compared to a pair of interactions from the same users.

- Two web pages from different web sites, compared to two pages from the same site.

- Two books by different authors, compared to two books by the same author.

- Two pictures of different cows, compared to two pictures of the same cow.

Clearly some of these pieces of data are not as independent as others, and using data with these types of strong relationships to both create and evaluate intelligence will lead to inaccurate evaluations. Randomly splitting data into training and testing sets does not always work in practice.

Two approaches to achieving independence in practice are to petition your data *by time* or *by identity*.

Partition data by time. That is, reserve the most recent several days of telemetry to evaluate, and use data from the prior days, weeks, or months to build intelligence. For example, if it is September 5, 2017 the intelligence creator might reserve data from September 2, 3, and 4 for testing, and use all the data from September 1, 2017 and earlier to produce intelligence.

If the problem is very stable (that is, does not have a time-changing component), this can be very effective. If the problem has a time-changing component, recent telemetry will be more useful than older telemetry, because it will more accurately represent the current state of the problem. In these cases you'll have to balance how much of the precious most-recent, most-relevant data to use for evaluation (vs training), and how far back in time to go when selecting training data.

Partition data by identity. That is, ensure that all interactions with a single identity end up in the same data partition. For example:

- All interactions with a particular user are either used to create intelligence or they are used to evaluate it.

- All interactions with the same web site are used to create intelligence or they are used to evaluate it.

- All sensor readings from the same house are used to create intelligence or they are used to evaluate it.

- All toasting events from the same toaster are used to create intelligence or they are used to evaluate it.

Most Intelligent Systems will partition by time and by at least one identity when selecting data to use for evaluation.

Evaluating for Sub-Populations

Sometimes it is critical that an intelligence does not systematically fail for critical sub-populations (gender, age, ethnicity, types of content, location, and so on.).

For example, imagine a speech recognition system that needs to work well for all English speakers. Over time users complain that it isn't working well for them. Upon investigation you discover many of the problems are focused in Hawaii—da pidgin stay too hard for fix, eh brah?

Ahem...

The problem could be bad enough that the product cannot sell in Hawaii—a major market. Something needs to be done!

To solve a problem, it must be measured (remember, verification first). So we need to update the evaluation procedure to measure accuracy specifically on the problematic sub-population. Every time the system evaluates a potential intelligence, it evaluates it in two ways:

- Once across all users.

- Once just for users who speak English with a pidgin-Hawaiian accent.

This evaluation procedure might find that the precision is 95% in general, but 75% for the members of the pidgin-speaking sub-population. And that is a pretty big discrepancy.

Evaluating accuracy on sub-populations presents some complications. First is *identifying if an interaction is part of the sub-population*. A new piece of telemetry arrives, including a context and an outcome. But now you need some extra information—you need to know if the context (for example the audio clip in the telemetry) is for a pidgin-Hawaiian speaker or not. Some approaches to this include:

1. **Identify interactions by hand**. By inspecting contexts by hand (listening to the audio clips) and finding several thousand from members of the target sub-population. This will probably be expensive, and difficult to repeat regularly, but it will usually work. The resulting set can be preserved long-term to evaluate accuracy on the sub-population (unless the problem changes very fast).

2. **Identify entities from the sub-population**. For example, flagging user as "pidgin speakers" or not. Every interaction from one of your flagged users can be used to evaluate the sub-population. When the context contains an identity (such as a user ID), this approach can be very valuable. But this isn't always available.

3. **Use a proxy for the sub-population**, like location. Everyone who is in Hawaii gets flagged as part of the sub-population whether they speak pidgin-Hawaiian or not. Not perfect, but sometimes it can be good enough, and sometimes you can do it automatically, saving a bunch of money and time.

A second complication to evaluating accuracy for sub-populations is in *getting enough evaluation data for each sub-population*. If the sub-population is small, a random sample of evaluation data might contain just a few examples of it. Two ways to deal with this are:

1. **Use bigger evaluation sets**. Set aside enough evaluation data so that the smallest important sub-population has enough representation to be evaluated (see the next section for more detail on the right amount of data).

2. **Up-sample sub-population members for evaluation**. Skew your systems so members of the sub-population are more likely to show up in telemetry and are more likely be used for evaluation instead of for intelligence creation. When doing this you have to

be sure to correct for the skew when reporting evaluation results. For example, if users from Hawaii are sampled twice as often in telemetry, then each interaction from Hawaii-based users gets half as much weight when estimating the overall accuracy compared to other users.

The Right Amount of Data

So how much data do you need to evaluate an intelligence? It depends—of course.

Recall that statistics can express how certain an answer is, for example: the precision of my intelligence is 92% plus or minus 4% (which means it is probably between 88% and 96%).

So how much data you need depends on how certain you need to be.

Assuming your data is very independent, the problem isn't changing too fast, and you aren't trying to optimize the last 0.1% out of a very hard problem (like speech recognition):

- Tens of data points is too small a number to evaluate intelligence.

- Hundreds of data points is a fine size for a sanity check, but really not enough.

- Thousands of data points is probably a fine size for most things.

- Tens of thousands of data points is probably overkill, but not crazy.

A starting point for choosing how much data to use to evaluate intelligence might be this:

1. Ensure you have thousands of recent data points reserved for evaluation.

2. Ensure you have hundreds of recent data points for each important sub-population.

3. Use the rest of your (reasonably recent) data for intelligence creation.

4. Unless you have ridiculous amounts of data, at which point simply reserve about 10% of your data to evaluate intelligence and use the rest for intelligence creation.

But you'll have to develop your own intuition in your setting (or use some statistics, if that is the way you like to think).

Comparing Intelligences

Now we know some metrics for evaluating intelligence performance and how to measure these metrics from data. But how can we tell one if intelligence is going to be more effective than another? Consider trying to determine whether the genre of a book is romance or not:

- One intelligence might have a precision of 80% with a recall of 20%.

- Another intelligence might have a precision of 50% with a recall of 40%.

Which is better? Well, it depends on the experience, and the broader objectives of the Intelligent System. For example, a system that is trying to find the next book for an avid romance reader might prefer a higher recall (so the user won't miss a single kiss).

Operating Points

One tool to help evaluate intelligences is to select an *operating point*. That is, a precision point or a recall point that works well with your intelligent experience. Every intelligence must hit the operating point, and then the one that performs best there is used. For example:

- In a funny-web-page detector the operating point might be set to 95% precision. All intelligence should be tuned to have the best recall possible at 95% precision.

- In a spam-filtering system the operating point might be set to 99% precision (because it is deleting users' email, and so must be very sure). All intelligences should strive to flag as much spam as possible, while keeping precision at or above 99%.

- In a smart-doorbell system the operating point might be set to 80% recall. All intelligences should be set to flag 80% of the times a person walks up to the door, and compete on reducing false positives under that constraint.

This reduces the number of variables by choosing one of the types of mistakes an intelligence might make and setting a target. We need an intelligence that is 90% precise to support our experience—now, Mr. or Mrs. Intelligence creator, go and produce the best recall you can at that precision.

When comparing two intelligences it's often convenient to compare them at an operating point. The one that is more precise at the target recall (or has higher recall at the target precision) is better.

Easy.

Curves

But sometimes operating points change. For example, maybe the experience needs to become more forceful and the operating point needs to move to a higher precision to support the change. Intelligences can be evaluated across a range of operating points. For example, by varying the threshold used to turn a probability into a classification, an intelligence might have these values:

- 91% precision at 40% recall.

- 87% precision at 45% recall.

- 85% precision at 50% recall.

- 80% precision at 55% recall.

And on and on. Note that any intelligence that can produce a probability or a score can be used this way (including many, many machine learning approaches).

Many heuristic intelligences and many "classification-only" machine learning techniques do not have the notion of a score or probability, and so they can only be evaluated at a single operating point (and not along a curve).

When comparing two intelligences that can make trade-offs between mistake types, you can compare them at any operating point. For example:

- At 91% precision, one model has 40% recall, and the other has 47% recall.

- At 87% precision, one model has 45% recall, and the other has 48% recall.

- At 85% precision, one model has 50% recall, and the other has 49% recall.

One model might be better at some types of trade-offs and worse at others. For example, one model might be better when you need high precision, but a second model (built using totally different approaches) might be better when you need high recall.

It is sometimes helpful to visualize the various trade-offs an intelligence can make. A *precision-recall curve* (PR curve) is a plot of all the possible trade-offs a model can make. On the x-axis is every possible recall (from 0% to 100%) and on the y-axis is the precision the model can achieve at the indicated recall.

By plotting two models on a single PR curve it is easy to see which is better in various ranges of operating points.

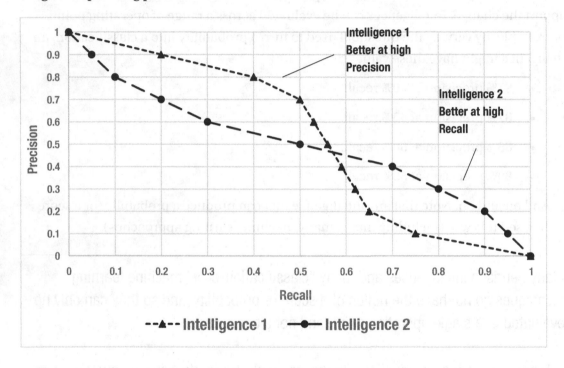

A similar concept is called a *receiver operating characteristic curve* (ROC curve). An ROC curve would have false positive rate on the x-axis and true positive rate on the y-axis.

Subjective Evaluations

Sometimes an intelligence looks good by the numbers, but it just...isn't... This can happen if you:

- Have a metric that is out of sync with the actual objective.

- Have a miscommunication between the experience and the intelligence.

- Have an important sub-population that you haven't identified yet.

- And more...

Because of this, it never hurts to look at some data, take a few steps back, and just think. Be a data-detective, take the user's point of view and imagine what your intelligence will create for them. Some things that can help with subjective evaluations include:

- Exploring the mistakes.

- Imagining the user experience.

- Finding the worst thing that could happen.

We'll discuss these in turn.

Exploring the Mistakes

Statistics (for example, those used to represent model quality with a precision and recall) are nice, they are neat; they summarize lots of things into simple little numbers that go up or down. They are critical to creating intelligence. But they can hide all sorts of problems. Every so often you need to go look at the data.

One useful technique is to take a random sample of 100 contexts where the intelligence was wrong and look at them. When looking at the mistakes, consider:

- How many of the mistakes would be hard for users to recover from?

- How many of the mistakes would make sense to the user (vs seeming pretty stupid)?

- Is there any structure to the mistakes? Any common properties? Maybe things that will turn into important sub-population-style problems in the future?

- Can you find any hints that there might be a bug in some part of the implementation or the evaluation process?

- Is there anything that could help improve the intelligence so it might stop making the mistakes? Some new information to add to the context? Some new type of feature?

In addition to inspecting random mistakes, it also helps to look at places where the intelligence was most-certain of the answer—but was wrong (a false positive where the model said it was 100% sure it was a positive, or a false negative where the model said the probability of positive was 0%). These places will often lead to bugs in the implementation or to flaws in the intelligence-creation process.

Imagining the User Experience

While looking at mistakes, imagine the user's experience. Visualize them coming to the context. Put yourself in their shoes as they encounter the mistake. What are they thinking? What is the chance they will notice the problem? What will it cost if they don't notice it? What will they have to do to recover if they do notice it?

This is also a good time to think of the aggregate experience the user will have. For example:

- How many mistakes will they see?

- How many positive interactions will they have between mistakes?

- How would they describe the types of mistakes the system makes, if they had to summarize them to a friend?

Putting yourself in your users' shoes will help you know whether the intelligence is good enough or needs more work. It can also help you come up with ideas for improving the intelligent experience.

Finding the Worst Thing

And imagine the worst thing that could happen. What types of mistakes could the system make that would really hurt its users or your business? For example:

- A system for promoting romance novels that classifies 100% of a particular romance-writer's books as non-romance. This writer stops getting promoted, stops getting sales, doesn't have the skills to figure out why or what to do, goes out of business. Their children get no presents for Christmas...

- All the web pages from a particular law firm get classified as funny. They aren't funny, but people start laughing anyway (because they are told the pages are funny). No one wants to hire a law firm full of clowns. The firm gets mad and sues your business—and they are a little bit pissed off and have nothing but time on their hands...

- The pellet griller controller's temperature sensor goes out. Because of this the intelligence always thinks the fire is not hot enough. It dumps fuel-pellet after fuel-pellet onto the fire, starting a huge, raging fire, but the intelligence still thinks it needs more...

These are a little bit silly, but the point is—be creative. Find really bad things for your users before they have to suffer through them for you, and use your discoveries to make better intelligence, or to influence the rest of the system (the experience and the implementation) to do better.

Summary

With intelligence, evaluation is creation.

There are three main components to accuracy: generalizing to new situation, making the right types of mistakes, and distributing mistakes well among different users/ contexts.

- Intelligence can be good at contexts it knows about, but fail at contexts it hasn't encountered before.

- Intelligence can make many types of mistakes (including false positives and false negatives).

- Intelligence can make random mistakes, or it can make mistakes that are focused on specific users and contexts—the focused mistakes can cause problems.

There are specific techniques for evaluating classification, regressions, probability estimations, and rankings. This chapter presented some simple ones and some concepts, but you can find more information if you need it.

Intelligence can be evaluated with data. Some data should be held aside and used exclusively to evaluate intelligence (and not to create it). This data should be totally independent of the data used to create the intelligence (it should come from different users, different time periods, and so on).

An operating point helps focus intelligence on the types of mistakes it needs to make to succeed in the broader system. A precision-recall curve is a way of understanding (and visualizing) how an intelligence operates across all possible operating points.

It is important to look at the mistakes your intelligence makes. Try to understand what is going on, but also take your user's point of view, and imagine the worst outcome they might experience.

For Thought...

After reading this chapter, you should be able to:

- Describe what it means for an intelligence to be accurate.

- Evaluate the quality of an intelligence across a wide range of practical criteria.

- Create useful quality goals for intelligence and progress toward the goals.

- Take a user's point of view and see the mistakes an intelligence makes through their eyes.

You should be able to answer questions like these:

- Describe three situations where an Intelligent System would need an intelligence with very high recall.

- For each of those three, describe a small change to the system's goals where it would instead need high precision.

- Select one of the Intelligent Systems mentioned in this chapter and describe a potential sub-population (not mentioned in the chapter) where bad mistakes might occur.

- Consider the system for classifying books into genres. Imagine you are going to examine 100 mistakes the system is making. Describe two different ways you might categorize the mistakes you examine to help gain intuition about what is going on. (For example, the mistake is on a short book vs a long book—oops! now you can't use book length in your answer. Sorry.)

CHAPTER 20

Machine Learning Intelligence

Machine learning is a powerful technique for producing intelligence for large, hard, open-ended, time-changing problems. It works by showing the computer lots (and lots) of examples of contexts and the desired outcomes. The computer produces models from these examples. And the models can be used to predict the outcome for future contexts.

Machine learning can be a huge force multiplier, allowing the computer to focus on things it is good at (like tuning every detail of a gigantic model so it works effectively in millions of contexts), while humans can focus on what they are good at (like picking a model representation that will generalize well and building systems and intelligent experiences that turn models into customer and business value). This chapter will give an overview of machine learning, introducing the steps and some of the key issues.

How Machine Learning Works

A machine learning algorithm is essentially a search procedure that looks for accurate models, using training data to evaluate the accuracy. Generally, machine learning algorithms do the following:

- Start with a simple model.

- Try slightly refined versions of the model (usually informed by the training data).

- Check to see if the refined versions are better (using the training data).

- And iterate (roughly) until their search procedure can't find better models.

245

© Geoff Hulten 2018
G. Hulten, *Building Intelligent Systems*, https://doi.org/10.1007/978-1-4842-3432-7_20

Sometimes the refinements take the form of adding more complexity to the model (for example by adding more if-then-else tests to a decision tree). Sometimes the refinements take the form of adjusting parameters in the model (for example, updating the weights in a linear model).

For example, recall that a decision tree represents intelligence with a tree. Each node contains an `if` condition, with one child for when the `if`-test is true and one child for when the `if`-test is false. Here is a sample decision tree for predicting how much money a movie will make:

```
If <won Academy Award>:
        Is True: If <has a top 10 star in the cast>:
                Is True: then $100,000,000.
                Is False: then $1,000,000.
        Is False: If <opened on labor day weekend>:
                Is True: then $50,000,000.
                Is False: then $1,000.
```

Machine learning for decision trees produces increasingly complex trees by adding if-tests until no further additions improve the model (or until the model reaches some complexity threshold). For example, a refinement of the sample decision tree for movie-success-prediction might be this:

```
If <won Academy Award>:
        Is True: If <has a top 10 star in the cast>:
                Is True: If <opened on labor day weekend>:
                        Is True: then $500,000,000.
                        Is False: Then $10,000,000.
                Is False: then $1,000,000.
        Is False: If <opened on labor day weekend>:
                Is True: then $50,000,000.
                Is False: then $1,000.
```

This model is a bit more complex, and possibly a bit more accurate.

A human could carry out the same process by hand, but machine learning algorithms automate the process and can consider millions of contexts and produce hundreds of thousands of small refinements in the time it would take a human to type "hello world."

Factors an intelligence creator must control when using machine learning include these:

- **Feature Engineering**: How the context is converted into features that the machine learning algorithm can add to its model. The features you select should be relevant to the target concept, and they should contain enough information to make good predictions.

- **Model structure complexity**: How big the model becomes. For example, the number of tests in the decision tree or the number of features in the linear model.

- **Model search complexity**: How many things the machine learning algorithm tries in its search. This is separate from (but related to) structure complexity. The more things the search tries, the more chance it has to find something that looks good by chance but doesn't generalize well.

- **Data size**: How much training data you have. The more good, diverse data available to guide the machine learning search, the more complex and accurate your models can become.

These elements must be balanced to get the best possible models out of a machine-learning process.

The Pros and Cons of Complexity

One of the key challenges in machine learning is controlling the complexity of the models it produces. They key tension is as follows:

1. You need very complex models to solve hard, big, open-ended problems.

2. The more complexity you allow, the more chances you have to learn a model that misunderstands the concept.

Your job as an applied machine learning practitioner is to balance these tensions. A machine learning process can fail in two ways:

- It might *underfit* the concept, in that its understanding of the concept might too simple.

- It might *overfit* the concept, in that its understanding of the concept is wrong; it happens to work in some contexts, but it just isn't right.

Let's look at an example. Consider building intelligence that examines books and classifies them by genre—sci-fi, romance, technical, thriller, historical fiction, that kind of thing.

You gather 1,000 books, hand label them with genres, and set about creating intelligence. The goal is to be able to take a new book (one that isn't part of the 1,000) and accurately predict its genre.

Underfitting

If the approach you take is too simple to understand what a "genre" is, you will underfit. It might go something like this:

1. For each genre, find the word that is most indicative of the genre:

   - For a romance, this "indicative word" might be "love."

   - For a sci-fi, it might be "laser."

   - And so on

2. When you get a new book, check to see if it contains any of the indicative words and label the book with the associated genre:

   - If the book contains the word "love," call it romance.

   - If the book contains the word "laser," call it sci-fi.

If every book has one and exactly one of the indicative words, this model might work well. But most books have lots of words. What should we do with a book that uses both the words "love" and "laser"? Should we call it a romance or a sci-fi? Whichever choice we make, we'll be making mistakes, because some sci-fi books talk about love; and some romance books talk about lasers.

This approach to modeling (checking for a single word per genre) is just too simple. This simple approach *underfits* the concept of genre, and will not generalize well.

Overfitting

Now imagine another approach—using a decision tree that is very, very complex, essentially encoding the exact string of words in the book into a tree form, for example:

1. If the first word of the book is "When"

2. And the second word of the book is "the"

3. And the third word of the book is "Earth"

4. And so on, exactly matching every word in the book...

5. Then the genre is "Science Fiction"

This model is very complex, and highly accurate (on the data used to create it), but it would not generalize to new books at all.

And it sounds silly, exactly memorizing every word in a book, but that is just an example. The fundamental problem is that machine learning almost always models problems in ways that don't match the underlying phenomenon. Because of this, parts of the model will *happen to* work on everything you know about, but *fail to* work in new situations. The technical term for this is *overfitting*, and avoiding overfitting is one of the key challenges in creating effective intelligence.

When an intelligence underfits or overfits a problem, the intelligence will not be accurate at dealing with new contexts; it will not generalize.

Balancing Complexity

A better model for our book genre example would be somewhere between the two. Maybe it would check for the presence of the 50 words that are most highly correlated with each genre:

*   For science fiction it might use laser, warp, star, beam, and so on.

*   For romance it might use... other words...

Then you could use a decision tree, a linear model, or a neural network to automatically learn how important each word is to each genre and how to combine and weight them to accurately predict genre.

This would be a better approach than the other two approaches. You can see that, I can see that. But machines? Maybe not.

And this 50-words approach still is not right. It will certainly underfit the concept of genre, because it doesn't take the context of the words into account, the sentences they were used in, the ideas they describe.

But that's OK. The point of machine learning isn't to find the "correct" representation for a problem. The point is to find the modeling process that produces the best generalization. You'll need to consciously choose how and where to underfit and to overfit to achieve that.

Feature Engineering

Contexts in an Intelligent System encode all the data relevant to an interaction. This might include information about the user, their history, what they are doing, the content they are interacting with, and more. Feature engineering is the process of turning contexts into a representation that machine learning can work with. This includes making the information have the right format for the model representation. It also includes making the information work well with the search procedure used to learn the model.

The rest of this section talks about ways of doing this, including these:

- Converting data to useful form.

- Helping the model use the data.

- Normalizing feature values.

- Exposing hidden information.

- Expanding the context.

- Eliminating misleading things.

Converting Data to Useable Format

In general, machine learning algorithms take a *training set* as input. A training set consists of a set of *training examples.* Each training example contains a set of variables that encode the context. These variables are called *features.* Each training example also contains an outcome variable, which indicates the correct answer for the context. This is often called the *label.*

Recall that outcomes are commonly one of the following: a category from a set of possibilities (for a classification); a number (for a regression); or a probability (maybe a number from 0 to 1.0, which represents probabilities from 0 to 100%, scaled down to the range of 0.0–1.0).

A training example can be thought of as a row in a spreadsheet. One column contains the label; the rest of the columns contain *feature values.* Each training example in a training set must have the same number of features, although some of the features values can be missing—machine learning is robust to that sort of thing.

For example, if you are trying to build a model for predicting a person's age from the types of music they listen to, you might have a feature representing the gender, the eye color, and the number of Journey songs they listened to last week (Figure 20-1). Some users might have listened to 7; some might have listened to 14. The training examples for these users will have the column (feature) for *Number of Journey Songs* filled in with the correct values (7 or 14, or whatever). But maybe your Dad won't tell you how many Journey songs he's listened to—maybe he's shy about that sort of thing. So your Dad's training example still has a column (feature) for number of Journey songs, but the value is undefined (or –1, or 0, or some other special code depending on your tools).

Age (The Label)	Gender	Number of Journey Songs	Eye Color	...
23	Male	7	Hazel	...
12	Female	0	Brown	...
47	Male	-1	Blue	...
...	...	...	...	...

Figure 20-1. *Example features*

Machine learning systems typically work two types of features:

- *Numerical features,* which are integers or floating-point numbers (like the number of Journey songs listened to last week).

- *Categorical features,* which are labels from a small set of values (like the gender of the person).

Most machine learning models and algorithms support both numerical and categorical features. But sometimes a bit of pre-processing is required. For example, internally neural networks work only with numerical features. To feed categorical features to a neural network you have to convert them to numerical values. One common way is to use a *one-hot encoding*, which turns a categorical feature into N numerical features, one for each possible categorical value. For example, to encode gender this way you would have the following conversions:

- Male -> [0, 1]

- Female -> [1, 0]

Figure 20-2 shows this encoding in a table.

Age (The Label)	Gender is Female	Gender is Male
23	0	1
12	1	0

Figure 20-2. *One-hot encoding of gender*

Helping your Model Use the Data

Feature engineering is the art of converting context into features that work well for your problem and with your model structure. For example, consider the model for predicting age from songs the person listens to. One approach would be to create one feature for every artist that ever recorded a song, ever. Each training example has tens of thousands of columns, one per artist, indicating how many songs from that artist the user listened to.

This representation contains lots and lots of information, which should help the machine learning process avoid underfitting. It is also complex. It requires the learning algorithm to consider many, many things in making its decision and might encourage overfitting. So is this the best way to go?

Maybe, depending on your tools and how much data you have. The only way to know is to try a bunch of different things, build models, evaluate them, and see.

Here are some other possible approaches to creating features for this problem:

- **Break down the songs by decade**: Have one feature per decade where the value is the number of songs from that decade the person listened to.

- **Separate popular songs from obscure songs**: Because maybe people growing up during a particular decade would have broader exposure to the music, so listening to obscure songs might give them away. So each decade would have two features, one for popular song-listens from that decade and one for obscure song-listens from that decade.

- **Break down the songs by genre**: One feature per genre where the value is the number of songs listened to from the genre.

- **Some combination between genre and decade and popularity**: Where you do all three of the previous methods simultaneously.

And once you get going you'll think of all sorts of ways to break down the problem, to extract information from the context (the songs the person listened to and the patterns of the listening) and partially digest it to expose the most relevant parts of it to the machine learning algorithm.

Think of it this way. If you happen to know there is a relevant concept called "genre" and you expose the concept in your features in a useful way, then the machine-learning algorithm can leverage it. It takes work on your part, but it can help produce models that better-match what is going on in your problem space by encoding your prior knowledge about the world.

Contrast this with giving raw data to the machine learning process (so the algorithm can figure it all out on its own). This can be extremely effective and efficient in terms of the effort you'll need to expend—if you have lots and lots and lots of data so you can avoid overfitting.

Different algorithms will do better with different approaches. Learn your tools, gain intuition, and make sure your evaluation framework makes it very easy to explore.

Normalizing

Sometimes numerical features have very different values. For example, age (which can be between 0 and about a hundred) and income (which can be between 0 and about a hundred million). Normalization is the process of changing numerical features so they are more comparable. Instead of saying a person is 45 with $75,000 income, you would say a person is 5% above the average age with 70% above the average income.

One standard way to perform normalization is to post-process all numerical features and replace them with their normalized value:

1. Subtract the mean value (shifting the mean to zero).

2. Divide by the standard deviation of the variable.

For example, in each training example replace:

$$Normalized\_Age = \frac{(Age - mean\_of\_all\_ages)}{Standard\_Deviation\_of\_all\_ages}$$

A learning algorithm can certainly work with un-normalized data. But doing the pre-processing can remove complexity, making the modeling task a bit easier so the algorithm can focus on other things. Sometimes it will help.

Exposing Hidden Information

Some features aren't useful in isolation. They need to be combined with other features to be helpful. Or (more commonly) some features are useful in isolation but become much more useful when combined with other features.

For example, if you are trying to build a model for the shipping cost of boxes you might have a feature for the height, the width, and the depth of the box. These are all useful. But an even more useful feature would be the total volume of the box.

Some machine-learning algorithms are able to discover these types of relationships on their own. Some aren't.

You can make *composite features* based on your intuition and understanding of the problem. Or you could try a bunch automatically, by creating new features by combining the initial features you come up with.

Expanding the Context

You might use external intelligence or lookups to add things to add information that isn't explicitly in the context. For example, you might do any of these:

- Look up the history of an identity in the context, like traffic to a web server.

- Look up the external temperature from a web service based on the location.

- Run a sentiment analysis program on a block of text and include the predicted sentiment as a feature.

Recall that you can only use these types of features if you can create them exactly the same way in the intelligence runtime and in the intelligence creation environment.

The next chapter will discuss more ways to organize intelligence, which will help you decide how to best leverage sources or legacy intelligences to expand the context.

Eliminating Misleading Things

Another approach is to delete features from your data.

WHAT!?!

Yeah. You just went through all this work to create features from contexts, now I'm saying you should delete some of them. Nuts!

But you may have created some really poor features (sorry). These can add complexity to the learning process without adding any value. For example, using a person's eye color to predict their age. Sure, every person has an eye color. Sure, it is in the context. But is it relevant to predicting how old the person is? Not really.

Including irrelevant features will reduce your ability to learn models that generalize well—because sometimes the irrelevant feature will look good by accident and trick your machine-learning algorithm into including it in the model.

Or you may have created way more features than your data will support. All of your features might be brilliant, but they can still lead to overfitting and hurt generalization.

So really you should:

1. Remove features that don't have any information about the outcome.

2. Select as many features as your data and model structure and search procedure will support, and select the best ones.

This is called *feature selection*. There are many techniques for performing feature selection. Here are some simple ones:

1. Measure the mutual information (think correlation) between each feature and the outcome feature, and remove the ones with lowest scores.

2. Train models with and without each feature, evaluate each model, and then remove features that don't help generalization accuracy.

3. Use your intuition and understanding of the problem to delete things that don't make sense.

These techniques can help debug your decisions about feature engineering, too. Create a new feature. Add it to the existing set of features. Train a new model—if the result isn't better, you might be on the wrong path. Stare at it for a while. Then stop and think. Maybe it will help you understand what is really going on.

Modeling

Modeling is the process of using machine learning algorithms to search for effective models. There are many ways an intelligence creator can assist this process:

- Deciding which features to use.

- Deciding which machine learning algorithms and model representations to use.

- Deciding what data to use as input for training.

- Controlling the model creation process.

But in all of these, the goal is to get the model that generalizes the best, has the best mistake profile, and will create the most value for your customers. This section talks about ways to do that, including common ways to adjust the modeling process to help.

Complexity Parameters

Every machine learning algorithm has parameters that let intelligence creators control how much complexity they add.

Some provide parameters that *control the size of the model the algorithm produces.* Some models add structure as they search, for example, adding nodes to a decision tree; adding trees to a random forest. These types of algorithms often have parameters that place hard limits on the amount of structure (number of nodes in the tree, or number of trees in the random forest). Many algorithms also have parameters for the minimum amount of training-set gain they need to see to continue the search.

When you are overfitting you should limit the size; when you are underfitting you should allow more size.

A machine learning algorithm might have parameters to *control the algorithm's search strategy.* The most common approach (by far) is to take greedy steps along the *gradient,* that is—to make the change that most improves the model's performance on the training data. Other options can include a bit of randomness, like random restarts (in case the search gets in a local maximum), or some lookahead.

Algorithms might also have parameters that control the size of steps taken in the search. That is, they find the gradient, and then they have to decide how far to move the model along the gradient. Smaller step size allows for more complexity. Some algorithms adaptively adjust the step size as the search continues, starting large to find a good general region, and then getting smaller to refine the model.

When you are overfitting you should use simpler search strategies; when you are underfitting you might try more complex search.

Some algorithms also use optimization algorithms directly; for example, by representing the problem as a matrix internally and using linear-algebra to find key properties. These algorithms might support various optimization options that trade off complexity.

You should basically always adjust these types of parameters based on the complexity of your problem and the amount of data you have. It's very, very rare to perform successful machine learning without tuning your algorithm's complexity parameters.

Identifying Overfitting

Recall that the goal of modeling is to build the model that is best at generalizing to new settings. One strategy to identify when a modeling process is starting to overfit is this:

1. Build a series of increasingly complex models on the training data, starting with brain-dead simple and progressing to crazy-complexity. If your features and model representation choices are reasonable, the more complex models should do better than the less complex ones—*on the training data.*

2. Now run these models on a hold-out test set. The brain-dead simple model should perform (roughly) worst. The next, slightly more complex model should perform (a bit) better. And the next more complex should perform better still. On and on—up to a point. Eventually, the more complex models will start to perform worse on the hold-out test data than the less complex ones. And the point where they start to perform worse is the point where your feature set, model search strategy, model representation, and available data are starting to get into trouble.

For example, see Figure 20-3.

Figure 20-3. *An example of overfitting*

If you start overfitting you'll need:

- Features that better match the problem.

- A model structure that better matches the problem.

- Less search to produce models.

- *Or more data!*

More data is the best way to avoid overfitting, allowing you to create more complex and accurate models. And thank goodness you've done all the work to set up an Intelligent System with a great intelligent experience and lots of users to produce data for you. This is where Intelligent Systems can shine, allowing your intelligence to grow as more users interact with it.

Summary

Machine learning produces intelligence automatically from data. It involves having a computer search for models that work well on your data (in a training set). One of the key tensions is between increased accuracy and poor generalization based on overfitting.

Feature engineering is a critical part of machine learning. It involves converting contexts into representations that your model can work with. But it also involves understanding your representation, your problem, and the machine learning algorithm's search strategy well enough to present data in the right way. Common techniques include normalization, feature selection, and composite features.

The modeling process involves a lot of iteration, picking the right machine learning algorithm, tuning the parameters, and squeezing the most value out of the available training data by carefully managing complexity.

For Thought...

After reading this chapter, you should:

- Have a good conceptual understanding of machine learning, including features, models, and the modeling process and the key tension between complexity and overfitting.

- Know the steps of applied machine learning and roughly how a generic machine learning algorithm would work.

- Be able to take a machine learning toolkit and produce usable models, and be prepared to take more steps in learning specific algorithms and techniques.

You should be able to answer questions like these:

Describe a context for an Intelligent System you haven't used in any previous answers to the questions in this book.

- Create ten numerical features for the context.

- Create five categorical features for the context.

- Propose one way to make the features less complex.

- Create an example of a model that uses your features to overfit—go crazy, find the limit of an outrageous overfitting.

- Bonus: Design a very simple machine learning algorithm. What is your representation? What is your search? What are your complexity parameters?

CHAPTER 21

Organizing Intelligence

In most large-scale systems, intelligence creation is a team activity. Multiple people can *work on the intelligence at the same time*, building various parts of it, or investigating different problem areas. Multiple people can also *work on the intelligence over time*, taking over for team members who've left, or revisiting an intelligence that used to work but has started having problems. Some examples of ways to organize intelligence include these:

- Using machine learned intelligence for most things, but using manual intelligence to override mistakes.

- Using one machine learned intelligence for users from France, and a different one for users from Japan.

- Using the output of heuristic intelligence as features into a machine learned intelligence.

This chapter discusses ways to organize intelligence and the process used to create intelligence to make it robust, and to allow many people to collaborate effectively.

Reasons to Organize Intelligence

There are many reasons you might want to move from building a single monolithic intelligence (for example, a single machine-learned model) to an organized intelligence:

- **Collaboration**: Large-scale intelligence construction is a collaborative activity. You may have 3, or 5, or 15 people working on the intelligence of a single Intelligent System. And if you do, you'll need to find ways to get them all working together, efficiently, instead of competing to be the owner of the one-intelligence-to-rule-them-all.

© Geoff Hulten 2018
G. Hulten, *Building Intelligent Systems*, https://doi.org/10.1007/978-1-4842-3432-7_21

- **Cleaning up mistakes**: Every intelligence will make mistakes, and correcting one mistake will often make more mistakes crop up in other places. Also, trying to get a machine-learning-based intelligence to stop making a particular mistake isn't easy—it often requires some experimentation and luck. Combining intelligences can provide quick mistake mitigation to backstop more complex intelligences.

- **Solving the easy part the easy way**: Sometimes part of a problem is easy, where a few heuristics can do a great (or perfect) job. In those cases, you could try to trick machine learning to learn a model that does something you already know how to do, or you could partition the problem and let heuristics solve the easy part while machine learning focuses on the harder parts of the problem.

- **Incorporating legacy intelligence**: There is a lot of intelligence in the world already. Existing intelligences and intelligence-creation processes can be quite valuable. You might want to incorporate them into an Intelligent System in a way that leverages their strength and helps them grow even more effective.

Properties of a Well-Organized Intelligence

Organizing intelligence can be difficult and problematic. Done wrong, the layers of intelligence come to depend on the idiosyncrasies of each other. Any change to one intelligence causes unintended (and hard to track) changes in other intelligences. You can end up with a situation where you know you need to make some changes, but you simply can't—just like spaghetti code, you can have spaghetti intelligence.

A well-organized intelligence will be all of the following:

- **Accurate**: The organization should not reduce the accuracy potential too much. It should be a good trade-off of short term cost (in terms of lower immediate accuracy) for long term gains (in terms of higher accuracy over the lifetime of the Intelligent System).

- **Easy to Grow**: It should be easy for anyone to have an insight, create some intelligence, and drop it into the system.

- **Loosely Coupled**: The ability for one intelligence to influence the behavior of other intelligences should be minimized. The interfaces between the intelligences should be clear, and the intelligences shouldn't use information about the inner working of one-another.

- **Comprehensible**: For every outcome that users have, the system should be able to pinpoint the intelligence (or intelligences) that were involved in the decision, and the number of intelligences involved in each decision/outcome should be minimized.

- **Measurable**: For every part of the intelligence, it should be possible to determine how much that part of the intelligence is benefiting users.

- **Suportive of the Team**: The organization strategy should work with the team. The organization should allow intelligence creators' successes to amplify one another's work. It should avoid putting goals in conflict or creating the need for participants to compete in unproductive ways.

Ways to Organize Intelligence

This section discusses a number of techniques for organizing intelligence and the process of creating intelligence, and evaluates them against the key properties of a well-organized intelligence. Most large Intelligent Systems will use multiple of these methods simultaneously. And there are many, many options—the techniques here are just a starting point:

- Decouple feature engineering

- Multiple model searches

- Chase mistakes

- Meta-models

- Model sequencing

- Partition contexts

- Overrides

The following sections will describe these approaches and rank them according to how well they meet the criteria for well-organized intelligence. This is a subjective scale that is attempting to highlight relative strength/weaknesses as follows:

++	A real strength
+	Better than average
	Average
-	Worse than average
- -	A challenge that will need attention

All of these methods are viable, and used in practice. But you should be prepared to mitigate the challenges inherent in the organization strategy you choose.

Decouple Feature Engineering

One approach to organizing intelligence creation is to separate the feature engineering tasks so that each intelligence creator has a clear part of the context to explore and to turn into features. For example, if trying to understand a web page:

- One intelligence creator can focus on the content of the page, using standard approaches like bag of words and n-grams to convert words into features.

- Another intelligence creator can focus on understanding the semantics of the text on the page, using parts of speech tagging, sentiment analysis, and so on.

- Another could look at the history of the web site, where it is hosted, who created it, and what else they created.

- Another could explore the images on the web page and try to create features from them.

- Another could look at the properties of the user.

Each participant needs to be able to inject their feature extraction code into the modeling process. They need to be able to tweak model parameters to take best advantage of the new work. They need to be able to deploy their work.

Some challenges of decoupled feature engineering include these:

- **Conflict on model building**: One type of model/set of parameters might work better for one type of feature than for another. Participants will need to balance trade-offs to grow the overall system, and not simply cannibalize value from existing feature sets.

- **Redundant features**: Multiple approaches to feature creation could leverage the same underlying information from the context. The resulting features may be very similar to each other. Intelligence creators may have conflict about how to remove the redundancies.

- **Instability in feature value**: When a new feature is presented to a machine learning algorithm it will usually change the direction of the model-building search, which can have wild impacts on the value of other features and on the types of mistakes the model makes. Adding a new feature may require some global understanding of the feature set/model and some work on other parts of the feature-creation code to keep everything in balance.

In summary, the approach of decoupling feature engineering is

- **Accurate: Average**

  There isn't much compromise in this approach, and intelligence creators can work in parallel to make gains.

- **Easy to grow: +**

  The act of adding a few new features to an existing model is conceptually easy. Not the absolute simplest, but quite good.

- **Loosely coupled: Average**

  Features can interact with each other, but as long as you aggressively remove redundancy, the coupling should not be a major problem.

- **Comprehensible: Average**

  When trying to debug an interaction there aren't good tools to pinpoint problematic features, and many model types make it particularly difficult. Sometimes you are left with "try removing features one at a time and retraining to see when the problem goes away."

- **Measurable:** –

  It's easy to measure improvement when the features are initially
  added. It isn't so easy to track the contribution of the features over
  time (for example, as the problem changes).

- **Supportive of the Team: Average**

  When there are clear boundaries in the context things can work well,
  but there are certainly plenty of ways to end up with conflict as to
  which features should be in and which should be out, particularly if
  there are any runtime constraints (CPU or RAM).

Multiple Model Searches

Another way to organize intelligence is to allow multiple creators to take a shot at the
model-building process. For example, maybe one team member is an expert with linear
models, while another is a master of neural networks. These practitioners can both try
to create the intelligence, using whatever they are most comfortable with, and the best
model wins.

Using multiple model searches can be effective when:

- You have intelligence creators who are experienced with different
  approaches.

- You are early in the process of building your Intelligent System and
  want to cast a wide net to see what approaches work best.

- You have a major change in your system (such as a big change in the
  problem or a big increase in usage) and want to reverify that you have
  selected the right modeling approach.

But using multiple model searches can result in redundant work and in conflicts,
because one approach will eventually win, and the others will lose.

The approach of multiple model searches is

- **Accurate:** -

  This approach makes it hard to leverage many intelligence creators
  over time. It should be used sparingly at critical parts of the
  intelligence creation process, such as when it is clear a change is
  needed.

- **Easy to grow: -**

  To ship a new intelligence you have to beat an old one. This means that new ideas need to be quite complete, and evaluated extensively before deploying.

- **Loosely coupled: Average**

  There is just one intelligence, so there isn't any particular coupling problem.

- **Comprehensible: Average**

  There is just one intelligence, so there isn't any particular comprehension problem.

- **Measurable: Average**

  There is just one intelligence, so there isn't any particular measurably problem.

- **Supportive of the Team: - -**

  This is a bit of a winner-take-all way of working, which means there are a modeling winner and a modeling loser. It also tends to promote wasted work—chasing a modeling idea that never pans out.

Chase Mistakes

Another approach is to treat intelligence problems like software bugs. Bugs can be assigned to intelligence creators, and they can go figure out whatever change they need to make to fix the problem. For example, if you're having trouble with a sub-population—say children—send someone to figure out what to add to the context, or what features to change, or what modeling to change to do better on children.

Intelligences will always make mistakes, so this approach could go on forever.

And one of the key problems with this approach is figuring out what mistakes are just sort of random mistakes, and which are systematic problems where a change in intelligence creation could help. When using this approach, it is very easy to fall into chasing the wrong problems, making everyone upset, and getting nowhere.

In my opinion, this approach should be used infrequently and only near the beginning of the project (when there are lots of legitimate bugs) or when there is a catastrophic issue.

The approach of chasing mistakes is

- **Accurate: -**

  Intelligence sometimes works this way (like with a sub-population problem), but it is easy to get drawn into chasing the wrong mistakes.

- **Easy to grow: -**

  Everyone needs to know everything to find and follow mistakes, develop productive changes, and deploy the fix. Also, this approach tends to lead to poor decisions about what problems to tackle.

- **Loosely coupled: Average**

  Doesn't really affect coupling.

- **Comprehensible: Average**

  Doesn't really affect comprehensibility.

- **Measurable: Average**

  Doesn't really affect measurability.

- **Supportive of the Team: -**

  This approach does not provide nice boundaries for people to work with. It is also easy to fix one mistake by causing another, and it won't always be clear that one fix caused the other mistake until much later. Done wrong, this approach can create a miserable work environment.

Meta-Models

The meta-model approach is to treat the predictions of the various intelligences in your system as features of a meta-intelligence. Every base intelligence runs and makes its decision, and then a meta-intelligence looks at all the proposed predictions and decides what the real output should be. Using meta-models can be

- Very accurate, because it brings together as many approaches as possible and learns which contexts each approach is effective in and which it struggles with.

- A great way to incorporate legacy intelligence. For example, when you find a new intelligence that is better than your original heuristics, you can throw away your heuristics... or you could use them as a feature in the new intelligence.

- A good way to get multiple intelligence creators working together. There are no constraints on what they can try. The meta-model will use the information they produce if it is valuable and ignore it if it isn't.

But meta-models can also be a bit of a nightmare to manage. Some complexities include these:

- The meta-intelligence and the base intelligences become tightly coupled, and changing any part of it might involve retraining and retuning all of it.

- If any piece of the system breaks (for example, one model starts behaving poorly) the whole system can break, and it can be very hard to track where and how problems are occurring.

If you want to use meta-models you will need approaches to control the complexity that interdependent models introduce, perhaps by

- Enforcing some structure about which models can change and when—for example, freezing the legacy intelligence and only changing it if you find severe problems.

- Building extra machinery to help retrain and retune all the intelligences that make up the system very, very easily.

In summary, the approach of using meta-models is

- **Accurate: ++**

  Short term, meta-models have the power to be the most accurate of the methods listed here. The cost of them is in the other areas. For raw accuracy, use meta-models.

- **Easy to grow: -**

  To ship a new intelligence you need to retrain the meta-intelligence, which risks instability in outcomes across the board. Careful testing is probably required.

- **Loosely coupled: - -**

  Changing any base intelligence usually requires retraining the meta-intelligence. Unintended changes in any of the intelligences (e.g. some change in one of the data sources it depends on) can affect the whole system, to the point of completely breaking it.

- **Comprehensible: - -**

  Every intelligence contributes to every decision. When there is a problem it can be extremely difficult (maybe impossible?) to track it down to a source.

- **Measurable: -**

  It is easy to measure a new intelligence when it is added to the system. It isn't so easy to track the contribution of the intelligences over time (for example, as the problem changes).

- **Supportive of the Team: Average**

  There can be conflict between intelligences, but they can be created independently. There may also be conflicts for resources when the intelligences need to run in a resource restrained environment.

Model Sequencing

Model sequencing is a restricted version of the meta-model approach in which the meta-model is constrained to be super-simple. In the sequencing approach, the models are put into order by the intelligence creator. Each model gets a chance to vote on the outcome. And the first model to vote with high confidence wins and gets to decide the answer.

This can be accomplished for classification by setting a default answer—if no one votes, the answer is "male"—and allowing each model to run with a high-precision operating point for the "female" answer. If any model is very sure it can give a high-precision "female" answer, then it does; if none of the models are certain, the default "male" is the return value.

Model sequencing has less accuracy potential than meta-models, which can combine all the votes simultaneously, but it is much easier to orchestrate and control.

The approach of model sequencing is

- **Accurate: Average**

  This approach trades off some potential accuracy for ease of management and growth.

- **Easy to grow: +**

  An intelligence creator can put a new model in the sequence (as long as it has high enough precision) without affecting any other part of the system.

- **Loosely coupled: ++**

  Models are completely uncoupled and are combined by a simple procedure that everyone can understand.

- **Comprehensible: ++**

  Every interaction can be traced to the piece of intelligence that decided the outcome, and each piece of intelligence can have a clear owner.

- **Measurable: Average**

  It is easy to measure how many positive and negative interactions each intelligence gives to users. The downside is that the first confident answer is taken, so other intelligences might not get all the credit (or blame) they deserve.

- **Supportive of the Team: +**

  Anyone can easily add value. Potential conflict points include what order to use to sequence of models and what precision threshold to demand. But telemetry should provide good data to use to make these decisions empirically, so they shouldn't make people argue—much.

273

Partition Contexts

Partitioning by contexts is another simple way to organize multiple intelligences. It works by defining some simple rules on the contexts that split them into partitions and then having one intelligence (or model sequence, meta-model, and so on) for each of the partitions. For example:

- One intelligence for servers in the US, one for all others.

- One intelligence for small web sites, one for large web sites, and one for websites that aggregate content.

- One intelligence for grey-scale images, one for color images.

- One intelligence for new users, one for users with a lot of history.

This approach has advantages by allowing you to use different types of intelligence on different parts of the problem, solve easy cases the easy way, and also control the incidence of mistakes made on various partitions. Of course, machine-learning algorithms can technically use this type of information internally—and probably pick better partitions with respect to raw accuracy—but manually partitioning can be very convenient for orchestration and organization.

The approach of partitioning is

- **Accurate: Average**

  This approach turns one problem into several problems. This doesn't have to affect accuracy, but it might, particularly as innovations in one area might not get ported to all the other areas where they could help.

- **Easy to grow: ++**

  An intelligence creator can define a specific partition and give intelligence that is tuned for it without affecting other partitions.

- **Loosely coupled: +**

  Models are completely uncoupled and are combined by an understandable procedure (unless the partitioning gets out of hand).

- **Comprehensible: +**

  Every interaction can be traced to the piece of intelligence that decided the outcome, and each piece of intelligence can have a clear owner.

- **Measurable: +**

  It is easy to measure how many positive and negative interactions each intelligence gives to users.

- **Supportive of the Team: +**

  Anyone can easily add value. Also, when one team member takes on a problem (by partitioning away something that was causing trouble for other models) it can be perceived as a favor: "I'm glad you took on those mistakes so my model can focus on adding this other value..."

Overrides

Overriding is an incredibly important concept for dealing with mistakes. The override structure for organizing intelligence works by having one blessed intelligence (usually created and maintained by humans) that can override all the other intelligences (usually created by machine learning)—no matter what they say.

One way this can be used is by hand-labeling specific contexts with specific outcomes. This can be used to spot-correct specific damaging mistakes. For example:

- This web page is not funny, no matter what any intelligence thinks.

- When all the toaster sensors read a specific combination, toast for 2 minutes (because we know exactly what product that is), no matter what the intelligence thinks.

Another way this can be used is by creating rules that can serve as guardrails, protecting from things that are obviously wrong. For example:

- If we've released 10 pellets into the pellet griller in the past 5 minutes, don't release any more, no matter what the intelligence says.

- If the web page has any of these 15 offensive words/phrases it is not funny, no matter what the intelligence says.

Override intelligence should be use extremely sparingly. It should not be trying to solve the problem; it should just be covering up for the worst mistakes.

The approach of using overrides is

- **Accurate: Average**

  As long as the overrides are used sparingly. They should be a backstop and not a complicated hand-crafted intelligence.

- **Easy to grow: ++**

  An intelligence creator (or an untrained person with some common sense) can define a context and specify an outcome. Tooling can help.

- **Loosely coupled: Average**

  The overrides are somewhat coupled to the mistakes they are correcting and might end up living in the system for longer than they are really needed. Over time they might turn into a bit of a maintenance problem if they aren't managed.

- **Comprehensible: +**

  Every interaction can be traced to any overrides that affected it. Intelligence creators might forget to check all the overrides when evaluating their new intelligence, though, so it can lead to a bit of confusion.

- **Measurable: +**

  It is easy to measure how many positive and negative interactions each override saved/gave to users.

- **Supportive of the Team: +**

  As long as overrides are used sparingly, they provide a simple way to make intelligence creators more productive. There is potential conflict between the intelligence creators and the people producing the overrides.

Summary

Most large Intelligent Systems do not have monolithic intelligences; they have organized collections of intelligence. Organizing intelligence allows multiple intelligence creators to collaborate effectively; to clean up mistakes cheaply, to use right types of intelligence to target the right part of the problem, and to incorporate legacy intelligence.

A well-organized intelligence will be: accurate, easy to grow, loosely coupled, compressible, measurable, and supportive of the team. It will, of course, also be accurate and sometimes organization needs to be sacrificed for accuracy.

There are many, many ways to organize intelligence. This chapter presented some of the prominent ones, but others are possible. Important organization techniques include: decoupling feature engineering; doing multiple model searches; chasing mistakes; meta-models; model sequencing; partitioning contexts; and overrides. Figure 21-1 shows the summary table.

Approach	Accurate	Ease of Growth	Loosely Coupled	Comprehensible	Measurable	Supports Team
Decouple Feature Engineering	Average	+	Average	Average	-	Average
Multiple Model Searches	-	-	Average	Average	Average	--
Chase Mistakes	-	-	Average	Average	Average	-
Meta-Models	++	-	--	--	-	Average
Model Sequencing	Average	+	++	++	Average	+
Partition Contexts	Average	++	+	+	+	+
Overrides	Average	++	Average	+	+	+

Figure 21-1. *Comparing approaches to intelligence organization*

For Thought…

After reading this chapter, you should:

- Understand what it takes to work on a large, complex Intelligent System, or with a team of intelligence creators.

- Be able to implement an intelligence architecture that allows the right intelligence to attack the right parts of your problem, and all participants to work together efficiently.

You should be able to answer questions like these:

- Describe an Intelligent System that does not need to have any intelligence organization—that is, it works with just a single model.

- What are some of the ways this might cause problems? What problems are most likely to occur?

- Design a simple intelligence organization plan that addresses the most likely problem.

PART V

Orchestrating Intelligent Systems

Chapters 22-26 discuss what it takes to run an Intelligent System. This includes taking control of all the parts of it, keeping its intelligence and experience balanced, growing it, debugging it, and ensuring that it is constantly achieving its goals.

This part discusses the tools needed, how to understand the mistakes an Intelligent System makes and what to do in response, and what to do if people start to abuse your system.

CHAPTER 22

Overview of Intelligence Orchestration

Intelligence orchestration is a bit like car racing. A whole team of people build a car, put all the latest technology into it, and get every aerodynamic wing, ballast, gear-ratio, and intake valve set perfectly—they make an awesome machine that can do things no other machine can do.

And then someone needs to get behind the wheel, take it onto the track, and win!

Intelligence orchestrators are those drivers. They take control of the Intelligent System and do what it takes to make it achieve its objectives. They use the intelligence-creation and management systems to produce the right intelligence at the right time and combine it in the most useful ways. They control the telemetry system, gathering the data needed to make the Intelligent System better. And they deal with all the mistakes and problems, balancing everything so that the Intelligent System is producing the most value it can for users and for your business. For example:

- If the intelligence gets better, the experience can be made more forceful to achieve better outcomes.

- If the problem changes, the intelligence management might need to be optimized to create and deploy intelligence differently or more quickly.

- If the system stops achieving its objectives, someone needs to figure out why and how to use the available systems to adapt.

This isn't easy.

Building Intelligent Systems and orchestrating them are very different activities. They require very different mindsets. And they are both absolutely critical to achieving success.

© Geoff Hulten 2018

G. Hulten, *Building Intelligent Systems*, https://doi.org/10.1007/978-1-4842-3432-7_22

This chapter describes what a well-orchestrated intelligence looks like and introduces the elements of orchestrating an Intelligent System, including the life-cycle, the orchestration environment, mistakes, and abuse.

Properties of a Well-Orchestrated Intelligence

A well-orchestrated Intelligent System will

- **Achieve its objectives reliably over time**: That is, it will grow, achieving its objectives better and better, until it reaches the point of diminishing return. Then it will stay there, producing value, as the user base, the problem, and the broader world change.

- **Have experience, intelligence, and objective in balance**: That is, the experience will be as forceful as the intelligence will support so the net benefit to users and business is maximized.

- **Have mistakes mitigated effectively**: So that high cost mistakes are understood, are detected quickly, and are corrected. And so lower cost mistakes are mitigated and do not put too large a burden on users or on the business.

- **Scale effectively over time**: So that the cost of maintaining the system in good shape (all the balancing, intelligence creation, mistake mitigations, and so on) scales favorably as the number of users and the complexity of intelligence scales. In particular, this means that machine learning and automation are used effectively.

- **Degrade slowly**: In that it should not require a bunch of manual effort to hold in a steady state. The effort of orchestration should be focused on finding opportunities, not scrambling to keep the wheels on the bus.

Why Orchestration Is Needed

Orchestrating an intelligence involves tuning it so that it produces the most value possible throughout its life cycle.

And right now you might be saying—wait, wait. I thought machine learning was supposed to tune the system throughout its life cycle. What is this? Some kind of joke?

Unfortunately, no. Artificial intelligence and machine learning will only get you so far. Orchestration is about taking those tools and putting them in the best situations so they can produce value—highlight their strengths, compensate for their weaknesses—and react as things change over time.

Orchestration might be needed because:

- The Intelligent System's objective changes.

- The Intelligent System's users change.

- The problem changes.

- The intelligence changes.

- The cost of running the Intelligent System changes.

- Someone tries to abuse your Intelligent System.

One or more of these will almost certainly happen during the life-cycle of your Intelligent System. By learning to identify them and adapt, you can turn these potential problems into opportunities.

Objective Changes

When your objectives change, your approach is going to have to change, too, and that means rebalancing the parts of your Intelligent System. You might want to change your objective when any of the following happen:

- **You understand the problem better:** As you work on something, you'll come to understand it better. You might realize that you set the wrong objective to begin with, and want to adapt. Recall that there are usually many layers of objectives. You might have a perfectly fine business objective, but come to understand that the user outcomes you are tracking are not contributing to the objective; or you might realize that the model properties you are optimizing aren't heading you in the right direction. In these cases you will want to change the objectives and adapt the system as needed to achieve the new goals.

283

- **You have solved the previous problem**: You may find that you have achieved what you wanted to achieve. The Intelligent System is running, users are getting good outcomes, your business is doing well. At this point you might want to set a higher goal: and adapt the system accordingly. Or you might want to figure out how to run the system more cheaply (which we will discuss later).

- **You realize the previous problem is too hard to solve**: Sometimes you have to admit defeat. You might have a system that has been running for a while, and just isn't getting to where you want it. You could shut down and find something else to do, or you could set an easier objective, change your approach, and build up more slowly.

- **You discover some new opportunity**: You might come up with a great idea. Something similar to your current Intelligent System, but just a little bit different. Maybe a totally new experience that can leverage the intelligence. Maybe a new type of intelligence you can build off of the existing telemetry. Sometimes chasing a new opportunity will require a new Intelligent System, but sometimes it can be accomplished with the systems you already have—by changing your objective.

Users Change

The user base of an Intelligent System will change over its life cycle, in various ways:

- **New users come**: And the new users might be very different than your existing users. For example, you might attract a new demographic, more casual users, users who exercise different parts of your Intelligent System, or users who just think differently. As your users change, you will have opportunities to learn and adapt. You will also have more telemetry to use to create intelligence and will have to deal with increasing sampling issues and intelligence-management complexity.

- **Usage patterns change**: As users learn to use your Intelligent System, they may change the way they interact with it. They will stop exploring how it works and become more focused on getting value out of it. Experienced users will benefit from different interactions than novice users, and so you might want your Intelligent System to change as your users become experts.

- **Users change their perception of the experience**: Over time users might become fatigued with forceful, frequent experiences. They might start to ignore them. They might get increasingly irritated with them. An experience that worked well for the first month of usage might be totally wrong in the long run—and it might need to change.

- **Users leave**: And as they do, you will probably cry a little bit. But life will go on. Depending on which users leave, you might have the opportunity to change the balance of intelligence and experience to optimizing for the users you still have. You will also have less telemetry, which might reduce your ability to create intelligence. Automated intelligence, especially, might need to be revisited—for example, to leverage more legacy data instead of relying on newer data.

Problem Changes

Intelligent Systems are for big, hard, open-ended, time-changing problems. These problems—by definition—change over time:

- **Time-changing problems are always changing**: And so the approaches and decisions you made in the past might not be right for the future. Sometimes a problem might be easy. At other times it might get very hard. Sometimes you will see lots of a particular type of context. At other times you won't see those contexts at all. As a problem changes, there is almost always opportunity to adapt and achieve better outcomes through orchestration.

- **Usage patterns changing**: As your users change how they interact with your Intelligent System, the problems you need to solve can become very different. For example, users behave differently in the summer than they do in the winter; they behave differently on weekends than weekdays; they interact with new content differently from how they interact with old content.

- **You start to solve the problem**: In that the intelligence starts to catch up with a big, hard problem. In this case you might want to revisit the approach. Maybe you will want to start locking in behavior instead of continuing to grow, using more intelligence caches (lookup tables) and investing less in telemetry.

Intelligence Changes

The intelligence of your system will change, when you do any of the following:

- **Get more data to build better intelligence**: As users interact with your system, the intelligence will have more data, and should get better. When intelligence is better, experiences can become more forceful.

- **Try a new approach that makes things a lot better**: Data unlocks possibilities. Some of the most powerful machine learning techniques aren't effective with "small" data, but become viable as users come to your Intelligent System and you get lots and lots of data. These types of changes can unlock all sorts of potential to try new experiences or target more aggressive objectives.

- **Try a new approach that makes things worse**: But you won't be perfect. Sometimes intelligence evolutions go backwards. This can be a particular problem with intelligence that is combined in aggressive ways (tightly coupled or hard to comprehend). It can also be a problem when you rely on lots of human intervention in intelligence creation (for example with heuristics)—sometimes the complexity just crosses a tipping point and a brittle intelligence-creation plan can't keep up. In these cases you might need to back off other parts of the system to give yourself time to get out of the problem.

Costs Change

Big systems will constantly need to balance costs and value. Changes to save money can require compromises in other places. Put another way, you might be able to change your experience or intelligence in ways that save a lot of money, while only reducing value to users or your business by a little. You might try to reduce either or both of these:

- **Telemetry costs**: By changing the way you sample data. This reduces your ability to measure your system and to create intelligence, but it might be a good trade-off. For example, you might collect less data when intelligence is not growing much, sample less from parts of the problem that are already solved, or remove parts of the contexts that aren't contributing to useful features (and better models).

- **Mistake costs**: Managing mistakes can be very expensive. We'll discuss dealing with mistakes more in Chapter 24. But you might find that you need to make fewer, less expensive mistakes, and you might need to change all sorts of things in the system to do this.

Abuse

Unfortunately, the Internet is full of trolls. Some of them will want to abuse your service because they think that is fun. Most of them will want to abuse your service (and your users) to make money—or to make it harder for you to make money. Left unmitigated, abuse can ruin an Intelligent System, making it such a cesspool of spam and risk that users abandon it.

In order to protect your Intelligent System from abuse you need to understand what abusers want, and then you need to do something to make your system less interesting to potential abusers. Sometimes a small tweak will do it. Sometimes combating abuse will become a major part of running your Intelligent System.

But abuse is a tough thing. Once one abuser figures out how to make a tenth of a penny per interaction off your users they will scale. They will tell their friends. Abuse can show up quickly and violently, so you need to be prepared to identify it and react.

We'll discuss Adversaries and Abuse in detail in Chapter 25.

The Orchestration Team

Orchestrating an Intelligent System requires a diverse set of skills. In particular, the team should:

- Be domain experts in the business of the Intelligent System, so they understand the objectives instinctively and know what users want from their interactions with the system.

- Understand experience and have the ability to look at interactions and decide how to make improvements in how intelligence is presented to users—what is easy, what is hard, what users will like, and what they won't.

- Understand how the implementation works so they know how to trace problems and have some ability to improve the implementation as needed.

- Be able to ask all sorts of questions of data and understand and communicate the answers.

- Know how to do applied machine learning, to be able to create new intelligence and inject it into the system when needed.

- Get satisfaction from making a system execute effectively day in and day out. Not everyone likes this—some people like to invent new things. Orchestration is about excellence in execution—taking some of the most awesome machines in the world, getting them on the race track, and winning.

An orchestration team could consist of a small number of jacks-of-all-trades. Or it could be made up of participants with diverse skills who work well together.

Summary

An Intelligent System needs to be orchestrated throughout its life cycle to succeed. Orchestration is the process of keeping all the parts of the system doing what they are good at, and supporting them where they run into problems. A well-orchestrated Intelligent System will do all of the following

- Achieve its objectives reliably over time.

- Have experience, intelligence, and objective in balance.

- Have mistakes mitigated effectively.

- Scale effectively over time.

- Degrade slowly.

One key activity of orchestration is rebalancing the intelligence and the experience over the Intelligent System's life-cycle. For example, making the experience more forceful when the intelligence improves; or making it less forceful when the problem gets harder.

Orchestration also involves dealing with mistakes, understanding what they are and when they are occurring, and then managing the system to make them less costly. Reasons orchestration would be needed include these:

- The objective changes.

- Your users change.

- The problem changes.

- Intelligence changes.

- The system needs to get cheaper to run.

Adversaries and abusers often become problems for successful Intelligent Systems. They must be identified and then convinced that your system isn't worth their time.

For Thought...

After reading this chapter, you should

- Understand what it means to orchestrate an Intelligent System and why it is needed.

- Know what type of team is needed to orchestrate an Intelligent System.

You should be able to answer questions like these:

- Imagine that a system's intelligence begins to behave poorly. What changes that aren't intelligence-based could be made to mitigate the issue?

- Pick two of your favorite Intelligent Systems. Which would be harder to orchestrate? Why?

- Describe the types of problems an Intelligent System might have if its user base doubles—and all the new users are from a different country than the previous users.

CHAPTER 23

The Intelligence Orchestration Environment

This chapter discusses some of the common activities that contribute to success at intelligence orchestration. These are examples of the types of things that can help get the most out of an Intelligent System, and include the following:

- **Monitoring the Success Criteria**: To know if the system is achieving its goal and if it is getting better or worse.

- **Inspecting Interactions**: To experience contexts as the user experienced them and see the outcomes the user achieved.

- **Balancing the Experience**: To make it more or less forceful, maybe changing the type of interactions, or maybe changing the prominence and frequency of the interactions.

- **Overriding the Intelligence**: To correct for infrequent bad mistakes, or to help optimize common contexts.

- **Creating New Intelligence**: Managing intelligence and producing new models to deal with emergent issues as the problem, the users, or the scope of data changes.

Intelligent Systems can benefit from all of these activities and more. Some of them can be built as part of the implementation—for example, a metrics dashboard that shows progress on the overall objectives. Some of them can be handled in a more ad hoc fashion—for example, querying raw telemetry sources to piece together interactions that are causing problems.

291

© Geoff Hulten 2018

G. Hulten, *Building Intelligent Systems*, https://doi.org/10.1007/978-1-4842-3432-7_23

This chapter discusses orchestration activities in turn and gives some ideas of how you can invest in each of the areas (if you choose to), along with some criteria for deciding how much to invest.

Monitor the Success Criteria

You must be able to tell if the system is achieving its goals. This involves taking telemetry, correcting for any sampling and outages, and producing an answer that all participants can understand—maybe a few numbers, maybe a graph. Recall that this can involve any of the following:

- **Business objectives and leading indicators**, particularly showing how they vary between users who heavily use the intelligent part of the system compared to those who don't.

- **User outcomes**, indicating how effective the Intelligent System is at helping users achieve the outcomes you want.

- **Model properties**, indicating how often the intelligence is correct or incorrect (no matter what outcomes the user ended up achieving).

Monitoring success criteria is critical. Every functional Intelligent System must have people who know how well it is doing, every day, and who care that it is achieving its goals.

Levels of investment for monitoring success criteria include:

- **Ad Hoc**: Where a few people have the skill to measure the systems objectives and produce answers, maybe by doing some cross checking of current telemetry settings, tweaking a script, and running it, copying into a presentation form (like a chart), and distributing.

- **Tool-Based**: Where anyone on the team can run a tool that creates and then runs the correct query and outputs the answer in a suitable presentation format.

- **Automated**: Where the system automatically runs the queries on a regular basis and archives the results somewhere that all participants can find them.

- **Alert-Based**: where the system automates the metrics, but also monitors them and provides alerts to participants when something happens, including

    - **A major degradation**: Where the metrics are far worse than they were at the previous measurement. For example, the number of errors went up 10% since yesterday.

    - **A significant, sustained erosion**: Where the metrics have trended gradually downward long enough to cross some threshold. For example, the fraction of users getting good outcomes has gone down 10% over the course of the last month.

- **Population Tracking**: Where the system tracks metrics for important sub-populations as well as the overall population. This might include people from a particular location, with a particular usage pattern, with a particular demographic, and so on.

Criterion for investing in monitoring success criteria:

- Almost every Intelligent System should implement alert-based monitoring for success criteria and for critical factors that contribute to success criteria.

Inspect Interactions

Inspecting an interaction involves gathering all the telemetry related to a particular user interaction and being able to see the interaction end-to-end. This includes the user's context at the time of the interaction, the version of the intelligence running; the answer given from the intelligence, the experience that resulted from this intelligence answer, how the user interacted with the experience, and what outcome the user eventually got.

This is important for debugging problems, for tracking mistakes, for understanding how users are perceiving the Intelligent System, and for getting intuition about user experience options.

Levels of investment for inspecting interactions include

- **Ad Hoc**: Where a few people have the skill to find all the parts of an interaction and understand how they relate. They may have to do some detective work to identify a specific interaction (or interactions with a certain property) and, based on sampling, may or may not be able to find any particular interaction.

- **Tool-based**: Where anyone on the team can identify an interaction (maybe by user and time) see the relevant telemetry. The tool may also support querying for specific types of interactions (maybe where the intelligence gave a particular answer and the user got a particular outcome) and inspect a sample of them.

- **Browser-based**: Where anyone on the team can find interactions as with the tool, but then experience the interaction as the user would have, seeing the experiences that the user saw, the buttons they clicked, the outcome they got, and so on.

- **Getting Data from Users**: Where orchestrators have the opportunity to flag specific types of interactions and ask users questions about their experience. For example, this can be done by specifying some contexts and then surveying a small fraction of users who hit that context. These surveys might ask the user to rate their experience, answer a question or two, or indicate if they had a good or bad outcome.

Criteria for investing in inspecting interactions:

- If you need a broad set of people to be able to understand interactions. For example, so experience creators and business stakeholders can understand how users are experiencing the Intelligent System.

- If you need to build support capabilities to help deal with mistakes. Tools can allow humans who aren't experts in the implementation to participate.

- If your system has high cost mistakes and you expect a major part of orchestration will involve regular mitigation, then you will want to invest in inspecting interactions.

Balance the Experience

As the problem, the user base, and the intelligence change, they create opportunities to optimize the experience. For example, when the intelligence is new—and poor quality—the experience might be very passive. Maybe it doesn't show up very often. Maybe it shows up in ways that are easy for users to ignore. But over time the intelligence will get better, and users might have more positive outcomes with more forceful experiences.

Levels of investment for balancing the experience include:

- **Ad Hoc:** Where changes to the experience are made by changing code and redeploying the software.

- **Parameter Updates:** Where the orchestrators can change parameters that affect the experience and push these parameters out relatively cheaply (maybe as intelligence is updated). Parameters might include:

    - The frequency of interaction.

    - The color or size of a prompt.

    - The text to use in the experience.

    - The threshold for automating an action.

    - And so on.

- **Experience Alternatives:** Where multiple experiences are created and orchestrators have the ability to switch between them (using something similar to a parameter). For example, maybe you create one version of the experience with a subtle prompt, one with a prominent prompt, and one that automates an action. Then the orchestrator can switch between these to achieve objectives over time.

- **Experience Language:** Where orchestrators can author and deploy experiences out of band of code changes, similar to separating intelligence from implementation. This might be accomplished by specifying experiences in scripts that clients download and execute. Or it might be more curtailed, so that non-engineers can safely make changes; for example an experience mark-up language with a restricted runtime.

Criteria for investing in balancing the experience:

- If your orchestration team does not have engineering resources, then you probably want to create some controls for the experience that are not engineering-based.

- If it takes a long time to deploy new client code to your customers, you might want to invest in ways to control experience through parameters and data-file updates.

- If you think you'll be changing your experience many times during the life-cycle of your Intelligent System, you might choose to invest in making it easy.

- If none of those are true—you do have engineering resources, you can easily deploy code, and you don't expect too many changes— it will almost certainly be cheaper to take an ad hoc approach to experience updates.

Override Intelligence

Overriding intelligence involves identifying a few contexts and hand-coding the answer the intelligence should provide for those contexts. You might want to do this to correct costly (or embarrassing) mistakes. You might want to do this to optimize a few common contexts. You might want to do this to support some business goal (like promote a piece of content in your system). You probably don't want to override intelligence too much— but when you need to, you are going to really need to.

Levels of investment for overriding intelligence include:

- **Ad Hoc**: Where you rely on intelligence creators to hack the learning process to try to achieve specific outcomes (which is very hard) or on engineers to hard-code the desired overrides into code and deploy them to customers.

- **As an Intelligence Feed**: Where overrides are treated as a very simple intelligence source that is deployed with the system's other intelligence and has the highest priority in the runtime. Perhaps represented using a data file in some simple format that maps contexts to outcomes.

- **Tool-based**: Where you treat overrides as an intelligence feed but create tooling to support orchestrators. Tools should include functions like these:

  - Ensuring the contexts to override are specified correctly.

  - Giving feedback on the prevalence of the specified contexts and the current intelligence responses for those contexts.

  - Giving feedback on existing overrides, including how many there are and how many users they are impacting.

  - Tracking who is doing overrides and how good/bad they turn out to be.

  - Managing expiration of overrides.

- **Browser-Based**: Where the tools are connected into a support suite, including ways to find interactions, view them, and override them all in one place.

Criteria for investing in overriding intelligence:

- If your system can make very costly mistakes you will almost certainly want to invest in overriding intelligence, possibly adding tracking and work-flow management around the tools discussed here. We'll discuss this in more detail in the next chapter, on dealing with mistakes.

- Overriding intelligence can also be a very helpful buffer for intelligence creators. The last thing you want is for lots of participants in the Intelligent System to use the product, find zillions of little problems, and file them as bugs against the intelligence creation process. Intelligence is going to make mistakes. Beating up the creators for every flaw will not help. Having the ability to override a few of the more important problems can make everyone feel better.

Create Intelligence

An important part of orchestration is controlling and creating the intelligence that drives your Intelligent System. This section recaps and summarizes some of the investments that can help intelligence grow and create impact.

Investments in creating intelligence might include:

- **Intelligent Management**: This will include the following:

    - Controlling when and how intelligence is updated and deployed.

    - Controlling what intelligence lives where.

    - Adding new sources of intelligence to the system.

    - Controlling how intelligence is combined.

- **Automated Machine-Learning Implementations**: These produce new intelligence on a regular basis with little or no oversight. These systems may provide controls to change the way intelligence is produced, including the following:

    - How much data to use.

    - Which data to use.

    - How often to run.

    - How much time to spend searching for good models.

    - Any other parameters of the modeling process that are tunable.

- **Intelligence Creation Environments**: Where the telemetry that captures context and outcome are gathered and made available for machine learning, including trying new machine-learning algorithms, new feature engineering, and incorporating new insights. These can produce ad hoc new intelligence, or find ways to improve existing automated learning systems.

- **Support for Feature Engineering**: Where intelligence creators are able to try new features very easily, and if they find ones that work are able to deploy them to customers quickly and safely.

- **Telemetry Systems**: Where the orchestrator can specify what data to collect and how much of it to collect.

Criteria for investing in creating intelligence:

Intelligence creation is kind of important for Intelligent Systems (it says so right in the name). You will probably end up investing in multiple systems to support this process.

The most effective tools tend to:

- Help prevent errors.

- Automate mundane tasks.

- Simplify multi-step processes.

- Leave some simple audit trail.

When building tools to support intelligence creation, avoid the temptation to mess with the core intelligence creation tools (like machine-learning systems or runtimes). These are generally standard and innovating with them is unlikely to be core to your value proposition.

Summary

This book has introduced many tools that are important for running an Intelligent System. This chapter summarized some of the key ones, including some criteria and approaches for investing in them. These include:

- Monitoring the success criteria

- Inspecting interactions

- Balancing the experience

- Overriding intelligence

- Creating intelligence

You can build a successful Intelligent System without investing much in these areas during implementation. But you'll probably need to do all of these activities during the system's life-cycle. You can decide how labor intensive you want to be, and how much you want to invest in tools.

And keep in mind that having good tools allows you to change your staffing over time from the system's creators (who often like to work on new exciting things) to skilled orchestrators (who need to be generalists with drive for sustained excellence). These are very different skill sets.

Often the orchestration environment will evolve as the Intelligent System does, with small investments being made where they add the most value over an extended period until no further investments make sense.

For Thought…

After reading this chapter, you should:

- Know the common tools needed to keep an Intelligent System healthy.

- Understand ways to start with low investment and scale the investment over time.

- Have some understanding of when the various tools will help, and which are most important for your Intelligent System.

You should be able to answer questions like these:
Imagine an important Intelligent System that exists today.

- Which area do you think it spends the most orchestration time: monitoring success, inspecting interactions, balancing the experience, overriding intelligence, or creating intelligence?

- What would you propose to reduce the cost in that area?

CHAPTER 24

Dealing with Mistakes

There will be mistakes. Humans make them. Artificial intelligences make them, too—and how. Mistakes can be irritating or they can be disastrous.

Every Intelligent System should have a strategy for identifying mistakes; for example, by monitoring critical metrics and giving users easy ways to report problems.

Every Intelligent System should also have a strategy for dealing with mistakes. Perhaps it's done by updating intelligence; perhaps by having humans override certain behaviors by hand; perhaps by offering a workaround or refund to affected users.

Some mistakes will be very hard to find. Some will be very hard to fix (without introducing new mistakes). And some will take a very long time to fix (hours to deploy a new model or months to come up with a new intelligence strategy).

This chapter will discuss ways of dealing with mistakes, including these topics:

- The types of mistakes the system might make (especially the bad ones).

- Reasons intelligences might make mistakes.

- Ways to mitigate mistakes.

Every orchestrator of an Intelligent System should embrace the reality of mistakes.

The Worst Thing That Could Happen

Ask yourself: What is the worst thing my Intelligent System could do?

- Maybe your Intelligent System will make minor mistakes, like flashing a light the user doesn't care about or playing a song they don't love.

- Maybe it could waste time and effort, automating something that a user has to undo, or causing your user to take their attention off the thing they actually care about and look at the thing the intelligence is making a mistake about.

© Geoff Hulten 2018
G. Hulten, *Building Intelligent Systems*, https://doi.org/10.1007/978-1-4842-3432-7_24

- Maybe it could cost your business money by deciding to spend a lot of CPU or bandwidth, or by accidentally hiding your best (and most profitable) content.

- Maybe it could put you at legal risk by taking an action that is against the law somewhere, or by shutting down a customer or a competitor's ability to do business, causing them damages you might end up being liable for.

- Maybe it could do irreparable harm by deleting things that are important, melting a furnace, or sending an offensive communication from one user to another.

- Maybe it could hurt someone—even get someone killed.

Most of the time, when you think about your system you are going to think about how amazing it will be, all the good it will do, all the people who will love it. You'll want to dismiss its problems; you'll even try to ignore them.

Don't.

Find the worst thing your system can do.

Then find the second worst.

Then the third worst.

Then get five other people to do the same thing. Embrace their ideas and accept them.

And then when you have fifteen really bad things your Intelligent System might do, ask yourself: is that OK?

Because these types of mistakes are going to happen, and they will be hard to find, and they will be hard to correct.

If the worst thing your system might do is too bad to contemplate, you might want to design a different system—one that couldn't do that bad thing, ever, no matter what the intelligence says. Maybe you make sure a human is part of the decision process. Maybe you use a less forceful experience. Maybe you find something completely different to do with your life...

Because the intelligence will make mistakes and, eventually, the worst thing will happen.

Ways Intelligence Can Break

An Intelligent System will make mistakes for many different reasons. Some of them are implementation or management problems; some of them are intelligence problems. This section discusses these potential problem sources, including the following:

- System outages
- Model outages
- Intelligence errors
- Intelligence degradation

The first step to fixing a mistake is understanding what is causing it.

System Outage

Sometimes computers crash. Sometimes the Internet is slow. Sometimes network cables get cut. Sometimes a system has subtle bugs in the way its systems interact. These are problems with the implementation or the operation of your Intelligent System, but they might show up the same way intelligence mistakes do, in user reports, escalations, and degrading metrics.

Isolating these types of problems can be difficult in large systems, particularly when intelligence is spread between clients (which are in different states of upgrade) and multiple servers (which can live in various data centers).

Catastrophic outages are usually easy to find—because everything tanks. But partial outages can be more subtle. For example, suppose 1% of your traffic is going to a particular server and the server bombs out in a crazy way. One percent of your users are getting a bad experience, and maybe they are reporting it, over and over... But that's just 1% of your user base. 99% of your users aren't bothered. Would you ever notice?

System outages should be rare, and they should be fixed immediately. If they become prevalent they will paralyze intelligence work—and they will be just plain bad for morale.

Model Outage

Related to system outages, a model outage is more an implementation problem than an intelligence problem—but it will have similar symptoms.

Model outages can occur when:

- A model file is corrupted in deployment.

- A model file goes out of sync with the code that turns contexts into features.

- The intelligence creation environment goes out of sync with the intelligence runtime environment.

- An intelligence goes out of sync with the experience.

These problems can be very hard to find—imagine if some feature code gets updated in the intelligence creation environment, but not in the intelligence runtime. Then when a new model (using the updated feature code) is pushed to the runtime (using the out-of-date feature code) it will be confused. It will get feature values it doesn't expect. It will make mistakes. Because of this, maybe the accuracy is 5% worse in the runtime than it is in the lab. All the testing in the lab shows that the intelligence is working fine, but users are getting a slightly worse experience.

Because these problems are so hard to find, every intelligence implementation should have checks and double-checks to make sure the intelligence-creation environment is in sync with the runtime environment, that everything is deployed correctly, and that all components are in sync.

Intelligence Errors

When the models that make up intelligence don't match the world perfectly (and they don't), there will be mistakes. Recall that creating intelligence is a balancing act between learning a very complex model that can represent the problem and learning a model that can generalize well to new contexts. There will always be gaps—places where the model isn't quite right.

And these gaps cause mistakes, mistakes that are hard to correct through intelligence creation. You can try another type of model, but that will make its own (new) types of mistakes. You can get more data, but that has diminishing returns. You can try more feature engineering—and it usually helps. But these types of mistakes will always exist.

They will appear a bit random. They will change over time (as the training data changes). They aren't easy to correct—it will require sustained effort, and it will get harder the further you go.

One additional challenge for intelligence errors is figuring out which part of the intelligence is responsible. When models are loosely coupled (for example, when they have an order of execution and the first model to make a statement about a context wins), it can be easy to determine exactly which model gave the incorrect answer. But when models are tightly coupled (for example, when the output of several models is combined using a complex heuristic or a meta-model), a mistake will be harder to track. If six models are each partially responsible for a mistake, where do you begin?

Intelligence Degradation

When an open-ended, time-changing problem changes, the intelligence you had yesterday will not be as good today. Changing problems compound generic intelligence errors because new mistakes will occur even when you don't change anything. Further, training data for the "new" problem will take time to accumulate, meaning you may need to wait to respond and you may never be able to get enough training data to learn any particular version of the problem well (by the time you learn it, it isn't relevant any more).

There are two main categories of change:

1. Where new contexts appear over time (or old ones disappear), in which case you will need to create new intelligence to work on the new contexts, but existing training data can still be used on old contexts.

2. Where the meaning of contexts changes over time, in which case existing training data can be misleading, and you'll need to focus on new telemetry to create effective intelligence.

One way to understand the degradation in your Intelligent System is to preserve old versions of your intelligence and run a spectrum of previous intelligences on current data—the intelligence from yesterday, from five days ago, from ten days ago, and so on. By looking at how mistakes change, you can gain intuition about the way your problem is evolving and use that intuition when choosing how to adapt.

Mitigating Mistakes

Random, low-cost mistakes are to be expected. But when mistakes spike, when they become systematic, or when they become risky or expensive, you might consider mitigations.

This section discusses the following approaches to mitigating errors:

- Investing in intelligence
- Balancing the experience
- Adjust intelligence management parameters
- Implementing guardrails
- Overriding errors

Invest in Intelligence

In a healthy Intelligent System, the intelligence will be constantly improving. One way to deal with mistakes is to wait for the intelligence to catch up.

In fact, almost every other approach to mitigating mistakes degrades the value of the Intelligent System for users who aren't having problems (by watering down the experience); or it adds complexity and maintenance cost in the long run (by adding manual tweaks that must be maintained). Because of this, improving the intelligence is a great way to deal with mistakes—when it is possible. The best ways to invest in improving intelligence with respect to mistakes are these:

1. Get more relevant telemetry or training data that contains the contexts where mistakes are occurring. This might allow the intelligence creation to start solving the problem with very little work.

2. Help intelligence creators prioritize the parts of the system they spend time on by categorizing mistakes into categories (by locals, age, user properties, and so on) and prioritizing the categories. Intelligence creators can then work on features and modeling that helps in those specific areas, maybe via partitioning and focused modeling.

3. Provide more resources to intelligence creation in terms of people and tools.

These investments will improve the overall quality of the intelligence and, over time, a rising tide will raise all ships.

And perhaps the worst way to invest in intelligence is to track intelligence errors as if they were software defects and hold intelligence creators accountable to fix them in order, one after the next, until there aren't any more. That's not the way it works. If there are errors that you absolutely must fix, then you should consider one of the other mitigation approaches discussed in this section.

Balance the Experience

If the errors are low-grade, random intelligence errors or are caused by intelligence degradation, they will be hard to solve. In these cases, you might want to rebalance the experience, making it less forceful and making the errors less costly.

If the problems are bad enough, you could consider essentially turning off an intelligent experience until you can get on top of the problem.

There are many chapters that discuss ways to balance intelligence and experience, so I won't cover them again here.

Adjust Intelligence Management Parameters

If the errors are because of degradation—that is, because new contexts are showing up quickly or old contexts are changing meaning—you might be able to address them by training and deploying new models faster.

You might also change what training data you use, for example by phasing out old training data more quickly (when contexts are changing meaning) or up-sampling new contexts in telemetry or training (which helps with changing meanings and when new contexts are appearing quickly).

These approaches are similar to investing in intelligence, but they can be more reactive. For example, when a holiday comes around, a bad weather event occurs, or a new product is launched your problem might change more-quickly than usual. An orchestrator might know this and tweak some knobs rather than waiting for intelligence creators to learn how to predict these types of events.

Implement Guardrails

Sometimes you encounter categories of mistakes that are just silly. Any human would look at the mistake and know that it couldn't possibly be right. For example:

- When the pellet griller is at 800 degrees you never want to add more fuel to the fire.

- When the user is 10 years old you never want to show them a horror movie.

In these cases, you could try to trick the intelligence creation algorithms to learn these things. You could gather focused training data. You could hound the intelligence creators. You could invest months of work...

Or you could implement a simple heuristic to override the intelligence when it is about to do something that is obviously crazy—a guardrail.

When using guardrails, make sure to:

1. Be conservative—only override obvious problems and don't get drawn into creating sophisticated intelligence by hand.

2. Revisit your decisions—by tracking the performance (and cost) of guardrails and removing or relaxing ones that become less important as intelligence improves or the problem changes.

Override Errors

Sometimes there is no way around it; your system will make expensive mistakes that can't be mitigated any other way, and you'll have to override these mistakes by hand. For example:

- You run a search engine and the top response for the query "games" is not about games.

- You run an anti-spam service and it is deleting all the mail from a legitimate business.

- You run an e-commerce site and it removed a product for "violating policy," but the product wasn't violating policy.

- You run a funny-webpage finder and it is marking the Internet's most popular joke site as not-funny.

When these mistakes are important enough, you might want to have special user experience to allow users to report problems. And you might want to have some processes around responding to these reports. For example, you might create a support group with the right tools and work-flows to examine every reported mistake within an hour, 24 hours a day, 7 days a week.

As with guardrails, make sure to use overriding sparingly and to track the quality and cost of overrides over time.

Summary

Mistakes are part of Intelligent Systems, and you should have a plan to measure mistakes and deal with them. Part of this plan is to understand what types of bad things your system can do. Be honest with yourself. Be creative in imagining problems.

In order to fix a problem, it's helpful to know what is causing the problem. Mistakes can occur when:

- A part of your Intelligent System has an outage.

- Your model is created, deployed, or interpreted incorrectly.

- Your intelligence isn't a perfect match for the problem (and it isn't).

- The problem or user-base changes.

Once you've found a problem you can mitigate it in a number of ways:

- By investing more in intelligence.

- By rebalancing the experience.

- By changing intelligent management parameters.

- By implementing guardrails.

- By overriding errors.

An active mistake mitigation plan can allow the rest of your Intelligent System to be more aggressive—and achieve more impact. Embracing mistakes, and being wise and efficient at mitigating them, is an important part of orchestrating an Intelligent System.

For Thought...

After reading this chapter, you should:

- Understand when and how mistakes put an Intelligent System at risk.

- Understand how to know if an Intelligent System is working, and to identify the common ways it might fail.

- Be able to mitigate mistakes using a collection of common approaches, and know when to use the various approaches.

You should be able to answer questions like these:

- What is the most widespread Intelligent System mistake you are aware of?

- What is the most expensive one?

- Design a system to address one of these two mistakes (the widespread one or the expensive one).

- Would it work for the other mistake? Why or why not?

CHAPTER 25

Adversaries and Abuse

Whenever you create something valuable, someone is going to try to make a buck off of it. Intelligent Systems are no different. If you spend energy, money, and time attract users, someone is going to try to make money off of those users. If you build a business that is putting pressure on a competitor, someone is going to try to make it harder for you to run that business.

These are some common ways that abuse can affect an Intelligent System:

- Abusers try to monetize your users, for example by spamming them.

- Abusers try to steal information about your system or users, to copy or sell.

- Abusers try to use your platform to host attacks on other systems.

- Abusers try to poison your system so it doesn't perform the way you want it to.

Some of these activities are illegal, but some of them aren't. And even when the activities are illegal, the global nature of the Internet makes it very hard to find the attackers and even harder to get them prosecuted.

Because of this, all successful Intelligent Systems need to be prepared to defend themselves from abuse.

This chapter explains the basics of abuse so you can understand the challenge, be ready to identify abuse when it happens to you, and have some tools to help make your Intelligent System harder to abuse.

© Geoff Hulten 2018
G. Hulten, *Building Intelligent Systems*, https://doi.org/10.1007/978-1-4842-3432-7_25

Abuse Is a Business

The first thing to know about abuse is that it is a business—a big business. The vast majority of people carrying out abuse are doing it to make money (although a bit of abuse is carried out for fun, for social justice, or to support espionage). Some of the ways abusers can make money include these:

- **Driving traffic**: By tricking your Intelligent System to show users things that the abuser wants them to see. This is essentially advertising. The abuser gets a deal with a web site, and gets paid for every user they direct from your site, to the target website. This is often called "spamming."

- **Compromising personal information**: Including social security numbers, contact information, passwords, banking information, and so on. Abusers can use this information directly or resell it for a quick profit to other abusers.

- **Compromising computers**: By tricking users to install bad things on their computers. Once they have bad things on a user's computer they can steal personal information or use the user's computer to launch further attacks. When abusers can use your Intelligent System to communicate with users, they can trick them to do all sorts of crazy things.

- **Boosting content**: By tricking your Intelligent System to behave in ways that they want. For example, an abuser might put fake reviews on a product in an e-commerce site to make the product more prominent and sell more.

- **Suppressing content**: By trying to hurt your Intelligent System or by trying to harm other users of your system. For example, an abuser might report a competitor's content as offensive.

- **Direct theft**: Of content, perhaps for resale, such as stealing digital content in an online game.

Abusers have created markets for all of these things (and more), so it is very easy for any abuser who finds a way to do things like these on your Intelligent System to turn that activity into money.

Abuse Scales

Imagine finding an activity you could do to earn a tenth of a penny. Click a button, select an option, click another button—Bam! A tenth of a penny shows up in your bank account.

It sounds pointless. You'd have to do that activity a thousand times just to make a dollar, a hundred thousand times to make a hundred dollars. What a waste of time!

But now imagine you can program a computer to do it for you, and the computer can do it a million times an hour, every hour, for the rest of eternity. This is Internet abuse. Generally Internet abuse involves an activity that is very unlikely to succeed (like tricking someone to give up a password), or is worth very little every time it does succeed (like sending traffic to a web site)—but the abuser does these-low value activities over and over and over and over. And they make good money doing it.

What this means is that you may not have an abuse problem one day. You think you're fine, but an abuser might be experimenting, trying different activities, measuring how often users fall for their scams or how much traffic they can produce off of your users, doing the math—and when they find math that is in their favor they can scale up, quickly.

It's easy for abuse to go from zero to disaster overnight.

Estimating Your Risk

Your Intelligent System will be interesting to abusers if any of the following are true:

- **It has a lot of users**: As your Intelligent System gets more popular it will have more users. This means abusers can scale their attacks further and make more money per attack. It's worth their time to experiment against your Intelligent System because if they find a way to make a tenth of a penny, they can make a lot of them.

- **Abusers can use your system to communicate with users**: Particularly if they can put a URL in their communication. The communication can be a message, an email, a web site, a picture, a review, a comment—anything. Abusers will find ways to make money off of communicating with users by tricking them and spamming them.

- **It interacts with user generated content**: And the intelligence plays any role in deciding which content to show, how to annotate or display the content, or how to order the content it does show. Influencing how users see content will influence which content they engage with—and how any associated money will flow. Abusers will try to get inside that kind of loop.

- **The mistakes it makes cost someone money**: Particularly if the costs can be directed toward specific parties. For example, when the smart toaster burns a particular brand of freezer-tart; or when your Intelligent System is putting the non-intelligent competitor out of business.

- **It does any other thing a dedicated mind can make money off of**: Think of abusers as smart hackers with dubious morals and a lot of time on their hands. Be prepared to be amazed by their ingenuity.

What an Abuse Problem Looks Like

When abusers scale an attack against your Intelligent System they will create odd usage patterns. You can often spot them in telemetry by looking for:

- Large groups of users who use your Intelligent System in degenerate ways (very focused on the parts that make them money).

- Contexts that see a spike in activity compared to normal usage.

- Contexts where the distribution of outcomes changes drastically (because abuser are sending you incorrect information).

- Patterns in user complaints and problem reports.

Abusers may try to blend in, but they will find it hard to match your legitimate user's activities, so you can usually spot their attacks if you spend the time looking for them. You can also usually spot their attacks retroactively by setting alerts for drastic changes. Traffic to a part of your system goes up by a huge amount? Someone should know. Complaints double? Someone should take a look.

Ways to Combat Abuse

If abuse does become a problem, some approaches include these:

- Add costs to your product.

- Becoming less interesting to abusers.

- Machine learning with an adversary.

- Get the adversary out of the loop.

Add Costs

You can stop abuse and make more money! Woo-hoo!

Well, charging more might scare away your legitimate customers too, so it may not be an option. But keep in mind that abuse is a business and your best way to stop abuse is to make it harder for abusers to profit.

For example, profiting from abuse will get harder if an abuser needs to:

- Pay 10 cents per account on your Intelligent System.

- Type in some squiggly characters for every review they want to leave.

- Buy a smart toaster for every attack they want to launch.

These costs are most effective when they impact abusers more than they impact legitimate users. For example, if each good user has to type in squiggly characters once, but the abuser needs to type them for every activity they do. Done right, adding cost might put abusers out of business, without legitimate users even knowing it is happening.

Becoming Less Interesting to Abusers

You could change your product to do less of the things abusers find interesting, for example:

- Removing or restricting communication channels.

- Storing less personal user information on the site and being careful about where you display it.

- Reducing the effect of user feedback on how content is presented.

These may also make your product less useful for real users, but sometimes a small tweak or two will make all the difference in breaking the ability of abusers to profit.

Machine Learning with an Adversary

You could use machine learning to identify abusive interactions and then delete them. I mean, by this point you're probably thinking: "This is a whole book about machine learning, so machine learning must be the right way to stop abuse, right?"

Unfortunately, not exactly. Machine learning is a fine tool, but abusers are very good at changing their attacks in patterns that fool machine learning. And it usually costs abusers much less to change an attack than it will cost you to chase them and change your machine learning.

You can do machine learning to combat abuse, and it will probably help, but I recommend you consider other options first and make sure you understand how machine learning can actually impact an abuser's business model before investing too much.

Get the Abuser out of the Loop

Whenever you identify abuse, you should block everything the abuser used to launch the attack, including the account, the toaster, the pellet griller, the funny web site, the sprinkler system—all of it. This will ensure abusers have to pay the most cost possible as they scale their attacks. The infrastructure they used to attack you yesterday is burned, so they need to go out and rebuild the infrastructure today.

Another option is to focus on creating your intelligence only from trusted users. Imagine you have 100,000 users who've been with you for years, using your intelligent service, producing telemetry and contexts with outcomes for training. These users are pretty safe—they aren't accounts created by abusers to launch some new attack. They are your customers. By restricting intelligence creation to "known good" users, you can often avoid abuse completely.

Summary

Whenever you create something valuable, abusers will come and try to benefit from your hard work, putting your users and your Intelligent System at risk.

The vast majority of abuse is done to make money. By understanding how abusers make money, you can control how interesting your Intelligent System is to them. Often the cheapest way to fight abuse is to make a few small tweaks in the way your system works so that abusers can't figure out how to make a reliable profit.

Abuse usually targets low-value activities that can be scaled dramatically—a tenth of a penny, one million times a day. You can often see abuse in telemetry as spikes of activity that doesn't match your regular usage patterns. You may not be able to stop it in real time doing this, but you can usually know if you are under attack.

Some practices can discourage abuse:

- Increase the cost of doing abuse.

- Change your Intelligent System to be less valuable to abusers.

- Use some machine learning (but be careful—abusers have the upper hand here).

- Trust your established users more than you trust new users, and delete all users who are involved in confirmed abuse.

For Thought...

After reading this chapter, you should:

- Know what abusers are and what they do.

- Be able to identify easy changes that will make your Intelligent System much less interesting to abusers.

You should be able to answer questions like these:
Consider your favorite Intelligent System.

- What is one simple change that would make it much more interesting to abusers?

- What is one simple change to make it less interesting to abusers?

CHAPTER 26

Approaching Your Own Intelligent System

Thank you for reading this book. I'm glad you got this far. You should now have the foundation to execute on your own Intelligent System project, knowing:

- **How to approach an Intelligent System**: What it is good for, when you need one, and how to set objectives for one.

- **How to design an intelligent experience**: One that achieves your objectives and produces data to help grow the intelligence.

- **What it takes to implement an Intelligent System**: How to execute, manage, and measure intelligence.

- **How to create intelligence**: Including many approaches, but particularly using machine learning.

- **How to orchestrate an Intelligent System**: To bring these parts together throughout its life-cycle and achieve the impact you want.

This chapter will review some key concepts and provide a checklist for approaching your own Intelligent System project.

An Intelligent System Checklist

Intelligent Systems are changing the world by closing the loop between users and intelligence. Intelligent Systems solve important problems, delight users, and help organizations achieve success.

To create an Intelligent System there are many things you'll need to do, and many decisions you'll need to make. This section collects the key ones of these into one place, with references to the chapters where you can find more detail to help.

319

© Geoff Hulten 2018
G. Hulten, *Building Intelligent Systems*, https://doi.org/10.1007/978-1-4842-3432-7_26

When approaching an Intelligent System project, I recommend that you consider the following steps.

Approach the Intelligent System Project

1. Begin by making sure an Intelligent System is right for you.

Intelligent Systems are useful when you expect to change the intelligence of your product many times through its life-cycle. This is commonly needed when your problem is:

* Large.

* Open-ended.

* Time-changing.

* Intrinsically hard.

In addition, an Intelligent System works best when:

* A partial solution is viable and interesting.

* You can use data from users interacting with the system to improve it.

* You can influence a properly scoped and meaningful objective.

* The problem justifies the effort of building the Intelligent System.

Chapter 2 discusses these topics in more details.

2. Define what success looks like for your Intelligent System.

Create consensus about what your Intelligent System should do in a way that *communicates the desired outcome*, is *achievable*, and is *measurable*.

Decide what *organizational objectives* (like profit or units sold) your Intelligent System will contribute to, and what *leading indicators* you can use to get quicker feedback that you're making progress toward them. Then decide what your Intelligent System should optimize for on a day-to-day basis by identifying the *user outcomes* you want to create and the *model properties* that will be critical for success.

Come up with a plan for measuring that your Intelligent System is achieving its objectives, which might include: *telemetry, waiting for outcomes to become clear, using human judgement,* and *asking users*.

And be prepared to invest over time to keep your goals healthy. Chapter 4 contains more detail on setting objectives for Intelligent Systems.

Plan for the Intelligent Experience

3. Decide how to present your system's intelligence to users to achieve your goals.

The goal of an intelligent experience is to:

- Present intelligence to the user.
- Achieve the system's objectives.
- Minimize intelligence flaws.
- Create data to help improve the intelligence.

To achieve these, you need to *balance the experience*, by trading off between:

- The *forcefulness* of the experience.
- The *frequency* of interactions.
- The *value of success*.
- The *cost of mistakes* (both discovering and recovering).
- The *quality of the intelligence*.

Balancing an intelligent experience is a process, in which an experience changes as intelligence changes. You'll probably want to leave yourself some room to iterate. Chapter 7 talks about balancing intelligent experiences.

And Chapter 8 talks about specific modes of interaction between users and intelligence that you can use as part of this balancing, including:

- *Automation*
- *Prompting*
- *Organizing*
- *Annotating*

You'll almost certainly use some hybrid of these in your Intelligent System, so that your experience is more forceful when the intelligence is certain, and less forceful when the intelligence is unsure.

4. Plan for getting data from your experience.

An ideal intelligent experience will produce data that lets you improve the Intelligent System (and specifically the intelligence itself) over time.

To be useful for improving the Intelligent System, data should:

- Include the *context, user action,* and *outcomes* of the interaction.

- Provide good *coverage* over the parts of your Intelligent System and problem space.

- Represent *real interactions* with your users.

- Be *unbiased*.

- Avoid *feedback loops*.

- Have sufficient *scale* to be useful for creating intelligence.

It is best—by far—when this data is created implicitly, as a natural byproduct of users interacting with your Intelligent System. But sometimes you need to ask for help understanding how your users perceive the outcomes they receive. Ways to get explicit user feedback include these:

- Letting users provide *ratings* in the experience (such as thumbs up, thumbs down, or 1-5 stars).

- Accepting *user reports* of bad outcomes in the experience (such as a "report as spam" button).

- Creating a support tier so users can *escalate* problems to get help (call a number for help).

- Prompting users to *classify* some of their outcomes you select.

But keep in mind that users won't want to spend a lot of time and attention helping you build your product. Focus on implicit data and use explicit data sparingly. See Chapter 9 for more detailed discussion.

5. Plan to verify your intelligent experience.

You'll need to plan how to:

- Know that the experience is functioning correctly (despite all the mistakes intelligence is making).

- Know that the experience is doing its part in helping your Intelligent System achieve its objectives.

To help with this, you might want to plan for tools to inspect the experience that will be generated for any particular context, and tools to help create and capture contexts where problems are occurring.

You might also want to plan for using different versions of the system's intelligence in these tools, including the live intelligence; a snapshot of intelligence and some simple or stub intelligence.

Chapter 10 covers verifying intelligent experiences.

Plan the Intelligent System Implementation

Implementing an Intelligent System requires all the work of implementing a traditional system, plus work to enable the system's intelligence. These additional components include:

- The *intelligence runtime*.

- *Intelligence management*.

- *Telemetry* to help grow the intelligence.

- An *intelligence creation environment*.

- Intelligence *orchestration* tools.

These components provide important capabilities for improving an Intelligent System over its lifetime. Building the right implementation can make it much easier to create intelligence and to orchestrate an Intelligent System.

Chapter 11 is an overview of the components of an Intelligent System implementation.

6. Decide where your intelligence will live.

Intelligence can be any of the following:

- *Static intelligence* in the product.

- *Client-side intelligence*.

- *Server-centric intelligence*.

- *Back-end (cached) intelligence*.

Or probably some hybrid of these. You should select where your intelligence will live based on the effect it will have on:

- The *latency in updating* your intelligence.

- The *latency in executing* your intelligence.

- The *cost of operating* your Intelligent System.

- Any requirement for *offline operation*.

Chapter 13 explores issues around where intelligence can live; it will have the detail you need to make good decisions for your application.

7. Design your intelligence runtime.

The intelligence runtime is where the intelligence is executed and experience is updated. An intelligence runtime must do all of the following:

- Gather the *context* you need so your intelligence can make good decisions.

- *Extract features* from context in a way that is safe, but that supports innovation in intelligence creation.

- Deal with *model* files and updating them.

- *Execute* models on contexts (and their features).

- Take the *results* and *predictions* produced by the execution and light up the intelligent experience.

Chapter 12 discusses intelligent runtimes in detail.

8. Plan for intelligence management.

Intelligence management is the system that lets you add new intelligence to the system and turn it on for users in a safe way. This includes:

- *Ingesting intelligence* into the Intelligent System.

- *Sanity-checking intelligence* to make sure it is not obviously harmful or corrupted.

- *Combining intelligence* when there are multiple intelligence sources, including making sure they are all in sync. Chapter 21 catalogs ways of combining intelligence.

- *Lighting up intelligence* for users in a controlled fashion. That can include silent intelligence, controlled rollout, flighting (A/B testing), and support for reversions.

Chapter 14 discusses methods for intelligence management.

9. Plan for telemetry.

Telemetry is a critical link to connect usage to intelligence, so an Intelligent System can improve over time. Telemetry is used to make sure the Intelligent System is working, understand the outcomes users are getting, and gather data to grow the intelligence.

There are more things in a system than you could possibly measure, so you need to make some decisions about how to control the scale of what you collect in telemetry. But you also need to provide tools to allow intelligence creators and orchestrators to adapt as the problem and Intelligent System change over time. These tools include:

- *Sampling* what is collected in telemetry, and turning up and down the sampling rate over time.

- *Summarizing* interactions in ways that allow the right information to be collected efficiently.

- Supporting *flexible targeting* of what contexts and users to sample telemetry from.

You can learn all about telemetry in Chapter 15.

Get Ready to Create Intelligence

Intelligence is the part of the Intelligent System that makes the hard decisions, mapping from contexts to predictions, powering the intelligent experience, creating value for users and for your organization.

10. Prepare to evaluate intelligence.

With intelligence, evaluation is creation. You should be able to easily measure any intelligence to know:

- How well it *generalizes*.

- The *types of mistakes* it makes.

- Its *mistake distribution*.

To do this you will need evaluation data, hopefully from telemetry. This should be sufficient data to get statistically meaningful evaluations. It should also be *independent* from training data (which is sometimes tricky) and it should allow you to understand how your intelligence performs on important *sub-populations*.

Chapter 19 (the longest chapter in the book) discusses evaluating intelligence—it's that important.

11. Decide how you will organize your intelligence.

Organizing intelligence creation is important when working with large Intelligent Systems where you need to:

- Have multiple people *collaboratively creating intelligence*.

- *Clean up mistakes* that intelligence is making.

- *Solve the easy part the easy way* and focus more complex techniques on the hard parts of the problem.

- *Incorporate legacy intelligence* or intelligence acquired from external sources.

Here are some common ways to organize intelligence creation:

- *Decouple feature engineering*.

- Perform *multiple model searches*.

- *Chase mistakes* (which is probably the worst).

- Use *meta-models*.

- Perform *model sequencing*.

- *Partition contexts*.

- Use *overrides*.

Chapter 21 discusses all of these ways to organize intelligence creation.

12. Set up your intelligence creation process.

Set it up to use all the tools and data available to create and improve intelligence. This involves:

- Choosing the *representation* for your intelligence (as described in Chapter 17).

- Creating *simple heuristics* as a baseline.

- Using *machine learning* to produce more sophisticated intelligence when needed (as described in Chapter 20).

- Helping *understand the tradeoffs* of various implementation options.

- And *assessing and iterating*, on and on, as long as the intelligence needs to improve.

Chapter 18 gives an example of the intelligence-creation process.

Orchestrate Your Intelligent System

You need to take control of all the tools discussed in this book and "drive the race car" to achieve success day in and day out for as long as the Intelligent System is relevant. A well-orchestrated Intelligent System will:

- *Achieve its objectives* reliably over time.

- Have experience, intelligence, and objective in *balance.*

- Have *mistakes mitigated* effectively.

- *Scale* effectively over time.

- *Degrade* slowly.

Orchestration is important for Intelligent Systems when:

- *Objectives change.*

- *Users change.*

- The *problem changes.*

- *Intelligence changes.*

- The system needs to be more *efficient.*

- *Abuse* happens.

Chapter 22 introduces orchestration in detail.

13. Plan for orchestration throughout your Intelligent System's lifecycle.

This planning includes how to get started and how to invest over time. In order to orchestrate an Intelligent System, you'll need to

- *Monitor the success criteria.*

- *Inspect interactions.*

- *Balance the experience.*

- *Override intelligence.*

- *Create intelligence.*

And more. Chapter 23 describes the activities of orchestration and how to decide when and where to invest.

14. Prepare to identify and deal with mistakes.

Because with intelligence, mistakes will come.
You will need to identify why your Intelligent System is getting mistakes, including:

- *System outages.*

- *Model outages.*

- *Intelligence errors.*

- *Intelligence degradation.*

And when you do identify a source of mistakes you must decide how to react:

- *Invest in intelligence.*

- *Balance the experience.*

- *Retrain intelligence* differently.

- Implement *guardrails.*

- *Override errors.*

By embracing errors, finding them proactively, and mitigating the damage they cause, you can free up lots of space for intelligence to shine. Chapter 24 explores mistakes.

15. Get ready for abuse.

If someone can figure out how to make a buck by abusing your users or your Intelligent System, they will.

You should be prepared to identify abuse quickly if it starts to occur, by looking for:

- Groups of users with *degenerate usage*.

- Contexts that see *spikes in activity*.

- Contexts that see drastic *changes in outcomes*.

- Patterns in complaints and *problem reports*.

If abuse does happen, keep in mind that it is a business. You don't need to block abuse. You just need to make the abuser think they can't make money and they'll stop. Here are some ways you can do this:

- Add costs to users of your Intelligent System.

- Become less interesting to abusers.

- Model with an adversary.

- Remove the adversary from the loop.

Chapter 25 is about adversaries and abuse.

16. Enjoy the ride.

And that's not the end, but the beginning. Intelligent Systems can live for many years. Attract users. Achieve objectives. Create intelligence. Orchestrate the thing—and win!

Summary

Thank you for reading this book. You should now have the knowledge to approach an Intelligent System project, with confidence.

I'm excited to see what you'll accomplish!

For Thought…

After reading this chapter, you should:

- Feel great about the work you put into reading this book.

- Be prepared to participate in Intelligent System projects, confident you have the knowledge to contribute.

You should be able to answer questions like these:

- What type of Intelligent System would you like to build?

- How will it change the world?

- What's stopping you? Get going!

Index

A

Abuse
  approaches
    abuser out of loop, 316
    add costs, 315
    less interesting for abusers, 315
    machine learning, adversary, 316
  business
    boosting content, 312
    computers, compromising, 312
    direct theft, 312
    driving traffic, 312
    personal information,
      compromising, 312
    scales (*see* Scales, abuse)
    suppressing content, 312
  telemetry, 314
  ways affecting intelligent
    system, 311
Anti-malware system, 101
APIs, 140
Automated experiences, 88

B

Balance intelligent experiences
  cost of mistakes, 81–83
  factors, 76
  forcefulness, 76–77
  frequency, 78–79

  intelligence quality, 83–84
  value of success, 79–81
Betting-based-interaction, 99
Big problems, 16

C

Changes in intelligence, 70
Checklist, intelligent system project
  approach, 320
  creation, 326
  implementation, 323–325
  orchestration, 327–329
  planning, 321–323
Conceptual machine learning
  algorithm, 30
  process, 31
Confidence interval, 28
Content processing, 188
Contexts
  forms, 187–188
  intelligence creation, 189
  runtime intelligence, 187–188
Cost of implementation, 188
Curves, 239–240

D

Data
  evaluation of intelligence, 237
  historical data, risks, 232

Data (*cont.*)
  partition data
    by identity, 234–235
    by time, 234
  test set, 233
Data experience
  betting-based-interaction, 99
  connecting to outcomes, 100
  context, actions and
      outcomes, 101–102
  data collection and labeling, 97
  feedback loops, 104
  machine learning systems, 97
  properties, 100
  real usage, 103
  scale, 105
  simple interactions, 98
  TeamMaker, 98–99
  traditional computer vision system, 97
  unbiased data, 103
  understanding outcomes
    escalations, 108
    implicit outcomes, 106
    reports, 107
    user classifications, 108–109
    user ratings, 107
Data models
  classification error, 30
  generalization error, 29
  regression error, 29
Data pitfalls
  bias, 178–179
  indirect value, 180
  privacy, 180–181
  rare events, 179
Data, working with, 25
  conceptual machine learning, 30
  data models, 29

  pitfalls, 31
    broken confidence intervals, 32
    data bias, 32
    inconclusive data, 33
    noise in data, 32
    out of date, 32
  questions, 27
  structured data, 25–27
Distribution of mistakes, 230

E

E-commerce store, 194
Evaluate intelligences
  curves, 239–240
  generalization, 226
  mistakes, 241–242
  offline, 225
  online, 225
  operating point, 238
  predictions
    probability, 231
    rankings, 231–232
    regressions, 230
  properties, 226
  subjective evaluations, 240–241
Execution
  change in right answer, 148
  conversion steps, 146
  intelligent experience occurring, 147
Experience
  levels, criteria, 295
  resources, 296

F

False negative, 67
False positive, 67

Feature selection, 257
Forceful automation, 94
Forcefulness, 76–77
Frequency, 78–79

G

Generalization, 226
Getting data
    bootstrap data, 214
        computer vision case, 214
        learning curve, 215
        methods, 214
    usage data
        ideal data-creation, 215
        options, 215

H

Horror-game system, 213
Human factor, intelligent mistakes, 70–71
Hybrid experiences, 93–94

I, J, K

Instability in intelligence, 139–140
Intelligence, 185
    changes in, 68, 70
    contexts, 187–189
    degradation, 305
    errors, 304–305
    management, 127
    pellet griller, example, 185, 187
    predictions, 190
        classifications, 190–191
        hybrids and combinations, 194
        probability estimation, 191–192
        rankings, 194
        regressions, 193

Intelligence creation, 209
    good creators, 221
        data debugging, 221
        intelligence-creation tools, 222
        math-first, 223
        verification-based approach, 222
    levels, criteria, 298
    process
        achieving accurate
          objectives, 209
        blink detector example, 210
        efficiency, reliability, 209
        implementation, working, 209
        meaningful intelligence, 209
        phases (*see* Phases, intelligence
          creation)
    tools, 299
Intelligence experience
    designing
        position consideration (*see* Position
          consideration)
        positioning patterns (*see* Patterns)
Intelligence implementation
    automatic updates, 125
    components
        environment, 128–129
        management, 127
        orchestration, 129
        runtime, 127
        simple/complex, 127
    creators, 125
    customers, 130
    plug-in, 124, 126
    user interaction, 123
    user-reported-offensive sites, 126
    web browsers, 124
    web page, 124
    web server, 123

Intelligence orchestration
    objectives, 281
    well-orchestrated intelligence
      (*see* Orchestration)
Intelligence representation, *see*
    Representation
Intelligence system goals, 35
    achievable, 36
    anti-phishing, 36–37
    desired outcome, 36
    layering, 43
    measurement, 36, 44–47
      A/B testing, 45
      ask the user, 46
      decoupling goals, 46–47
      hand labeling, 45
      processing, 47
    types, 38
      leading indicators, 39–41
      model properties, 42–43
      organizational objectives, 38–39
      user outcomes, 41–42
Intelligence telemetry pipeline, 128
Intelligent experiences, 53
    data collection, 60–61
    data creation, growth, 59
    effectiveness, 62
    failure reasons, 111
    flaws, 58–59
    intended (*see* Intended
      experience)
    mistakes (*see* Mistakes)
    objectives, 53
      achieving system objectives, 53
      data creation, 53
      flaws, 53
      user presentation, 53
    options, 61

system's objectives (*see* Intelligent
    system, objectives)
users (*see* Users)
verification, tools, 112, 113
Intelligent interaction modes
    annotation, 92–93
    automated experience, 87–88
    hybrid experiences, 93–94
    organization, 90–91
    prompting action, 89–90
Intelligent systems, 3–4, 6
    building, 11
    checklist (*see* Checklist, intelligent
      system project)
    creation, 11
    data to toast, 7–8
    deciding, telemetry, 11
    elements, 4, 22
      creation, 5
      experience, 4
      implementation, 5
      objective, 4
      orchestration, 5
    heuristic, 8
    Internet toaster, 6–7
    monitor, key metrics, 11
    objectives
      design, 57–58
      home lighting automation, 57
      ineffective experience, 58
      leading indicators, 57
      minimize power amount, 58
      model properties, 57
      organizational outcomes, 57
      sensors, 58
      user outcomes, 57
    sensors, 8–10
    toast, machine learning, 10

Intended experience
  errors, 115
  user, 112
  working with context
    computer vision system, 113
    crafting, 113
    Intelligent System state, 112
    producing manual contexts, 113
    recorded usage, 113
    work-flow, 114
  working with intelligence, 114–115
Interactions levels, criteria, 294
Intrinsically hard problems, 18
Iris detector, 213

L

Learning curve, 215
Lighting up intelligence
  controlled rollout, 165–166
  flighting, 166–167
  silent intelligence, 164
  single deployment, 163

M

Machine learning
  algorithms, 192, 245
  complexity, 247–248
  complexity parameters, 258
  data size, 247
  feature
    creation, 254
    engineering, 247, 250, 253
    selection, 257
    values, 251–252
  genre, 254
  hidden information, 255

if-test, 246
irrelevant features
    elimination, 256–257
label, 251
modeling, 257
model search complexity, 247
model structure complexity, 247
normalization, 255
overfitting, 249, 259–260
training set, 251
underfits, 249
Management
  complexity, 158
  deploying the intelligence, 157
  frequency, 159
  human systems, 159
  lighting up intelligence, 157, 162
  sanity checking, 157, 160
  turning off intelligence, 157, 167
Mistakes, 58–59
  accuracy, 64
  artificial intelligence, 65
  changes, 68, 70, 72
  crazy mistakes, 65–66
  decisions, 63
  degradation, 305
  effectiveness, 64
  human factor, 70–71
  inaccurate intelligences, 65
  inconvenient, 63
  intelligence errors, 304–305
  interactions, 64
  machine learning, 65
  mitigation, 65
    balancing the experience, 307
    guardrails implementation, 308
    intelligence management
      parameters, 307

Mistakes (*cont.*)
    investing in intelligence, 306
    override errors, 308–309
  model outages, 304
  perfection, 65
  pre-conceived notions, 66
  system outages, 303
  types, 64, 66–68, 301–302
  verification, 117
Model outages, 304
Model types
  decision tree
    forests, 205
    tree structure, 203–204
  linear model, working, 202
  neural networks
    components, 206
    tasks, 206
Monitoring success criteria
  goals, 292
  levels, 292

N

Normalization, 255

O

One-hot encoding, 252
Open-ended problems, 16
Operating point, 238
Operation
  distribution costs, bandwidth
    charges, 148
  execution costs, 149
  offline, 149
Orchestration, 129
  abuse, 287

costs change
  mistake costs, 287
  telemetry costs, 287
importance, 282
intelligence changes, 286
objectives
  opportunities, 284
  solving previous problems, 284
  understanding problems, 283
problems
  changing usage patterns, 285
  solving problems, 286
  time-changing problems, 285
properties, 282
team skills, 288
types
  balancing experience, 291
  creating intelligence, 291
  experience, 295
  inspecting interactions, 291
  intelligence creation, 298
  intelligence overriding, 291, 296
  interactions, 293
  metrics dashboard, 291
  monitoring success criteria, 291
  telemetry sources, 291
user changes
  changing perception, 285
  changing usage patterns, 284
  new users, 284
Organizing intelligence
  comparing approaches, 277
  monolithic intelligences, 277
  properties
    accuracy, 264
    comprehensible, 265
    growth, 264
    loose coupling, 265

measurable, 265
    spaghetti code, 264
    team cohesion, 265
  reasons, 263
    cleaning mistakes, 264
    collaboration, 263
    incorporating legacy, 264
    problem solving, 264
  technique (*see* Techniques)
Overriding intelligence
  levels, criteria, 296
  work-flow management, 297

P, Q

Passive experience, 77
Patterns
  back-end (cached) intelligence
    advantages, 153
    disadvantages, 154
  client-side intelligence
    disadvantages, 152
    extraction code, 151
    reasons, challenges, 151
  hybrid intelligence, 154
  server-centric intelligence,
    advantages, 152
  static intelligence
    advantages, 150
    disadvantages, 150
Phases, intelligence creation
  data collection, 213
    methods (*see* Getting data)
  defining success, 212
  environments
    low light sensor, 211
    questions, 211
  evaluation steps, 216

heuristic creation
    computer vision, techniques, 218
    methods, 217
  iteration, 220
  machine learning, 218
    complex artificial neural
      networks, 218
    machine-learning toolkit, 218
    standard approach, 218
  maturity creation
    maturity spectrum, 221
    sanity-check, 221
    stages, 220
  tradeoffs, flexibility, 219
Position consideration
  latency
    executing, 146
    updating, 144
  operation (*see* Operation)
Precision-recall curve (PR curve), 240
Predictions
  probability, 231
  rankings, 231–232
  regressions, 230
Probability, 231
Problem changing, 145
Properties, intelligence, 226
  client-side summarization, 177
  flexible targeting, 177
  sampling, 175–176
  server-side summarization, 177

R

Rankings, 231–232
Receiver operating characteristic curve
    (ROC curve), 240
Regressions, 230

Representation
    code
        heuristic-based intelligence, 199
        human-based intelligence, 199
        runtime performance, 199
        working, 198
    criteria, human, computer
        creation, 197, 198
    data files, 198
    hand-labeling specific contexts, 197
    lookup tables, uses, 199
    machine learning, 197
    models
        encode intelligence, 201
        types (*see* Model types)
Runtime
    context, 134–135
    effective, 141
    execution, 138
    feature extraction, 135–137
    information gathering, 133
    instability, intelligence, 139–140
    intelligence action, 133
    intelligence APIs, 140
    models, 137

S

Sanity checking
    compatibility, 160–161
    mistakes, 162
    runtime constraints, 161
Scales, abuse
    Internet, 313
    risk
        communication, 313
        dedicated mind, hackers, 314
        mistakes, 314

        user generated content, 314
        users, 313
Sub-populations evaluation, 235–237
System outages, 303

T

TeamMaker, 98–99
Techniques
    decoupled feature engineering
        approach, 267
        challenges, 267
    meta-model
        approaches, 271
        uses, 270
    mistakes
        approaches, 270
        software bugs, 269
    model sequencing
        approaches, 273
        uses, 272
    multiple model searches
        approach, 268
        effectiveness, 268
    overriding
        approaches, 276
        guardrails, 275
        hand-labeling specific contexts, 275
        uses, 275
    partition contexts
        advantages, 274
        approaches, 274
        working, 274
    subjective scale, 266
Telemetry system
    data collection, intelligence, 174–175
    data pitfalls, 178
    functions, 172

outcomes, 173–174

properties, 175

Testing set, 233

Time-changing problems, 17

Toaster-controlling
        algorithm, 10

Toasters rule or update, 9

Types of mistakes, 227–229

U

Users

annotation, 54, 56

attention, 89

automation, 54

challenges, 54

functioning, 55

light-automation, 55

light sensor reading, 55

motion sensor data, 55

organize, 54

predictions, 54

prompt, 54, 56

smart-home product, 55

system's goals, 56

Uses, intelligent system

cost effective approaches, 21

data usage, 19

elements, 22

partial system, 19

system interface, 20–21

types, 16

    big problems, 16

    intrinsically hard problems, 18

    open-ended problems, 17

    time-changing problems, 17

V, W, X, Y, Z

Verification, intelligent experiences

continual, 117

effectiveness, 116

high-quality, 116

Internet, 116

measures, 116

mistakes, 117

poor-quality, 116

scam sites, 116

users, 118

warning, 116

Get the eBook for only $5!

Why limit yourself?

With most of our titles available in both PDF and ePUB format, you can access your content wherever and however you wish—on your PC, phone, tablet, or reader.

Since you've purchased this print book, we are happy to offer you the eBook for just $5.

To learn more, go to http://www.apress.com/companion or contact support@apress.com.

Apress®

Printed in the United States
By Bookmasters